The Story of the Space Shuttle

Springer
London
Berlin
Heidelberg
New York
Hong Kong
Milan
Paris
Tokyo

The Story of the Space Shuttle

Published in association with
Praxis Publishing

David M Harland
Space Historian
Kelvinbridge
Glasgow
UK

SPRINGER–PRAXIS BOOKS IN ASTRONOMY AND SPACE SCIENCES
SUBJECT *ADVISORY EDITOR*: John Mason B.Sc., M.Sc., Ph.D.

ISBN 1-85233-793-1 Springer-Verlag Berlin Heidelberg New York

Springer-Verlag is a part of Springer Science + Business Media (*springeronline.com*)

British Library Cataloguing-in-Publication Data
Harland, David M. (David Michael), 1955–
The story of the space shuttle. – (Springer-Praxis books in astronomy and space sciences)
1. Space shuttles 2. Astronautics
I. Title
629.4′41

ISBN 1-85233-793-1

Library of Congress Cataloging-in-Publication Data
Harland, David M. (David Michael), 1955–
The story of the space shuttle / David M. Harland.
p. cm.
Includes bibliographical references and index.
ISBN 1-85233-793-1
1. Space shuttles–United States. I. Title.

TL795.5.H363 2004
629.44′1′0973–dc22

2004041821

Project Copy Editor: Alex Whyte
Cover design: Jim Wilkie
Typesetting: BookEns Ltd, Royston, Herts., UK

Printed in the United States of America on acid-free paper

Columbia lifts off.

In memory of

Dick Scobee
Mike Smith
Judy Resnik
Greg Jarvis
Ron McNair
Ellison Onizuka
and
Christa McAuliffe

who perished attempting to reach space,
and of

Vladimir Komarov
Georgi Dobrovolksy
Viktor Patsayev
Vladislav Volkov
Rick Husband
William McCool
Dave Brown
Kalpana Chawla
Mike Anderson
Laurel Clark
and
Ilan Ramon

who perished attempting to return to Earth.

Other books by David M Harland

The Mir Space Station – a precursor to space colonisation
The Space Shuttle – rôles, missions and accomplishments
Exploring the Moon – the Apollo expeditions
Jupiter Odyssey – the story of NASA's Galileo mission
The Earth in Context – a guide to the Solar System
Mission to Saturn – Cassini and the Huygens probe
The Big Bang – a view from the 21st century

with John E Catchpole
Creating the International Space Station

with Sy Liebergot
Apollo EECOM – Journey of a lifetime

with Ben Evans
NASA's Voyager Missions – exploring the outer Solar System and beyond

with Paolo Ulivi
Lunar Exploration – human pioneers and robotic surveyors

I have decided today that the United States should proceed at once with the development of an entirely new type of space transportation system designed to help transform the space frontier of the 1970s into familiar territory, easily accessible for human endeavour in the 1980s and '90s. This system will center on a space vehicle that can Shuttle repeatedly from Earth to orbit and back. It will revolutionise transportation into near space, by routinising it.
Richard Nixon, President of the United States of America, 5 January 1972

From a pilot's standpoint, you could not ask for a more superb flying machine.
Bob Crippen, Pilot, STS-1

The Space Shuttle did more than prove our technological abilities, it raised our expectations once more; it started us dreaming again ...
Ronald Reagan, President of the United States of America,
in his Address to a Joint Session of Congress, 28 April 1981

Politicians are a strange bunch of critters.
John Young, Commander, STS-1, upon hearing that Ronald Reagan, having first praised the Shuttle as a "brave adventure", promptly cut NASA's budget

... obviously a major malfunction.
Stephen Nesbitt, NASA Public Affairs Officer, 28 February 1986

We are on a true space 'ship' now, making her way above any Earthly boundary.
William Shepherd, first commander of the ISS, 18 March 2001

The Columbia is lost. There are no survivors.
George W Bush, President of the United States of America,
addressing the nation, 1 February 2003

Table of Contents

List of Figures

List of Tables

Foreword

I was an astronaut from 1967 to 1997. I was involved with the birth of the concept of a winged spacecraft during 1969, the year Man landed on the Moon; I flew on the Space Shuttle six times; and after 35 years I am still involved in the program. That spaceship has been my calling for three decades, and through this book I have relived those wonderful years in vivid detail.

In *The Story of the Space Shuttle*, David Harland puts me there, just as he will put you there. You do not only read the words, you experience the events to your core. The text is eminently readable; it is coherent and flows across the eyes and through the mind. The events are expressed clearly, concisely and with precision, and their descriptions are logically and internally consistent. I learned about lots that I did not know, and what I did know about the Shuttle is here presented in a wonderfully weaved historical context. Despite the amount of detail and the in-depth treatment of scientific and technical issues, it is a very comfortable read. It explains the issues and expresses them so well that one does not struggle with the language, the concepts or the technical terms. He always provides enough background to enable you to understand the foundational principles of the points he makes. The details of incidents are not just spectacular; they are extraordinary. I lived through many of them and can appreciate the fidelity with which they have been captured. There is a beauty in their detail, and in the depth of history and understanding that they reveal.

Although painstaking attention to detail is the foundation of this encyclopaedic work, it is not at the expense of lucidity. The reader gets no sense of drowning in technical terms, and the text is refreshingly free of the acronyms that make some NASA documents so impenetrable. In addition to *who*, *what* and *when* there is *how* and *why*. The book presents the 'big picture' in a relevant historical perspective. It reveals causes and effects, currents and consequences. In telling how the Space Shuttle evolved, and why it worked out the way it did, Harland not only shows indirectly how programs generically work in this developmental system, he also explains the operating principles constituting to human spaceflight in this era. Harland knows the spacecraft hardware and understands the inherent technologies, and although this book is full of technical issues, it is nevertheless a comfortable read as he is succinct in its explanation.

The history of the Space Shuttle is much more of a technical history than was the case of Apollo, which was the simplest and most efficient way to get a specific job done. The Space Shuttle is technically complex, very vulnerable, and very difficult – and these are the essential elements of its history. However, while Harland appreciates this landscape, he puts a background on the canvas prior to painting the picture. He also understands the science that has been achieved. As with the technology, he gives the reader enough background and explanation to grasp what was being attempted, and follows up by explaining the significance of the results to spaceflight and other related fields.

The history of a space program is not just the evolution of a technical system; it is also a *psychosocial history*. While addressing the details, the technologies, the science and the program management, Harland never forgets that the *human* is the base for all of this, and that it is the individual who provides the impetus for space exploration.

As the title of this book indicates, Harland presents the history of the Space Shuttle as a *story*. The book reads as much like an unfolding drama as any book on history. It pulls you along. I flew on every Space Shuttle and they all took me to space and brought me home. On my first space mission I flew on Challenger on her maiden voyage, and on my final mission I flew on Columbia. I am pulled and propelled forward by the stories of Challenger, Columbia, and the other ships in the fleet, and by the unfolding drama of *their* lives. I had first-hand experience of what happened to them, but it is the dramatic revelation of their stories that leads me on irresistibly. In this history, the text is not only of linguistic fact but is also of imagistic truth: I can *see* it, *hear* it, *taste* it, *feel* it, and I end up *knowing* it. In his prose Harland has a wonderful control of historical perspective – a sense of the momentum – and in the historical trajectory one can feel the flow of the past, through the present and into the future. He understands the currents, the forces, and the sociopolitical pushes, pulls and squeezes that produced the story of the Space Shuttle. To reiterate, he explains not only *who*, *what* and *when*, but also *how* and *why*, and in this book he shows how a space program evolves in the current developmental paradigm.

Mercury, Gemini, Apollo, Skylab and Apollo Soyuz were built on a vision; in each case, the historical reality for the most part followed a preconceived plan. In the current era there either is *no* real plan, or the real plan is not evolving as intended! The Space Shuttle and the International Space Station had no strong vision to drive their rationale or steer their trajectory. Their stories did not follow detailed scripts, so they have evolved in the progression of events. The space programs of the current era – which are known only by their designations of X-30, X-33, X-34, X-37, X-38, X-40 and X-43 – were *not* the children of strong, clear and logical visions. Nor did they come into the world with clear purposes, strong mandates, logical plans or a project management that would nurture them to maturity. Consequently, one by one, they have all passed into obscurity, with barely a notice.

The Story of the Space Shuttle is not only a wonderful, readable history, it is also a lesson in the creation and development of plans, programs, spaceships and missions. While it deals with particular events, it also addresses the operating principles that drive the evolution and operation of space programs in the current era – the great, the good, the bad and the terrible. The book also highlights the factors that are essential to the development of any future space program. At its most fundamental level, it is about the human quest: visions, leadership, plans, processes and execution. It is about turning dreams into reality.

Story Musgrave

Author's Preface

By 1981 John Young, 51 years of age, was NASA's most experienced astronaut. In 1965, on Gemini 3, he became one of the first men to ride the Titan missile. The following year, on Gemini 10, he flew a very complex orbital rendezvous mission. In 1969 he flew in lunar orbit on Apollo 10, and in 1972, in command of Apollo 16, he landed on the Moon. Although most of the moonwalkers retired soon after returning to Earth, Young stayed on with the agency, became Chief of the Astronaut Office, and took the lead in preparing for the Space Shuttle. Nevertheless, as he rode the bus out to Launch Complex 39 at the Kennedy Space Center on 12 April 1981 – which was, coincidentally, the 20th anniversary of Yuri Gagarin's pioneering orbit of the Earth – his vast experience could not have fully prepared him for what he was about to do. Before being entrusted with astronauts, all previous rockets had been 'man rated' in a series of tests that identified the flaws that made them blow up. But the Shuttle was different: it required a crew to fly it. Its first test flight would therefore be the riskiest space mission ever attempted. Failure could strike any component of its incredibly complex system at any time. This was *flight test* in the classic sense of the pilot who straps into a new aircraft and takes it up *to see if it will fly*. One million people had gathered to watch, most of them camped at the Space Center, but some on cruise liners anchored offshore, and millions more watched on television, all with a niggling half-expectation that it would blow up. The tension mounted as the clock ran down to its fateful moment. *No one* – not even Young and his co-pilot, 'rookie' astronaut Bob Crippen – really, truly, knew what to expect.

As Columbia lifted off and soared into the sky on a spectacular pillar of fire and red smoke, Young's heart maintained a steady 70 beats per minute. Crippen's soared and he yelled ecstatically, "What a ride!" Two days later, as Columbia began its final approach to the vast dry lake at Edwards Air Force Base, Crippen could not contain himself, "What a way to come to California!" During the roll-out, Young, ever the cool test pilot, casually enquired of the control tower, "Do you want me to put it in the hangar?" After a *textbook* first flight, NASA pronounced its new spacecraft to be "a magnificent flying machine". Amazingly, the agency's aim was to make such hair-raising missions routine!

Incredibly, by January 1986, with 24 missions successfully achieved, launches had become so routine that they attracted no media interest. But then Challenger was lost 73 seconds after lift-off, within sight of the small group of spectators, many of whom were relatives of the dead crew. After redesigning the flawed components that had directly caused this catastrophic failure, flights resumed in September 1988, and the Shuttle demonstrated its versatility by flying a remarkable series of missions, in some cases visiting the Mir space station owned by the Russians. A decade after its 'return to flight', the Shuttle began the assembly of the International Space Station, a collaborative venture by over a dozen nations, including the Russians. As Columbia returned from an independent science mission on 1 February 2003, it was lost during re-entry, a mere 15 minutes from its scheduled landing.

While it is true that the Shuttle has killed more spacefarers than any other launch vehicle, this must be considered in context. During the Mercury project, astronauts rode two Redstone and four Atlas missiles, for Gemini they rode 10 Titan II missiles and in Apollo spacecraft they rode five Saturn IB and 10 Saturn V rockets without a single loss – although in 1967 one Apollo crew was killed when a fire broke out in their cabin during a test on the ground. Of the 113 Shuttle missions mounted, all but two were successful. And during those missions the Shuttle showed itself to be an extremely versatile vehicle for satellite deployment, astronomical and solar research, environmental monitoring and investigations of how materials and biological systems react to microgravity. The Shuttle's most spectacular rôle has been as a platform for spacewalkers. The lesson of early spacewalking tests by the Gemini missions was that to work effectively a human being must be anchored. The Robotic Manipulator System on the Shuttle has enabled astronauts to conduct a wide variety of tasks, ranging from the brute force spinning up of a communication satellite to the delicate servicing of the Hubble Space Telescope. When NASA envisaged a Shuttle, it was in the context of assembling a space station. It took a long time, but this is finally underway. The design of the station is intimately linked to the design of the Shuttle. To assess one necessarily involves considering its relationship to the other.

The oft-cited 'failing' of the Shuttle has been its inability to reduce the cost of sending payloads aloft to the promised $100 per kilogram (a figure of $25,000 would be more realistic) but with hindsight this was never a serious possibility. This must not cloud an assessment of the Shuttle's successes, however.

This book tells the story of the Space Shuttle, relating its various types of mission and the tasks that its astronauts, particularly when spacewalking, have undertaken. In the early 1970s when the Shuttle was conceived, few would have believed possible what it has since achieved.

David M Harland
Kelvinbridge, Glasgow
March 2004

Acknowledgements

I would like to thank Neville Kidger, Roeluf Schuiling, Craig Covault, John Catchpole and Keith Wilson for their reportage in the aerospace press; Rich Orloff for supplying press kits; Brian Lawrence for assistance with the Mission Log; David Portree for information concerning EVAs; Phillip Clark for his Satellite Digest; Bill Harwood of CBS News for his timeline of Columbia's re-entry; Dwayne Day of the Space Policy Institute at the George Washington University in Washington D.C. for assistance with Department of Defense missions; Robert Dempsey of the Johnson Space Center for miscellaneous snippets of information; Mike Gentry of the JSC Media Resource Center for supplying pictures; Story Musgrave for the Foreword; Brian Harvey, Alasdair Downes, Sarah Dougan, Philip Baker, Flo McGuire, Alex Williams, Manfred Wiesinger, David Shayler, David Woods, Kipp Teague and Ken Glover for miscellaneous assistance; Alex Whyte for proof-reading; and, of course, Clive Horwood of Praxis for his continuing support over the years.

1

Origins

GRANDIOSE IDEAS

In March 1952, *Collier's*, the mass-market magazine published in New York, included a short article by Wernher von Braun entitled *Across The Last Frontier*. As a rocket pioneer, von Braun had built the V-2 during the Second World War to serve as a long-range ballistic weapon, but he always regarded it as the means of achieving orbital flight, and advanced his vision of the human colonisation of space in this and other articles that were published over the next two years. However, it was Chesley Bonestell's paintings of winged space planes, enormous rotating space stations and ungainly interplanetary ships that captured the imagination of a generation that was awed by the incredible pace of technological advancement in jet engines, rockets and atomic energy resulting from the war.

The V-2 could rise to the fringe of space on a ballistic trajectory, but always fell back to Earth because it did not have nearly enough energy to achieve the speed of 8 kilometres per second required to go into orbit. The Bell X-1 rocket-powered aircraft broke the 'sound barrier' during a high-altitude soaring arc in 1947, but was confined to the atmosphere. At that time, atomic power was widely seen as the solution to any problem requiring a vast supply of energy, but the prospect of an atomic rocket was rather remote. In fact, the idea of 'spaceships' was thought to be so far beyond what was feasible that it belonged to the realm of science fiction, if not pure fantasy. Indeed, in 1956, Professor Richard van der Riet Woolley, the Astronomer Royal in England, dismissed space travel as "utter bilge". The launch of Sputnik on 4 October 1957 transformed the mood, however, and overnight von Braun's articles were seen to have been more prophetic than fantastic.

By the end of the 1950s – continuing the hybrid aircraft evolution pioneered by the Bell X-1 – the high-performance X-15 rocket plane was generally regarded as the best route to orbital flight. At this time, the US Air Force planned the X-20 as a new hybrid, combining a spaceplane with a vertically launched rocket. Dubbed the DynaSoar, the X-20 was to soar hypersonically on the fringe of the atmosphere, circling the globe to undertake reconnaissance and deliver a nuclear weapon from far beyond the reach of Soviet Union's air defences. While the X-15 and the X-20 were launched differently, both aircraft were designed to land on a runway, be refurbished and relaunched. It would take time to develop the technology and scale it to match von Braun's vision, but the door was now open, and the logic was impeccable: the spaceplane would build the orbital station that would serve as a base for mounting expeditions to destinations beyond the Earth. However, in the rapid transition from fantasy to reality there was no time to pursue the most logical evolutionary route. In reacting to the shock of Sputnik, on 1 October 1958 America created the National Aeronautics and Space Administration (NASA), which promptly set out to place a human being into orbit ahead of the Soviets. It was a race! As there was no time to upgrade the X-15 sufficiently to attain orbit and survive re-entry, and as the X-20 was still being designed, the only option was to develop a less sophisticated ballistic-return capsule. NASA saw its Mercury 'man-in-space-soonest' project as a 'crash' response to a national crisis, and the inelegant capsule was an expedient that would serve until a spaceplane became operational. The way to the planets was still via von Braun's space station and, left to itself after Mercury, NASA may well have pursued this strategy. But on 12 April 1961 the Soviets launched Yuri Gagarin on an orbit of the Earth. A few weeks after NASA sent Alan Shepard on a suborbital hop in early May, President John F. Kennedy decided that it was time to "take longer strides" and set his nation the staggering goal of achieving a lunar landing before the decade was out. As this was to be another 'crash' programme, the 'spaceplane and space station' scenario was impractical. However, the direct route to the Moon would require a rocket with unprecedented lifting power. This would have to be chemically propelled because atomic power was not yet an option, and so von Braun turned his attention to the development of the Saturn V. When 'Old Glory' was planted on the lunar surface on 21 July 1969, NASA and von Braun had met Kennedy's challenge.

COST–BENEFIT

Yet even as NASA basked in Apollo's success and set out to exploit its expendable rockets and ballistic capsules with a multifaceted Apollo Applications programme, the new Nixon administration, eager to reduce federal spending, told NASA to devise a more cost-effective technology that would consolidate its leadership in space.

The most expensive part of a trip into space is the energy required to attain low orbit, and most of the Saturn V rocket's prodigious energy was expended in climbing out of the Earth's deep 'gravity well'. All subsequent manoeuvres – even setting off for the Moon – were relatively inexpensive. Cutting the cost of spaceflight therefore meant reducing the cost of achieving orbit. In developing its response to Nixon's demand, NASA proposed a 'modular' approach, employing a reusable spaceplane with the initial mission of assembling a 12-person station in low orbit. In the fullness of time, this would act as a base from which to send an expedition to Mars. NASA regarded the station and its transportation infrastructure as an *integrated* system, but when the plan was costed it became apparent that the agency would have to choose either the space station or the spaceplane, as it could not afford both.

The use of existing, expendable, rockets to build the space station would be the simpler task, and it would form a logical follow-on to the Skylab orbital workshop, scheduled for 1973, which was the only part of the Apollo Applications programme to survive Nixon's budgetary axe. It would take longer to develop the innovative spaceplane, but, because it was to reduce the cost of access to orbit, developing the transportation infrastructure would ensure the programme's long-term future. If the 'Space Shuttle' was to achieve a low operating cost, it would require almost total reusability and a very high flight rate. In fact, the cost–benefit analysis used to show how the Shuttle would recoup its development costs envisaged *weekly* flights. Such a high rate of missions would impose an unprecedented operational strain on NASA, which was accustomed to expeditionary missions at intervals of several months. In effect, to operate the Shuttle cost-effectively, it would have to run like a commercial airline. Would there be sufficient work to sustain such a flight rate? The only option was to phase out the expendable rockets and assign *all* payloads to the Shuttle. This *Shuttle-only* policy was formally recognised by the designation of the Shuttle as the National Space Transportation System.

In fact, the annual satellite launch rates during the Shuttle's formative years were broadly supportive of the case for a weekly flight rate. The launches attributed to the United States included all services provided for Intelsat and foreign institutions. When the Federal Communications Commission opened up the market for satellite broadcasting in 1972, it was expected that the resulting proliferation of commercial satellites would not only inflate the launch rate but also create a market in which the Shuttle could generate revenue.

Table 1.1 Annual satellite launch rates

	1969	1970	1971	1972	1973	Total
US	38	29	32	27	23	149
USSR	68	79	81	70	83	381
Total	106	108	113	97	106	530

Source: Royal Aircraft Establishment.

WAYS AND MEANS

NASA had selected a ballistic capsule for Mercury because it was a more effective solution than a spaceplane given the limited mass that a contemporary rocket could place in orbit. As aerodynamic vehicles, the streamlined X-15 and X-20 had higher lift-to-drag ratios, and the development of an orbital equivalent would have required a heavyweight rocket. This was not considered to be a serious issue as larger rockets were under development. A more serious issue, however, for which there was no ready solution, was the insulation of the skin of the craft from the thermal stress of hypersonic re-entry. A conical capsule that penetrated the atmosphere blunt-end forward would be much more manageable because (a) the area in need of thermal protection could be minimised and (b) a capsule that would fly only once could use an ablative shield – by flaking off into the slipstream, this carried away the heat and prevented it from penetrating the structure. The conic angle was defined to ensure that the superheated plasma that curled around from the base did not come into contact with the side of the capsule. On the other hand, the capsule had such a low lift-to-drag ratio that it would be incapable of flying and its 'pilot' would be obliged to make an ignominious parachute descent. It was from this logic that the world drew the distinction between an aircraft and a spacecraft. Yet, soon after Apollo 11 landed on the Moon, the spaceplane was announced as the way forward! What caused this about-turn? Previously, a spaceplane had been seen as the obvious way forward because it built on the ever-faster, ever-higher trend in post-war aircraft development, but the race to be first to send an astronaut into orbit forced Mercury to use a ballistic capsule because (a) the available rockets could lift it, (b) a person could fit into it, and (c) it could be shielded for re-entry. Gemini and Apollo, driven by another deadline, had used the same technology for the same reasons. In pursuing Apollo, however, a truly enormous rocket with a prodigious lifting power had been built, and this had opened up options for spacecraft development. But could this new rocket launch a spaceplane?

The Air Force's X-20 had been cancelled long before it could begin flight testing. The X-15 continued to investigate hypersonic flight through to 1968, but no attempt was made to upgrade it for orbital flight – a step that would have required fitting it with a substantial booster and some form of thermal protection system. NASA had studied the subsonic descent characteristics of a number of lifting-bodies, so-called because they derived a high lift-to-drag ratio from the squat shape of their wingless airframes. As with the X-15, they were dropped from a B-52, and since they could fly and manage their energy, they could land on a runway. NASA had not fired a lifting-body on a sub-orbital trajectory to assess its *hypersonic* characteristics – a lack of data that was a matter of serious concern – but it was natural for it to use its lifting-body experience as a starting point in designing the Shuttle. A large, high lift-to-drag aircraft can accommodate the thermal stresses of flying hypersonically more easily than a smaller one because the heating rate and the total heat load are functions of mass and size. As the size and payload of a vehicle increases, its surface area increases more rapidly than its mass. Although some form of reusable thermal protection still had to be devised for the Shuttle, the thermal stress diminished as the airframe was scaled up. The early overriding mass constraint that had inhibited spaceplane development had been alleviated by the availability of powerful rocket engines, such as those that powered the Saturn V. A spaceplane was therefore now an option, but was it a requirement? Might a ballistic system be scaled up? Although a capsule was ideally suited to sending small crews on brief missions, it could not carry much cargo on the return trip. The later Apollos carried a battery of cameras but, because these instruments were in the service module, they had to be jettisoned at the end of the mission. Apollo was used to fly to and return from Skylab, but this was feasible only because the station carried all the apparatus that would be required for the research programme. Each crew could ferry up a certain amount of cargo – such as spare parts and perishables – but these items had to be small and lightweight because the shape of the capsule was defined by the plasma sheath that formed on re-entry. The ballistic capsule was adequate to transport crew to and from orbit, to staff a space station, and for brief trips to the Moon, but it was not a platform for scientific instruments and offered no potential as a cargo carrier. NASA intended the Shuttle to have a payload bay large enough to carry a satellite, a battery of instruments, a research laboratory or a module for a space station. Such a 'space truck' would have to be a scaled up lifting-body vehicle. This dictated that the spacecraft developed at great expense to fly to the Moon be written off as a technological *cul-de-sac*, and an essentially new – yet, ironically, rather familiar – line of development be pursued instead. Despite the existence of powerful rocket motors and the availability of data on lifting bodies, the configuration of the Shuttle was essentially open. The design was sensitive to a wide range of factors: an aerodynamic control system would be used for the first time; non-ablative thermal insulation had yet to be developed; no high-performance liquid-propellant rocket engine had been designed to be reusable, and when designed, it would have to be throttleable across a wide range of thrust to accommodate the particular operational requirements of the Shuttle. By this time, on the other hand, NASA and its primary contractors had an extensive knowledge base. All the major aerospace companies submitted proposals, either singly or as consortia – with such an enormous contract on offer they could not afford to elect out and the long-term significance of being selected out was unthinkable. A wide variety of possible configurations were offered, incorporating varying degrees of reusability, but the one aspect on which all agreed was that a single-stage-to-orbit vehicle was impracticable. However, a composite vehicle would complicate the aerodynamics during launch and require a means of separating the elements safely in normal circumstances and in an emergency.

One major factor guiding the design was the need to meet the requirements of the Air Force, as NASA had had to secure the support of the Department of Defense in order to make a sceptical Congress accept the Shuttle. The Air Force – which was responsible for launching classified satellites – had reluctantly agreed to abandon its various expendable rockets, but imposed a stiff penalty. In addition to requiring that the Shuttle be capable of flying at short notice and sufficiently often to deal with its full range of mission scenarios, it insisted that the Shuttle be sufficiently voluminous and powerful to accommodate the largest satellite that it was ever likely to produce. Specifically, it demanded that the Shuttle be able to launch a satellite that was twice as heavy and three times as bulky as would fit on the Titan III, the most powerful rocket in its inventory at that time. This meant that the orbiter had to be much larger than NASA had envisaged – having an 18-metre-long payload bay with a capacity of 30 tonnes. Also, because the Air Force insisted on being able to mount single-orbit missions, the orbiter had to be capable of manoeuvring extensively during re-entry, using hypersonic soaring to fly 2000 kilometres off the orbital ground track in order to counter the Earth's rotation and return to the launch site. This required the vehicle to have a large delta-shaped wing, in order to hit the atmosphere at a low angle of attack and then fly a steep banking turn at extreme altitude. NASA had envisaged an orbiter having short, stubby, straight wings that would hit the atmosphere at a high angle of attack, take most of the heat on its belly, and make a direct approach. The stubby wing would have given the orbiter excellent subsonic glide characteristics, but a delta wing is a high-speed configuration that has little to recommend it in subsonic flight. Consequently, an orbiter designed to suit the Air Force would not just need to endure much greater structural and thermal loads during re-entry, its poor subsonic characteristics would oblige it to make a very steep descent and touch down at high speed.

The preferred option was a fully-reusable vertically launched stack with a large fly-back booster and a small orbiter – both of which would be piloted – mated either belly-to-belly or piggyback fashion. They would separate at altitude and the booster would return to land on a runway at the launch site while its charge flew on to orbit. But the more the fly-back booster was studied, the more certain it became that the separation would have to occur at a higher altitude and at a greater speed than had been planned, which turned the booster into a hypersonic aircraft of unprecedented size, and its development cost began to compare with that of a massive hypersonic airliner – which the industry had previously dismissed as impractical. Deleting the fly-back booster meant yielding total reusability, which in turn meant increasing the overhead when the Shuttle was operational, but the prospective budget permitted no alternative. NASA considered mounting the orbiter on the first stage of a Saturn V, and Boeing, its manufacturer, offered to make its rocket reusable, but this idea was rejected. Meanwhile, the orbiter itself was a cause for concern. How would the cryogenic tanks be integrated into its structure? How would residual propellant be vented in orbit? How would the tanks be rendered inert?

After much consideration, it was decided to use an external tank that would be jettisoned. As the tank could not be discarded until the orbiter's engines shut down, and that would not occur until the orbiter had reached orbital velocity, it was evident that the tank would not be reusable. The provision of a tank for each mission would add considerably to the operational overheads. Given that the tank was external, the decision to use strap-on boosters for thrust-augmentation at launch was a logical step. For simplicity, augmentation motors were usually solid rockets. The Air Force had developed a powerful solid motor to augment its Titan II, thereby creating the Titan III, but even this would not have the thrust to augment an orbiter enlarged to match the Air Force's 30-tonne payload requirement, therefore a more powerful motor was ordered. These motors (of which there would be two per stack) would parachute into the ocean, be recovered and reused. Their refurbishment might take a considerable time in relation to the high flight-rate, and be costly, but this was a step towards reusability. Some concern was expressed that the use of solid rockets represented an unacceptable risk because they could not be shut down, but as the Air Force had intended to launch its Manned Orbiting Laboratory on a Titan III with two astronauts riding on top in a Gemini capsule – a project that had been cancelled only in 1969 – the use of solid rockets was of no great concern to the Air Force.

In March 1972, Richard Nixon formally ordered NASA to develop the National Space Transportation System, and the configuration for the Shuttle's stack was revealed. It would obviously take at least five years to develop such an innovative vehicle, but it was confidently expected that the Shuttle would become operational in time to refurbish Skylab to serve as an interim orbital base until a modular space station could be built. However, the full magnitude of the technology, systems and integration problems soon became evident, with almost every aspect of the Shuttle's development and operation posing serious issues:

- the stresses that the stack would endure at launch;
- throttling down the engines to cope with the aerodynamic pressure associated with flying through the sound barrier;
- releasing the solid rockets;
- the greatest acceleration load that the remainder of the stack should endure as it continued to climb;
- the hypersonic aerodynamics of manoeuvring during re-entry, and the use of the flight control systems;
- a radiative thermal protection system; and last, but by no means least,
- devising and facilitating abort options for the various phases of the ascent.

THE STACK

The Shuttle comprises the delta-winged orbiter, the external tank, and two boosters, all bolted together in a configuration referred to as the stack.

With a dry mass of about 65 tonnes, a wingspan of 24 metres, and a 37-metre-long airframe, the orbiter is dimensionally similar to a DC9 airliner. It does not look like an airliner, however, as its profile is dominated by the engines clustered beneath its vertical tail assembly. When the three Space Shuttle Main Engine (SSME) units are running at 100 per cent of their rated power, they each deliver 170,000 kilograms of thrust. Uniquely for such powerful rocket engines, they can be throttled down to 65 per cent, and up to 109 per cent. The force is imparted to a thrust frame in the aft bay which distributes it partly through the lower attachment points to the external components, and partly to the orbiter's mid-structure that forms the 18-metre-long, 4.6-metre-diameter cylindrical payload bay. Three Auxiliary Power Units (APU) in the aft bay provide power to gimbal the SSMEs during the ascent, and to drive the aerodynamic control surfaces during the descent.

During periods of high activity – such as ascent and descent – when most of the orbiter's systems are operating and both primary and backup systems are active, excess heat is shed by a flash evaporator system. The payload bay doors have to be opened within approximately an hour of attaining orbit to expose the radiators on their inner surfaces to enable excess heat from the vehicle's systems to be dissipated, and the flight must be curtailed if the doors fail to open. In space, a trio of fuel cells combine cryogenic oxygen and hydrogen in such a way as to generate electricity and yield water as a by-product.

Two Orbital Manoeuvring System (OMS) engines are mounted in pods on each side of the base of the tail. At 2,670 kilograms of thrust, these engines can be fired singly or in tandem. They are gimballed to deliver their force through the orbiter's centre of mass, even when fired singly. The OMS is the primary system for making the de-orbit burn. The orbiter has smaller Reaction Control System (RCS) engines for attitude control. There are 14 thrusters clustered in the nose and 12 in each of the OMS pods. Most produce 400 kilograms of thrust and are used for general translational and rotational manoeuvres, but six that deliver 11 kilograms of thrust provide fine control during proximity operations. All of these engines burn monomethyl hydrazine in nitrogen tetroxide, a hypergolic mixture that eliminates the need for an igniter.

As it would have been impractical to coat the entire surface of the orbiter with an ablative to shield its aluminium skin from the heat of re-entry, a ceramic material was developed that is able to radiate surface heat so efficiently that a thickness of only a few centimetres was required. Although it sounds deceptively trivial to describe the tiles of this Thermal Protection System (TPS) as silica coated with borosilicate, their development required a significant advance in materials science. Furthermore, as they are so lightweight, they impose only one-tenth of the mass penalty that a traditional ablative. Several forms of Reusable Surface Insulation (RSI) were manufactured, differing mainly in the additives employed, to cater for the range of thermal stresses expected. Areas required to endure temperatures of 650 °C are protected by white tiles, and those required to withstand 1,200 °C are protected by black tiles. The nose and the leading edges of the wings, which have to endure 1,600 °C, are coated by a Reinforced Carbon–Carbon (RCC) laminate of graphite cloth impregnated with resin. Applying the tiles posed a problem in itself. The orbiter's surface flexes in response to dynamic loads, but as the rigid ceramic tiles were brittle they had to be glued on. A strain-isolating pad was first bonded to the metal to absorb the flexure, and the tile was stuck to the pad. It took some time to perfect this bonding process, but once the problems were resolved, the tiles were able to withstand both the orbiter's structural dynamics and the aerodynamic loads of the airflow.

The Solid Rocket Booster (SRB) is a 46-metre-long, 3.7-metre-diameter cylinder. It has a conical nose cap containing a parachute, and a flared skirt at the rear, around the nozzle. The metal casing weighs 83 tonnes, but this rises to 586 tonnes when it is filled with the atomised aluminium powder that is burned along with ammonium perchlorate oxidiser. The solid propellant does not actually fill the motor, it forms an annular column within the casing, shaped to control the combustion in order to vary the thrust according to a predefined profile. Each of the Shuttle's two SRBs delivers 1,300 tonnes of thrust. An auxiliary power system gimbals the nozzles to vector the thrust to steer the stack along a specific trajectory up through the atmosphere.

The External Tank (ET) is essentially an 8.4-metre-diameter cylinder 42 metres in length that consists of two tanks linked by an intertank ring. The hydrogen tank forms the lower two-thirds of its length, and the oxygen tank is at the top. Although its dry mass is just 30 tonnes, it is loaded for launch with 700 tonnes of propellant in the ratio of 6 parts oxygen to 1 part hydrogen by weight. The outer surface of the ET is covered with cork and a thick coat of foam insulation to prevent the build-up of a 'hot spot' that might induce a convection current in its cryogenic contents (the oxygen is at −180 °C and the hydrogen is at −253 °C). The propellants flow through pipes running down the outside of the hydrogen tank and into the orbiter's aft bay, whereupon lightweight high-pressure turbopumps feed the orbiter's engines.

At its peak thrust, the orbiter draws propellant at the rate of 500 kilograms per second. The ET is the main load-bearing element of the stack, and the orbiter and the SRBs are bolted to it. The ET is mated to the orbiter by the attachment points of the aft bay and by a bipod on the intertank ring to the orbiter's nose. An SRB is fastened to the circumferential support ring at the base of the ET by an attachment at the lowest field-joint, just above its nozzle segment, and to the intertank ring by an attachment near the top of the motor casing. The SRBs are jettisoned by pyrotechnic systems in their fixtures.

LAUNCH FACILITIES

Rather than build new facilities – as had been done in the past when a new rocket was introduced – NASA opted to modify its Saturn V facilities for the Shuttle. The Vehicle Assembly Building (VAB), the 160-metre-per-side cube that had been built to house up to four Saturn Vs, was re-equipped to prepare two Shuttles. Assembly begins by erecting the two SRBs on the Mobile Launch Platform (MLP). This had extra vents cut into it to pass the SRB efflux through to the flame trench. First the nozzle segment is bolted to the platform, the three motor segments are then added, and finally the nose is put on top. The segments are mated by an annular tang-and-clevis mechanism and a ring of 180 bolts. Once both SRBs are in place, the ET is mounted between them. The assembly is now ready for the orbiter. This is prepared in the Orbiter Processing Facility (OPF), a new building alongside the VAB that can 'turn around' two orbiters simultaneously. Payloads can be exchanged and an orbiter can be stripped down and rebuilt, as necessary, to refurbish it for flight, but all processing must be done in a horizontal configuration. Once ready, the orbiter is towed into the VAB, where a crane lifts it, swings it vertical, and then mates it with its ET.

Shuttle payloads are stored in Operations and Checkout (O&C) Building several kilometres south of the VAB. This also has living facilities for astronauts preparing for a mission. They are driven to Launch Complex 39 on the Merritt Island Launch Area in an Astrobus four hours before a launch is due. The Crawler–Transporter (CT), affectionately known as 'the tortoise', is unchanged. It drives into the VAB, retrieves the loaded MLP, and moves down the Crawlerway at a top speed of 1 kilometre per hour. Unloaded, the crawler weighs 2,750 tonnes, and is twice that when fully loaded. Once across the Banana River causeway, it either continues east for pad 39A, or turns left and heads north to pad 39B. Each metre of its 6-kilometre journey consumes a litre of fuel. At the pad, the crawler offloads the MLP onto the concrete pad structure with an accuracy of a few centimetres, and withdraws. Although the Shuttle stack is only 56 metres tall, it is fully 24 metres from the tip of the orbiter's tail to the opposite side of the ET. The lower third of the 120-metre tall tower that serviced the Saturn V was discarded, and instead of ferrying the remainder back and forth on the MLP it was mounted on the pad as the Fixed Service Structure (FSS). The swing arm with the 'white room' at its end was retained for access to the orbiter's side hatch. A new facility, the Rotating Service Structure (RSS), was added to protect the stack from the elements during its pad check-out. This incorporates a clean room to enable payloads to be installed in the payload bay after the Shuttle has been checked, and shortly before launch it swings 120 degrees away from the vehicle on an arcuate rail track. A sound-suppression system floods the pad with water to absorb the low-frequency acoustic shock that accompanies SRB ignition in order to prevent this from reflecting off the concrete, deflecting the orbiter's flight control surfaces and damaging their actuators. Nevertheless, at launch the sound level at the periphery of the complex rises almost instantaneously to 160 decibels. The Launch Control Complex (LCC) alongside the VAB, with its independent firing rooms, was modified to deal with the Shuttle's launch process. As soon as the Shuttle clears the tower, flight control is 'handed over' to the Johnson Space Center in Houston.

A completely new facility for a spacecraft was the 5-kilometre-long runway of the Shuttle Landing Facility (SLF) for landings at the Kennedy Space Center – if the weather in Florida precludes a landing there, a Shuttle will land at Edwards Air Force Base in California. In the event of an abort, an orbiter will try an emergency descent to any suitably equipped commercial airport.

GLIDE TRIALS

Although the orbiter is designed for a gliding descent, its delta wing is shaped to permit dynamic soaring during the hypersonic phase of the re-entry. As a result, the orbiter makes a much faster and steeper final approach than a conventional aircraft – ten times steeper than an airliner, in fact – and does not level out until a few seconds before landing.

To give the astronauts an opportunity to practice flying this unique glider, a Boeing 747 known as the Shuttle Carrier Aircraft (SCA) was adapted to carry an orbiter on its back. This took the orbiter Enterprise to high altitude and released it, thus enabling it to rehearse the subsonic phase of the return from orbit. This phase of the programme, called the Approach and Landing Test (ALT), was conducted by

Table 1.2 Enterprise approach and landing tests – captive flights

#	Date	Crew
1	18 Feb 1977	Unoccupied
2	22 Feb 1977	Unoccupied
3	25 Feb 1977	Unoccupied
4	28 Feb 1977	Unoccupied
5	2 Mar 1977	Unoccupied
6	18 Jun 1977	Haise and Fullerton
7	28 Jun 1977	Engle and Truly
8	26 Jul 1977	Haise and Fullerton

Table 1.3 Enterprise approach and landing tests – free flights

#	Date	Crew	Duration (min)	Altitude (m)	Landing (km/h)
1	12 Aug 1977	Haise and Fullerton	5.3	7,265	340
2	13 Sep 1977	Engle and Truly	5.5	7,878	360
3	23 Sep 1977	Haise and Fullerton	5.6	7,484	354
4	12 Oct 1977	Engle and Truly	2.6	6,788	368
5	26 Oct 1977	Haise and Fullerton	2.0	5,760	352

Fred Haise, Gordon Fullerton, Joe Engle and Dick Truly, of whom only Haise had flown in space. His flight had been on Apollo 13, which had frustrated his one and only chance to walk on the Moon. Enterprise was delivered on 31 January 1977. The programme started with a series of captive flights in which the inert orbiter remained on the back of the 747 to determine the dynamics of the combination in all phases of the SCA's flight profile. Haise and Fullerton first rode in the captive orbiter on 18 June 1977, and made their first gliding descent to Edwards on 12 August. On their final flight, on 26 October, they landed on the concrete runway.

The ALT trials were completed on schedule, but the first launch of Orbital Flight Test (OFT) was repeatedly delayed by development problems with the SSME (the lightweight, extremely high-pressure engine had a tendency to misfire) and the TPS (the 30,000 individually shaped tiles had to be put on manually, and a skilled worker could apply only two or three per week), slipping progressively from 1978 to 1981. However, it is not the purpose of this book to dwell on the protracted development of the Shuttle; rather, the focus is on its subsequent use.

A Boeing 747 was modified to serve as the 'Shuttle Carrier Aircraft', both to enable the orbiter Enterprise to make the Approach and Landing Tests (in this case with a tail cone over its engines) and later to the ferry operational orbiters across the country.

An orbital view of the Kennedy Space Center showing the Shuttle Landing Facility, the Operations and Checkout Building, the Vehicle Assembly Building and the crawlerway to the two pads of Launch Complex 39.

The crane in the Vehicle Assembly Building prepares an orbiter for mating with its External Tank and Solid Rocket Boosters, which have already been installed on the Mobile Launch Platform.

An overhead view of a stacked Shuttle. Notice the exhaust vents cut into the Mobile Launch Platform.

A crawler transports a stacked Shuttle.

A view of the Vehicle Assembly Building with the squat Launch Control Center by its side, with the crawlerway leading to the pads.

Driving up the ramp to one of the pads of Launch Complex 39. (Notice the Fixed Service Structure and the Rotating Service Structure.)

The crawler delivers the MLP.

After the crawler has withdrawn.

2

The early years

TEST FLIGHTS

In March 1978, John Young, a veteran of two Gemini and two Apollo missions, and one of the dozen men to have walked on the Moon, was named as the commander of the first orbital test flight of the Space Transportation System (STS). Although Bob Crippen was a 'rookie' in terms of flight experience, he was assigned to accompany Young because his field of expertise was the orbiter's computer system. Neither man had participated in the ALT phase of the Shuttle's development, as they both had to concentrate on the preparations to launch Columbia, the first spaceworthy orbiter of the fleet.

As it stood on the pad, the Saturn V had the majestic elegance of a monument. The Shuttle stack, in contrast, resembled a bird that was eager to take to the air. On 10 April 1981, the countdown for STS-1 incorporated a number of 'planned holds' to allow time to resolve minor problems. The last of these holds was at T–9 (that is, nine minutes of the process remained, although it was possibly much longer in real time). When the count resumed, the pace picked up. At T–31 seconds, when control passed to the orbiter, a timing fault in the General Purpose Computer (GPC) caused the sequence to be halted, leading soon after to the launch being 'scrubbed'. As the Shuttle was a fly-by-wire vehicle, its computer system was so important that it had been made multiply-redundant. Although one processor was sufficient to control the orbiter, four identical units ran in parallel to guarantee a level of redundancy even if two of the four computers malfunctioned. They were all set to run the same program at critical phases of the mission, compare their output and 'vote down' any unit that yielded a spurious result. To guard against a generic fault, a fifth processor, built by a different manufacturer and programmed independently, served in a 'supervisory' capacity. The launch had been abandoned because a timing error had prevented the various machines from interfacing properly. As this was a straightforward problem to fix, the count was recycled and the launch was reset for 12 April – a date which, incidentally, marked the twentieth anniversary of Yuri Gagarin's pioneering orbital mission. As the clock ran through the final few seconds in the count, no one really knew what to expect. There was a distinct chance that the vehicle would be lost in a catastrophic failure. After prolonged testing, the Space Shuttle Main Engine (SSME) had proved sufficiently reliable for operational use, and Columbia's engines had been fired for 20 seconds during the Flight Readiness Firing (FRF) on 20 February 1981. However, the SRBs – the most powerful solid rockets ever built – had been tested while mounted horizontally on a frame, rather than standing upright, and had never been launched on test flights, so the first time that their casings would be subjected to the stress of flight would be during the launch of STS-1. This mode of engineering development, called 'all up' testing, made the Shuttle the first US rocket to carry a crew on its first test flight. It was a risky strategy.

At T–31 seconds, the orbiter's internal sequencer successfully took command. Starting at T–6.6 seconds, it ignited the trio of SSMEs at 120-millisecond intervals, ran them up to full power, and assessed their performance for any indicator heading out of tolerance. If the ultra-high-performance hydrogen-burning engines survived the first few seconds, they could be expected to operate smoothly for the ascent to orbit. Although the orbiter's engines had ignited, it did not represent a commitment to launch, because the stack remained bolted to the pad. Young and Crippen felt the orbiter vibrate as the SSMEs started, and they felt the pitching motion as the thrust pushed the stack forward on its pedestal. Known as 'the twang', this oscillation had an amplitude of approximately 2 metres, and a period of several seconds, which meant that the stack had yet to return to vertical when the count reached T=0 and the command was sent to start the SRBs. On this signal, an igniter fired at the top of each motor, blasting a jet of extremely hot flame down the length of the motor's core to initiate solid-propellant combustion and in an instant the SRBs more than tripled the thrust. This signal also fired the pyrotechnics in the massive bolts that clamped the SRBs to the MLP to release the stack, because, one way or another, the Shuttle *was* about to leave the pad. In contrast to the slow and stately manner in which the Saturn V lifted off, within seconds the Shuttle had cleared the tower. Nothing had prepared the astronauts for the 'feel' of the ascent. Young was a man of few words, but Crippen, riding a rocket for the first time, exuberantly exclaimed: "What a ride!"

The stack weighed 2,000 tonnes, which was 1,000 tonnes lighter than a Saturn V. As the Shuttle's 3,100 tonnes of thrust was only marginally less than that of the Saturn V, the acceleration was much more pronounced. As soon as the Shuttle had cleared the tower, it rolled onto the desired azimuth, and as it climbed it pitched back to head out over the Atlantic. As it accelerated towards the speed of sound, and the aerodynamic stresses on the stack reached their peak, the SSMEs throttled back to 65 per cent and the SRBs – by virtue of the configuration of their propellant – reduced their thrust. This phase lasted for some 20 seconds after which the thrust was rapidly piled back on. The SRBs exhausted their propellant after 2 minutes, and were jettisoned at an altitude of 45 kilometres. After following a ballistic arc, the boosters made a parachute descent and splashed down in the ocean, to be recovered, refurbished, and reused. The SRB-phase of the ascent had been essentially a vertical climb out of the atmosphere. As the orbiter continued to ascend, it pitched further back to start to increase its horizontal speed. As it consumed the ET's propellant, the Shuttle rapidly accelerated. When it reached 3 *g*, the SSMEs were progressively throttled to preclude imposing excessive structural loads on the vehicle. Eight and a half minutes after launch, the SSMEs were shut down. After Main Engine Cut-Off (MECO), pyrotechnics jettisoned the spent ET, which re-entered the atmosphere over the Pacific and was destroyed. Columbia briefly fired its Orbital Manoeuvring System (OMS) engines to avoid the same fate.

Prior to launch, Young is said to have opined that there was only a 50/50 chance of achieving orbit and returning to Earth without a serious mishap. The ALT flights had proved the orbiter's aerodynamics in the subsonic phase of the descent profile, but as the only way to assess its hypersonic flight characteristics was by computer extrapolation of wind tunnel data, the plasma-induced radio blackout that would accompany Columbia's re-entry promised to be the most worrisome part of the flight for the flight controllers. Their anxiety was heightened when the payload bay doors were opened to expose the radiator that each carried on its inner surface, and it became evident that several of the tiles that were to protect the orbiter from the thermal stress of re-entry had shaken loose from the OMS pods on each side of the vertical tail housing. It was later discovered that although the pad had been flooded with water in order to suppress the reflection of the acoustic shock accompanying engine ignition, there had been a miscalculation and the shock had knocked off some of the tiles. The most crucial areas were the nose, the leading edges of the wings and the belly. It was impossible to tell whether these sections had been damaged. Even a few missing tiles could spell danger, because they might peel off like a zipper to expose a long strip of the aluminium skin beneath, in which case the vehicle would be unlikely to survive re-entry. Columbia reportedly turned to face its belly towards an imaging reconnaissance satellite to enable this risk to be assessed, but this historic image has not yet been released.

With Columbia in its in-orbit configuration, Young and Crippen gave a televised tour of their split-level cabin. They portrayed the cavernous payload bay, which on this occasion was almost empty, tracked the Earth drifting by, and showed Crippen, in space at last, enjoying some weightless acrobatics. What struck the viewers most, however, was that the commander of the world's first true spaceship monitored his flight deck controls wearing spectacles.

After two full days in space, Columbia was oriented tail-forward for the de-orbit manoeuvre which was performed above the Indian Ocean by the OMS engines. This caused the trajectory to dip into the atmosphere on its next low pass, half a world away. Meanwhile, Columbia pointed forward and raised its nose so that the thermal stresses of re-entry would be borne by its nose, the leading edge of its wing and its belly. The re-entry profile was straightforward – no attempt was made to evaluate cross-range performance. As the vehicle dug deep into the atmosphere at Mach 25, Young and Crippen, through the flight deck's large wrap-around windows, watched the sky turning from the intense black of space to a fluorescent pink as the tenuous air of the upper atmosphere was compressed and became an ionised shockwave that blacked out all radio communication for 11 minutes. Throughout the descent, the 3-axis accelerometers of the Aerodynamic Coefficient Identification Package recorded the orbiter's hypersonic and supersonic flight characteristics in order to verify the predictions of the computer models. On reaching subsonic speed, Columbia flew the ALT-proven gliding descent and made a perfect landing on the dry lake at Edwards.

Crippen was as enthusiastic as ever: "What a way to come to California!" After Columbia had come to rest, the fleet of recovery vehicles set about safing the orbiter. Only after the aft bay had been vented of propellant fumes were the crew allowed to disembark. As he bounded down the stairway, Young, the laconic veteran, displayed a broad grin and punched his clenched fist into the air. "From a pilot's standpoint," Crippen concluded, "you could not ask for a more superb flying machine."

It had been hoped to launch the second flight in September but the date slipped, and on 9 October a spillage while pumping propellant into the orbiter's forward Reaction Control System (RCS) meant that the adjacent tiles had to be replaced. The next countdown on 4 November was scrubbed by a faulty fuel cell. Auxiliary Power Unit (APU) problems prompted the abandonment of the attempt on 11 November, but things went smoothly the next day and the age of the reusable spaceship became a reality. This time Columbia was flown by Joe Engle and Dick Truly, and when they opened the bay doors they were delighted to see that there was no sign of damage to the tiles. The upgraded sound-suppression system had reduced the shockwave to an acceptable level. This time, the bay had much more than an instrumentation package. NASA's Office of Space and Terrestrial Applications had supplied a pallet of instruments to study the Earth, including a carbon monoxide atmospheric monitor and a prototype of a terrain-mapping radar. However, the most intriguing item was the Remote Manipulator System (RMS). Many tasks on future flights would require this 16-metre-long, triple-jointed Canadian-built 'arm', which was remotely controlled from the aft flight deck. It had both automatic and manual modes of operation. Although it suffered a few teething problems, the faults were rectified and it was successfully put through its paces. It had been difficult to test on Earth because it was designed for use in space and could not actually support its own weight in terrestrial gravity. In return for supplying the arm, Canada's own astronauts were to be allowed to fly on the Shuttle. An ambitious five-day flight had been intended, but this was cut to two days when a fuel cell had to be shut down. This technology had been developed for Gemini, then used for Apollo. It combined cryogenic hydrogen and oxygen in a catalyst to make electricity, with water (which, if not required, was periodically vented overboard) as a by-product. If illuminated by the Sun, such a water dump was a spectacular sight. Although there were three fuel cells, because they were the only source of power the mission rules required the curtailment of a flight if one failed. In this case, one had become flooded with water, and the early return prompted the cancellation of the plan to test the new spacesuit by having Engle don it and depressurise the airlock.

It was originally intended to make six test flights before declaring the Shuttle to be 'operational', but even before Columbia made its maiden flight this test phase had been reduced to four flights, and when STS-2 returned NASA announced that the remaining tests would take place in March and June 1982. When STS-3 launched on the appointed date, it was taken to be an omen that the programme would indeed be able to stick to a formal schedule. Fred Haise had retired, so Jack Lousma and Gordon Fullerton flew Columbia for its third outing. They used the arm to lift the 160-kilogram drum-shaped Plasma Diagnostics Package off its bay mount and held it out to measure the physical properties of the ionosphere in the immediate vicinity

Columbia is launched for STS-1.

John Young at the controls of Columbia on STS-1.

Columbia touches down on the dry lake at Edwards Air Force Base.

After rolling to a halt on the dry lake, Columbia awaits the attention of the recovery vehicles.

Columbia is prepared for relaunch as STS-2.

of the orbiter. A fault in the arm pre-empted the plan to sweep the far larger Induced Environment Contamination Monitor around the bay to assess the extent to which outgassing 'polluted' the environment in which instruments carried in the bay would be called upon to operate, and it was limited to making *in-situ* measurements. It was while studying the extent to which the orbiter interacted with the ionosphere that the effect promptly dubbed 'Shuttle glow' was first noted. To assess the vehicle's thermal characteristics, Columbia was held in various orientations with respect to the Sun in order to verify that the thermal protection tiles obviated the requirement to establish a 'barbecue' roll – as had been done with earlier spacecraft – to even out thermal stresses. This indifference to solar heating meant that an orbiter could safely spend extended periods in arbitrary orientations, serving as a science platform.

The excitement on STS-3 came at the end of the mission. Half-way through the planned seven-day flight, it became clear that the weather at Edwards would have degraded to such an extent that it would not be practicable to land there. The Kennedy Space Center's Shuttle Landing Facility (SLF) was not a viable option, as it was a 100-metre-wide strip of concrete, and at this stage in the test programme the vast expanse of a dry lake was considered to be essential for safety. The only other immediately available desert site was at White Sands in New Mexico. A landing at White Sands would pose no problem for the orbiter, but there was no recovery equipment there to tend to it, so, during the final days of the mission, two special trains transported the recovery crew's vehicles from Edwards to Northrup Field at White Sands. It was a hectic time on the ground, but in space life aboard Columbia had become routine, and studies of plant and insect adaptation to microgravity were underway. Unfortunately, an hour before the scheduled de-orbit time, a sandstorm swept across White Sands, and John Young, flying a chase plane overhead, ordered a 24-hour postponement. Columbia landed without incident the next day. It would not be the last mission to be extended to wait out poor weather. This time, a high-flying C-141 Starlifter aircraft equipped with a thermal imaging system was able to monitor Columbia as it began its re-entry, and this data confirmed the predicted distribution of thermal stress across the orbiter's belly.

Ken Mattingly and Hank Hartsfield took Columbia up for the final test flight exactly on schedule. The event was marred only by the fact that faulty parachutes prevented the SRBs from being recovered. With missions seemingly routine, the Press focused its attention on the 'classified' payload carried in the payload bay for the Department of Defense. Although it was not revealed until much later, the cover on the aperture of the CIRRIS missile-tracking telescope failed to release. However, this time the arm was able to assess the extent to which outgassing polluted the bay using the Induced Environment Contamination Monitor, and Mattingly was able to test the airlock in order to clear the way for the spacewalk assigned to the next flight. A series of Programmed Test Inputs (PTI) during re-entry initiated the process of 'stretching the envelope' by assessing the orbiter's hypersonic characteristics. The nose was pitched down to reduce the angle of attack in order to assess the increased thermal stress. For the first time, Columbia landed on the concrete at Edwards. On this Independence Day, half a million people had driven to the high desert to witness the conclusion of the test series, among them President Ronald Reagan, who declared the National Space Transportation System to be operational.

Table 2.1 Orbital test flights

Flight	Orbiter	Launch	Return	Objective
STS-1	Columbia	12 Apr 1981	14 Apr	Instrumentation
STS-2	Columbia	12 Nov 1981	14 Nov	Instrumentation; mapping radar
STS-3	Columbia	22 Mar 1982	30 Mar	Instrumentation; science experiments
STS-4	Columbia	27 Jun 1982	4 Jul	Instrumentation; missile tracking

This decision was criticised by some as premature. They pointed out that a new aircraft normally makes hundreds of flights prior to entering service. Others likened the Shuttle to the X-15 rocket plane with which NASA had investigated hypersonic flight. Although the X-15 had made almost 200 flights, it was a research aircraft and had no operational rôle. The Shuttle, on the other hand, was specifically intended to provide a transportation service to and from low orbit and, although it was a great advance in the state of the art, for that reason it could not possibly make hundreds – or even dozens – of test flights. A fairer comparison would be the Saturn V rocket, which sent astronauts out to orbit the Moon on only its third flight. The decision to declare the Shuttle operational was political, and this cleared the way for commercial operations. The main effect on operations was the increase in the crew complement, and the resultant removal of the ejection seats that had offered the two-man crews a means of escape during the initial phase of the ascent. From now on, in the event of a problem during the ascent, the astronauts would be committed to accompany their vehicle to its fate. On the other hand, NASA had investigated abort options to give the orbiter the best possible chance of survival, so as Reagan's words echoed across the desert, NASA was looking forward to increasing the pace of operations to fulfil its promise of providing routine and cheap access to low orbit.

THE DREAMTIME

The last two test flights had lifted off on schedule, and NASA had high hopes that it would be able to build up the flight rate and expand its operations by pursuing independent programmes. The flight schedule – or the 'manifest' as NASA preferred to refer to it in order to convey the impression of a commercial transport operation – interleaved these many strands of activity.

It had been intended to send the first Tracking and Data Relay System (TDRS) satellite up to geostationary orbit on the first operational flight, but a commercial satellite deployment mission had been given priority for Columbia's fifth flight. As expected, it took off exactly on time, on 11 November 1982. In addition to the flight crew of Vance Brand and Bob Overmyer, mission specialists Joe Allen and Bill Lenoir rode in removable seats at the rear of the flight deck. The first spacewalk of the programme had to be cancelled because the new extravehicular suits were faulty, but the main objective of dispatching a pair of HS-376 commercial communications satellites was achieved without incident. Each satellite was successfully inserted into

Table 2.2 Shuttle manifest, c. mid-1982

Flight	Date	Orbiter	Objective
STS-5	Nov 1982	Columbia	SBS, Anik
STS-6	Jan 1983	Challenger	TDRS-A
STS-7	Apr 1983	Challenger	SPAS-1; Palapa; Anik; OSTA-2
STS-8	Jul 1983	Challenger	TDRS-B; Insat
STS-9	Sep 1983	Columbia	Spacelab 1
STS-10	Nov 1983	Challenger	Classified; cancelled
STS-11	Dec 1983	Columbia	LDEF; SolarMax repair
STS-12	Jan 1984	Discovery	TDRS-C; Palapa-B2
STS-13	Mar 1984	Challenger	Classified
STS-14	Apr 1984	Columbia	OAST; Anik; RCA
STS-15	May 1984	Discovery	TDRS-D; Westar
STS-16	Jun 1984	Columbia	SBS; Leasat
STS-17	Jul 1984	Challenger	Spacenet; Westar
STS-18	Aug 1984	Discovery	Arabsat; Telstar; Leasat
STS-19	Sep 1984	Columbia	Spacelab 3
STS-20	Oct 1984	Challenger	OSTA-4; RCA; Intelsat
STS-21	Nov 1984	Columbia	Spacelab 2
STS-22	Nov 1984	Discovery	Classified
STS-23	Jan 1985	Challenger	HST; LDEF retrieval
STS-24	Jan 1985	Columbia	OSS; Arabsat; Intelsat
STS-25	Feb 1985	Discovery	Classified
STS-26	Apr 1985	Atlantis	Spacelab D1
STS-27	Apr 1985	Columbia	Classified
STS-28	May 1985	Challenger	Telstar; RCA; GStar; Leasat
STS-29	Jun 1985	Atlantis	OAST; Satcom; Spacenet
STS-30	Jul 1985	Columbia	Classified

geostationary transfer orbit by a small Payload-Assist Module (PAM) and used its own motor for the circularisation burn to take up its assigned geostationary position. It had been planned to attempt a crosswind landing at Edwards, but the salt flat was waterlogged and a normal landing was made on the runway instead. The landing was so smooth that Overmyer was prompted to ask the pilot of a chase plane to verify that the gear was running along the concrete. After this remarkable mission, Columbia was sent back to Rockwell's factory near Los Angeles to be refitted, and a new orbiter – Challenger – was readied for service.

STS-6 was set for 20 January 1983, but the FRF on 19 December revealed a hydrogen leak in one of the SSMEs. The follow-up firing on 25 January revealed leaks in all three engines, so the launch slipped to April. This double delay was a disappointing disruption of a programme that had begun to seem routine. However, once the engine problems had been overcome, Challenger was dispatched without incident. To begin the process of 'stretching the envelope' of engine performance to satisfy the Air Force's demand that the Shuttle deliver the promised mass-to-orbit capacity, after maximum aerodynamic pressure (known as Max-Q) as the stack went supersonic, the SSMEs were throttled up to 104 per cent of their nominal thrust.[1] It was argued by critics that this risked the crew's lives, but incremental development was standard practice in the aerospace industry, and for a craft as expensive to operate as the Shuttle there was no viable alternative. The primary objective was to deploy the first of the TDRS satellites. Since the PAM was limited to 1.2 tonnes of payload, the 2.2-tonne TDRS required the more powerful two-stage Inertial Upper Stage (IUS). The folded-up package was mounted on the IUS, and this stack was carried lengthwise in the bay. The annular tilt-table of the deployment system raised the 17-tonne stack and spring-ejected it across the top of the orbiter's cabin. The IUS fired its big engine on time to enter geostationary transfer orbit, but when the second-stage motor attempted to make the circularisation burn it malfunctioned, stranding its payload in a useless orbit. Although the satellite was eventually able to manoeuvre onto its assigned station, the problem with the IUS threw the programme into chaos. Meanwhile, Story Musgrave and Donald Peterson made NASA's first spacewalk since the heady days of Skylab. During re-entry, Challenger made further manoeuvres to explore the orbiter's hypersonic flight characteristics.

The IUS had been developed by the Air Force specifically to deliver its large Shuttle-deployed satellites to geostationary orbit, and it was intended to be the most reliable inter-orbit transfer stage ever made. Its flight-control systems were multiply-redundant, yet had been crippled by a straightforward mechanical fault. It could not be used until the problem had been rectified, but that did not mean that the Shuttle was out of business. Grounding the IUS simply meant that the payloads that needed it had to be pushed down the manifest, which in turn provided slots for commercial and scientific payloads. While this complicated the process of payload integration, NASA soon proclaimed one of its strengths to be dynamic scheduling. It had been hoped to fly Challenger again in June and in August – the latter to deploy the second TDRS satellite – and then, with two relays in service, Columbia would resume flying in September with the first Spacelab, after which it would carry the first satellite for the Air Force in October. The second TDRS and the classified satellite, which would also require an IUS, were immediately withdrawn from the manifest, and Spacelab was placed on hold, pending the rescue of TDRS-A. The Spacelab science activities would be degraded without two relays, but NASA was reluctant to cancel this key mission, with its international crew. It was difficult to increase the flight rate as long as there was only one serviceable orbiter, as this was determined by the turnaround time, and although a management team had been created to streamline this process, there was only so much slack in the system. Nevertheless, allowing for the delays in its launch as STS-6, Challenger was launched on STS-7 more or less on schedule. It successfully deployed a pair of commercial satellites and a German-built free-flyer. The retrieval of this Shuttle Pallet Applications Satellite (SPAS) involved a rehearsal of the rendezvous that was later to be used to retrieve the ailing SolarMax satellite. In commanding this mission, Bob Crippen became the first astronaut to make two

[1] The 100 per cent level was simply the benchmark in the specification, not the engine's maximum performance. The goal was to work up to 109 per cent, because this would be required to lift heavy military payloads into polar orbit.

The first view of an orbiter in space was provided by the SPAS free-flyer when it took this picture of Challenger during the STS-7 mission.

flights on the Shuttle. He had hoped to make the first landing at the Kennedy Space Center, but weather constraints forced him to divert to California.

With Spacelab on hold while TDRS-A slowly manoeuvred towards its station, it was decided to fly STS-8 on schedule, with the Payload Flight Test Article instead of TDRS-B. After the successful deployment of an Indian satellite, the RMS hoisted this ballasted dumb-bell framework of approximately the size and mass of a large satellite, and swung it about to assess the stresses on the arm's joints. Although the arm performed satisfactorily with this 3.3-tonne mass, it flexed alarmingly when the orbiter manoeuvred. The orbital requirements of the Indian satellite had necessitated the first night launch, which was spectacular, and this, in turn, mandated the first landing in darkness, which was assigned to Edwards for the increased margin of safety. After the SRBs had been recovered NASA made a shocking discovery: a wide segment of the 75-millimetre-thick insulation that lined the aft nozzle of one of the boosters had been almost completely eroded by the exhaust plume. This was the first significant damage to an SRB. This was attributed to a faulty batch of material. In fact, this had been the first use of an uprated booster that, by being 4 per cent more powerful, facilitated a 1.4-tonne increase in the Shuttle's payload. The erosion of the SRBs was to become a matter of some concern, not least because there was so much variation in the damage.

As soon as it became clear that the solitary TDRS satellite would be on-line in October, John Young was given the go-ahead to fly the STS-9 Spacelab mission, but Columbia's launch was slipped by a month while the SRB damage was studied. The European Space Agency's pressurised laboratory module, with its external pallet of instruments, was the heaviest payload to date and took up the entire bay. It was a mission of firsts:

- the first to fly payload specialists;
- the first to fly a foreign astronaut (Ulf Merbold, representing the European Space Agency, was a German);
- the first to use an orbit inclined at 57 degrees to the equator for better latitude coverage;
- the first to operate a shift system to facilitate 24-hour operations; and
- the first to exploit the orbiter's maximum duration.

In fact, the low rate at which the cryogenics were consumed allowed the flight to be extended for a day, and there was therefore very little margin when Young prepared Columbia for its de-orbit burn. He subsequently admitted that he "turned to jelly" when, after firing an RCS thruster to align the vehicle, a shock propagated through the orbiter's structure and the primary GPC dropped off-line. Another processor promptly took over, but it too shut down. This was the programme's first in-flight crisis. There was concern that there might be a generic fault in the GPC system, and it took seven hours to reset the computers for the de-orbit burn. The radio blackout during the initial phase of re-entry was the tensest moment in the programme since Young had flown Columbia back after its first test flight. The flight controllers were undoubtedly greatly relieved to see it make a textbook landing on the dry lake, not just because of the in-orbit problem but also because the laboratory was the heaviest payload to be returned so far, and its presence moved the vehicle's centre of mass significantly forward and affected its landing characteristics. What no one was aware of was that while Columbia had been lining up for its approach, a hydrazine fire had flared in the aft compartment containing one of the APUs – the electrical generators that provided the power to operate the aerodynamic control surfaces. The fire was discovered by the recovery team while venting fumes from the aft bay. An analysis revealed that it was due to microscopic débris contaminating the circuitry. Although an orbiter required only one APU, it would be doomed if it were to be denied power to move its control surfaces at such a critical time. Fortunately, the APUs had been physically isolated in order to minimise the risk of all three units being destroyed. It had nevertheless been a narrow escape.

Even without the second TDRS relay, the 10-day Spacelab had downlinked more science data than Skylab had during its six months of occupancy. It had been thought that a complete TDRS network was a prerequisite to such missions, but recertifying the IUS was clearly going to be a protracted process. The network would probably not be able to be expanded until 1985. This was bad news for the Air Force, because its most modern satellites had been specifically designed for the Shuttle and were too large to be off-loaded onto its rapidly depleting stock of Titan III rockets.

The long-term future for the Shuttle programme received a boost in January 1984 when Ronald Reagan ordered NASA to build a space station "within a decade". This was exactly what NASA had been hoping for. It not only gave the Shuttle a new strategic mission – the construction of the orbital facility – but it also provided a destination: a space community to serve. The space station, however, was a project for the future; the overriding objective for 1984 was to increase the flight rate.

So many flights had been deleted from the manifest that NASA abandoned the original straightforward numerical series for identifying flights. Instead, it introduced a scheme based on the Financial Year (FY) to which its costs were attributed. Flights were identified by an alphabetic designator starting with the final digit of the FY. To account for flights from California, launches from the Kennedy Space Center were designated by '1' and those from Vandenberg Air Force Base by '2'. The sequence ran through each year, starting at 'A'. This new scheme was introduced for FY84, which ran from October 1983 to September 1984. Although STS-9 had flown after the start of FY84, it fell within the FY83 budget. However, when the Air Force decided to cancel its flight rather than assign it a non-IUS payload, the 41A label was rendered void, so the new scheme got off to an unfortunate start, and not just by missing out its first designator; there were problems in space too.

In February 1984, STS-41B released two commercial satellites but both PAM stages fizzled out within seconds and stranded their payloads in low orbit. Then the balloon released to serve as a radar target to rehearse rendezvous procedures in preparation for the SolarMax repair inflated so rapidly that it ripped itself to shreds. Next, a fault in the RMS prevented the release of the SPAS – the first free-flyer to be refurbished and reflown. However, the public imagination was caught when Bruce McCandless flew 100 metres from Challenger to test the Manned Manoeuvring Unit

Although no one was aware of it, as Columbia landed after STS-9 with Spacelab 1, there was a fire raging in its aft compartment.

The Westar 6 communications satellite is deployed by STS-41B.

(MMU). This was the first time an astronaut had, if only temporarily, become an independent satellite. Challenger landed at the Kennedy Space Center. This historic roll-out was marred by a brake seizure as the vehicle tried to remain on the centreline of the 100-metre-wide concrete strip; however, returning to Florida would reduce the turnaround time by a week. The spectacular failure of the PAM rocket stages stalled the commercial satellite aspect of the manifest. However, although the lightweight and heavyweight stages were both now grounded for recertification, NASA advanced another strand of its multifarious manifest.

In April, Bob Crippen took Challenger up to rendezvous with SolarMax to carry out the main feature of the early part of the programme: the repair of a satellite. To reach SolarMax, the Shuttle had to climb to 560 kilometres, the highest operating altitude to date. Its launch marked the first time that a Shuttle flew a direct ascent trajectory. In previous cases, the Shuttle had settled into a low initial orbit and then used a series of OMS burns to raise its altitude to about 300 kilometres. In this case, however, the OMS engines were fired immediately after the ET was released in order to yield a high initial apogee, marginally below SolarMax's orbit, at which point, around 45 minutes after launch, the OMS-2 burn was used to circularise at that level. This ascent profile enabled the LDEF to be deployed into a high orbit, safe from decay for the year or so it was intended to remain in space, and made the rendezvous with SolarMax a straightforward operation. By now all the aspects of the SolarMax repair that could be tested had been rehearsed in space, so it was expected that the recovery would go smoothly. However, the apparatus that was specifically built to mate with a grapple-fixture on the satellite refused to engage, and in attempting to grasp the satellite manually George Nelson induced a rotation that sent it spinning out of control. Without sunlight on its solar panels, the satellite switched over to its battery, but because this could only support a few hours of operation it began to look as if SolarMax would be killed off rather than repaired. However, the ground controllers were able to stabilise the satellite, and Challenger returned two days later. This time Terry Hart used the RMS to grab the satellite and set it down on a tilt-table at the rear of the bay. From this point, the repair operation went smoothly, and the Public Affairs Office slipped into overdrive. Crippen had hoped, finally, to be able to land in Florida, but unacceptable weather there forced a diversion. His earlier enthusiastic remark, "What a way to come to California!" was haunting him.

Discovery's debut mission as STS-41D – the twelfth flight of the programme – started as an exercise in frustration. In fact, it took three countdowns to get the new orbiter off the ground. After a perfect FRF on 2 June, NASA set the launch for 22 June, postponed it to 25 June, and scrubbed the third attempt at T–9 minutes due to a GPC fault. The next day, the count ran smoothly, down to the point at which the SSMEs were to ignite at 120-millisecond intervals. The hydraulically activated fuel valve of the first engine failed to open, but before the misfire was diagnosed, enabling the GPC to intervene, the second engine had fired and had already reached 20 per cent thrust. It was immediately ordered to shut down and ignition of the third engine was inhibited. The flame fizzled out and the billowing cloud of steam rapidly dispersed, leaving the Shuttle stranded on the pad. This was the first time a launch had been scrubbed following engine-start (an event that NASA dubbed a post-ignition abort). Although spectacular, this was not an unprecedented situation, because the orbiter was in essentially the same state as after a FRF test. Forty-five minutes later, Hank Hartsfield's dejected crew disembarked. It subsequently emerged that a very hot colourless flame had persisted at the base of the stack for several minutes as free hydrogen burned off. As the engines that had fired had to be refurbished, the launch had to be pushed back by a month. The FY84 schedule had slipped so far behind that NASA had already cancelled mission 41E; it now cancelled 41F and added one of that mission's commercial satellites to Discovery's manifest, making the revised 41D the first Shuttle to be assigned a trio of satellites. The attempt to launch Discovery on 29 August was foiled by a timing fault in the orbiter's master event controller, but the next day, after a brief delay for a private aircraft to clear the restricted airspace around the Kennedy Space Center, 41D was finally dispatched. In orbit, everything went well. Two of the satellites released rode the recertified PAM stage and, to everyone's relief, were successfully inserted into geostationary transfer orbit. The third satellite, an HS-381 built specifically for the Shuttle, incorporated its own kick-motor. Immediately after the de-orbit burn, the active APU developed a hydraulic leak, so the backup, which was already running in standby, was used to drive the aerodynamic surfaces during re-entry. The descent was routine until the nosewheel made contact with the Edwards dry lake, at which point the orbiter slewed to the right. Hartsfield applied the rudder to straighten up and run parallel to the centreline. When he finally applied the brakes, the orbiter rocked back and forth with a frequency of several times a second, prompting him to suspect that one of the brakes was repeatedly locking and releasing. It was found that the shock absorber of the right gear had lost nitrogen pressure and the 'flat' mechanism had repeatedly bottomed out, causing that side of the vehicle to bounce. The fault had developed precisely as the wing passed through horizontal to lower the nose gear onto the runway, just as the pressure on the wing transitioned from up-lift to down-force. This roll out, the trickiest yet, gave cause for concern, because although the runway at the Kennedy Space Center was long, it offered little room for lateral manoeuvring.

The cancellation of 41E and 41F, and 41D's successful deployment of three satellites, meant that the thirteenth flight, 41G, effectively restored the year's schedule. With two orbiters in service, it was also possible to accelerate the pace: 36 days after Discovery's landing, Challenger was launched at the first attempt. For this science mission, the initial crew was augmented at the 'last minute' by two payload specialists and their names had to be sewn in an arc around the lower edge of the already prepared crew patch. Although Crippen had been assigned command of 41G, this had been in the expectation that he would have a year following 41C to prepare his new crew, but 41G had been advanced by six months and much of the responsibility for preparing this mission fell to his rookie pilot, Jon McBride. The main event was the spacewalk during which the procedure that was to be used to replenish the Landsat 4 Earth-resources satellite was rehearsed. Much of the flight was spent on radar mapping, but this became an exercise in frustration because the orbiter's TDRS antenna, which was a dish on a short boom projecting over the starboard sill just aft of the cabin, ceased to follow the relay satellite. Since the data

Judy Resnik sends a poignant message on STS-41D.

could not be relayed in real time, the only option was to lock the antenna in position and repeatedly reorient the orbiter, first to scan the surface and then to point the stuck antenna at the satellite to download each successive strip of data. As a result, instead of the planned 50 hours, only 9 hours of data could be returned in the time assigned to the experiment. When Crippen finally managed to land in Florida, the CapCom teased him with the retort, "But Crip, the beer's been sent to Edwards!"

For the first flight of FY85, flight 51A, NASA set out on an ambitious rescue mission. After it had dispatched another two communications satellites, Discovery was to retrieve the two HS-376s that had been stranded in low orbit in February when their PAM stages had failed. This would be a much more difficult task than SolarMax, not because there were two satellites, but because these satellites had not been designed to be manipulated in orbit. Nevertheless, after discarding a handling bar that would not fit, both satellites were securely stowed in the bay. It was a *tour de force* for spacewalkers Joe Allen and Dale Gardner.

The next flight was shrouded in mystery. This first Department of Defense mission was so classified that even the imminence of its launch was not announced until the clock picked up at the end of the T–9 minute hold; the secret had obviously been kept because, for once, there were no Soviet 'trawlers' offshore. With the exception of the 8-minute climb to orbit, all communications were encrypted. The Air Force reluctantly let NASA release occasional, rather bland, progress reports. No video downlink was released. Over the years, the media had grown accustomed to NASA's openness, so the 'closed door', which generated resentment, became the focus of attention. In the revised FY85 manifest, 51B had been pushed back, so this military flight, the fifteenth of the programme, was 51C. Upon the recertification of the IUS, the Air Force had exercised its authority and requisitioned a schedule slot in the interest of national security. In effect, this was the 'STS-10' mission. The crew was the same as had originally been assigned, but with the late addition of Gary Payton, who became the first 'spaceflight engineer' to fly; his rôle was to supervise the activities involving the classified payload. The Air Force intended to have one of its own officers on each classified mission in order to avoid revealing data about the payload to the NASA astronauts, even if they had previously been in the military. When the IUS's first-stage underperformed, the controller burned the aft-pointing thrusters to make up the shortfall in velocity to allow the second stage to deliver the secret satellite. With its satellite away, Discovery landed. Lasting only three days, it was the shortest flight since the Shuttle had been made operational, but it reflected the Air Force ethos: get airborne, carry out the mission, and return to base with the minimum of distractions – in other words, operational efficiency.

As NASA was equally keen to exploit the recertification of the IUS, it assigned the second TDRS satellite to the next flight and scheduled it for 7 March. A week before launch was due, however, an exhaustive analysis of communications faults by TDRS-A raised suspicions of a flaw in the design. TDRS-B was unloaded to allow it to be thoroughly inspected, and STS-51E was rolled back to the VAB. The relentless launch process immediately switched its attention to STS-51D, which was set for April. After a hold for the weather, when Discovery finally lifted off the window had only 55 seconds remaining. It deployed two communications satellites, but one, an

HS-381 Leasat, failed to activate as it left the bay. Every effort to kick-start the mechanism using the RMS failed and the slowly spinning hulk had to be abandoned. However, the real drama unfolded at the end of the mission during the first attempt to land on the SLF in a crosswind. Development was underway to install nosewheel steering, but although the mechanism was in place, the control system had not yet been certified and Karol Bobko had to apply differential braking to keep the vehicle on the centreline. The crosswind was just 8 knots, but the differential loads stressed the main gear so much that one of the 275-psi tyres blew out, another was badly eroded, and one of the brakes seized. Luckily, Discovery rolled to a halt a few seconds later. If the blow-out had occurred earlier, the orbiter could all too easily have veered into the scrub adjacent to the strip. With hindsight, it was clear that expecting to land on such a narrow runway in a crosswind prior to perfecting the nosewheel steering system had been inviting disaster. The orbiter's brakes had been causing concern for some time as they had shown much greater wear than expected. There were four brake units on each main gear. Weight had been a key factor in their design. A carbon-coated beryllium stator disk was developed to combine low mass with high efficiency, but its size meant that it could not sustain heavy braking. NASA demoted the SLF to emergency-only status, and ordered future missions to land at Edwards. Until the nosewheel steering could be introduced, no crosswinds would be allowed; this constraint was viable because there were a dozen runways on the dry lake, at all azimuths and there was room to run wide. As events transpired, however, it would be five years before another Shuttle landed in Florida.

The countdown for the seventeenth Shuttle flight – the postponed 51B mission – was held up for two minutes to check that an oxygen drain valve was properly set, so Challenger launched a mere 10 days after Discovery's return. This was the first flight to attract a group of protesters as the Spacelab carried two squirrel monkeys, which Animal Rights considered to be cruel. Despite this added interest, 51B was the first Shuttle launch not to be broadcast live by the television networks. The only excitement during the ascent was the shutdown of one of the three APUs. The crew operated two shifts to run materials and life science experiments, and the landing at Edwards was absolutely textbook. By mid-1985, the Shuttle programme had reached the point where it was unlikely to raise any great interest unless it offered something new and dramatic. As missions had become routine, the Shuttle was no longer news. In fact, it is doubtful if anyone was aware that the missions were being flown out of their manifested order.

When STS-51G was launched in June it carried its own pair of satellites and the one left from the cancelled 41E, and all three were dispatched without incident. In addition, an astronomical free-flyer was released and retrieved by the RMS. This Shuttle-Pointed Autonomous Research Tool for Astronomy (SPARTAN) was conceived in the mid-1970s as a platform to be released to conduct observations and then retrieved and brought back to Earth. It would carry instruments that normally would have been developed for sounding rockets. To a community accustomed to observations lasting only a few minutes, a two-day run represented a considerable improvement. The project was managed jointly by the Naval Research Laboratory and the Goddard Space Flight Center. It had its own battery for power, thermal

STS-51D deploys the Leasat 3 communications satellite.

control and data storage, and on being oriented by the RMS and released, it undertook programmed observations, re-orienting itself as necessary with a pointing accuracy of 3 minutes of arc. On this first outing, as SPARTAN-101, it had an X-ray telescope. Although designed to be reusable, this 100-class bus was given to the Smithsonian Air and Space Museum in Washington because NASA had a much more capable bus ready. The only excitement of the mission was after landing, towards the end of the roll-out, when the left main gear sank into a soft patch of the dry lake and entrenched itself.

A delay in integrating the telescope package for STS-51F resulted in Spacelab 2 flying after the life sciences Spacelab 3. To avoid a recurrence of this embarrassing situation, it was decided to split the Spacelab programme into a number of themes, name each appropriately, and pursue them in parallel. Although the European Space Agency's preference was for multidisciplinary missions such as Spacelab 1, this change reflected NASA's desire to fly more focused missions to avoid the conflicting requirements of a mixed payload. Spacelab 2 was also the first to utilise an all-pallet configuration, and its Instrument Pointing System (IPS) carried three instruments for solar studies. The launch attempt on 12 July was scrubbed at T–3 seconds when a hydrogen coolant valve failed to close and the SSMEs – all three of which had ignited – had to be shut down. When launched two weeks later, it became the first flight to execute an abort during the ascent. After jettisoning its SRBs, Challenger continued to climb under its own power, but the centre engine began to overheat. The flight controllers monitoring the situation throughout the 104 per cent phase were relieved to see that it did not stray into the red line before it throttled back to 65 per cent for the final phase of the ascent. However, it continued to overheat and, at T+350 seconds, the GPC shut it down. By this point, with two engines, the Shuttle would have sufficient energy to achieve low orbit, so an Abort To Orbit (ATO) was ordered. This involved mission commander Gordon Fullerton turning the selector to 'ATO' and pushing a button to enact his selection. The GPC throttled up the two remaining engines to 91 per cent, and added 70 seconds to the burn to compensate. However, no sooner had the abort been initiated than a temperature rise was spotted in one of the remaining engines. At this point the controllers began to suspect a faulty sensor rather than a genuine problem with the engine, and they recommended that the crew intervene to inhibit the GPC from shutting down this engine. Unable to reach orbit on one engine, Challenger would have had to invoke a Transatlantic Abort Landing (TAL) which, in this case, was an emergency landing at Zaragoza in Spain. The final 220-kilometre circular orbit was not ideal for the observational programme planned for the telescopes on the palletised Spacelab, but any orbit was better than none, and the observing plan was hastily revised. NASA undoubtedly took comfort from the fact that the abort procedure had worked, even if in this case the term abort appeared to be overly dramatic. Initially, the IPS encountered problems because its star trackers proved to be rather too sensitive and could not identify assigned guide stars, but once reprogrammed it produced the desired pointing accuracy of 1 second of arc. Although a polarimeter was disabled after only a few hours by a fault in its power supply, it mysteriously restarted on the final day of the mission and the astronauts described its 'white light' view of the solar disk as "stunning".

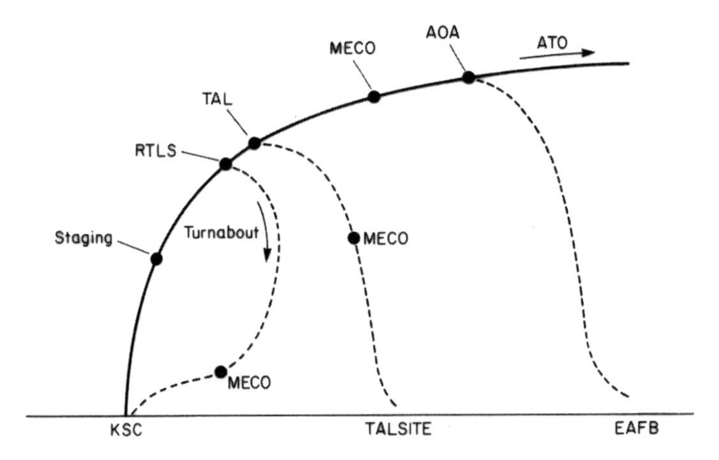

The sequence of abort options for the launch phase of a Shuttle mission: Return To Launch Site (RTLS), Transatlantic Abort Landing (TAL), Abort Once Around (AOA) and Abort To Orbit (ATO).

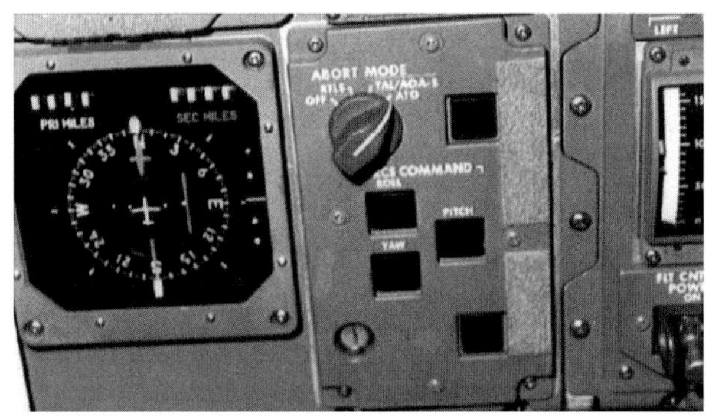

SSME problems during Challenger's ascent for the STS-51F mission prompted an Abort To Orbit.

Precisely three weeks after Challenger landed, Discovery began the programme's twentieth mission. In addition to deploying another three communications satellites, STS-51I rendezvoused with the Leasat that had been abandoned in April, and Bill Fisher and James van Hoften spacewalked to bypass its failed timer, thus allowing it to boost itself up to geostationary orbit. This rescue, undertaken as a 'bonus' at the end of an already full satellite-deployment mission, demonstrated another aspect of the way in which the Shuttle programme was maturing. Failed satellites abandoned by one flight could be retrieved, repaired and sent on their way by a later flight – an advance that provided a tremendous operational flexibility and vindicated those who had argued that repair work would be one of the Shuttle's rôles. And impromptu EVAs made crew training a dynamic process accommodating substantial last-minute tasks. Overall, NASA was proving to be quite adept at what it had inappropriately dubbed 'routine' operations.

The four-day mission of STS-51J in October – which marked Atlantis's debut – deployed an IUS with two strategic communications satellites for the Department of Defense and was "highly successful". Three weeks later, Challenger launched with a record crew of eight astronauts and a Spacelab that had been co-sponsored by West Germany. This was flight STS-61A. Orbiter propellant had been traded for payload mass, but despite exploiting gravity-gradient stability there was insufficient margin in the cryogenic consumables to facilitate an extension beyond the nominal mission, and Challenger came home after seven days. During the dry lake roll-out, the newly activated nosewheel steering was tested. In the absence of any crosswind, Hank Hartsfield steered Challenger off to one side of the centreline and then back again. The Spacelab flights had tested prototypes of the kind of scientific apparatus that could be used on the space station. The next flight, STS-61B, set off in November to test how that station should be assembled. After another trio of communications satellites had been deployed, Jerry Ross and Sherwood Spring went out and built and stripped down girder structures, their activities being recorded stereoscopically to facilitate a computer study of the efficacy of different assembly methods, in order to resolve a debate as to whether spacewalking astronauts would be able to assemble the space station manually. In parallel with the ongoing deployment of satellites, the various strands crucial to the future of the programme were being drawn together.

It had been hoped to mark Columbia's return to service on 18 December, but a variety of issues consumed all the slack built into the countdown and the launch was postponed 24 hours, and this was scrubbed at T–14 seconds by a hydraulics fault on one of the SRBs. The attempt on 6 January 1986 ended at T–31 seconds when a valve in an SSME malfunctioned. A hold at T–31 second is frustrating because the APUs cannot be kept running for longer than five minutes, and if a fault cannot be rectified in that time the launch must be scrubbed. The count on 7 January had to be abandoned due to unacceptable weather at the Spanish and West African TAL sites, and poor weather at the launch site precluded a launch on 10 January. Since the launch procedure did not call for the crew to enter the orbiter until after the ET had been loaded with cryogenics, it made sense, once they were on board, for them to sit out unacceptable weather until the window closed. On a flight to rendezvous with a satellite, the window might only be a few minutes long, but on a less time-critical mission it could last several hours. With each launch postponement, Robert 'Hoot' Gibson's crew dutifully powered down the orbiter and disembarked. They finally left the ground, without a hitch, on 12 January. Columbia had only one satellite in its bay; a second satellite, a Leasat, had been withdrawn for inspection after the loss of its predecessor and replaced by the GAS Bridge Assembly. If the launch had been delayed once more it is likely that this mission would have been cancelled and its payload added to a later flight, because Columbia had an appointment with Halley's comet in March. As it was, 61C was recalled a day early in order to make a start on this turnaround, and thus became the first flight to be cut short in the absence of any fault. With the nosewheel steering certified, Columbia had been assigned a SLF landing. However, this effort to save time was foiled by the weather, not just on the recall day but also on the originally assigned day, so after an enforced day's extension Gibson landed at Edwards, which automatically slipped the turnaround by a week.

As part of Columbia's refit, various sensors had been installed, and during this re-entry a camera mounted in a streamlined pod at the top of the tail took a series of images to record the evolution of the thermal stresses across the upper surface. Lightweight blankets were to be fitted in areas that could be shown not to require the full protection of the tiles. Various manoeuvres were performed to further refine computer models of hypersonic flight. A sensor mounted in the nosewheel bay noted the chemical composition of the plasma that came into contact with the skin of the vehicle during the early phase of the re-entry, while another set of sensors recorded the stresses on the nose cap and the leading edge of the wing, both of which bore the worst of the heating. These sensors had been conceived when the Shuttle was under development, but Columbia's refit had been the first opportunity to install them. Once the orbiter's hypersonic characteristics were fully understood, and the 'turnabout' manoeuvre involved in a Return-To-Landing-Site (RTLS) was certified, it would be possible to relax the payload-mass constraints. This was all part of the ongoing process of stretching the envelope to make the Shuttle more flexible.

Table 2.3 The first twenty operational Shuttle flights

Flight	Orbiter	Launch	Return	Objective
STS-5	Columbia	11 Nov 1982	16 Nov	Communications satellites
STS-6	Challenger	4 Apr 1983	9 Apr	TDRS satellite
STS-7	Challenger	18 Apr 1983	24 Apr	Communications satellites
STS-8	Challenger	30 Aug 1983	5 Sep	Communications satellite
STS-9	Columbia	28 Nov 1983	8 Dec	First Spacelab
STS-41B	Challenger	3 Feb 1984	11 Feb	Communications satellites
STS-41C	Challenger	6 Apr 1984	13 Apr	Deploy LDEF and fix SolarMax
STS-41D	Discovery	30 Aug 1984	5 Sep	Communications satellites
STS-41G	Challenger	5 Oct 1984	13 Oct	Environmental satellite
STS-51A	Discovery	8 Nov 1984	16 Nov	Release and retrieve satellites
STS-51C	Discovery	24 Jan 1985	27 Jan	Classified
STS-51D	Discovery	12 Apr 1985	19 Apr	Communications satellites
STS-51B	Challenger	29 Apr 1985	6 May	Microgravity Spacelab
STS-51G	Discovery	17 Jun 1985	24 Jun	Communications satellites
STS-51F	Challenger	29 Jul 1985	6 Aug	Astronomical Spacelab
STS-51I	Discovery	27 Aug 1985	3 Sep	Communications satellites
STS-51J	Atlantis	3 Oct 1985	7 Oct	Classified
STS-61A	Challenger	30 Oct 1985	6 Nov	German-sponsored Spacelab
STS-61B	Atlantis	27 Nov 1985	3 Dec	Communications satellites
STS-61C	Columbia	12 Jan 1986	18 Jan	Communications satellite

3

The Challenger accident

A MAJOR MALFUNCTION

As the 24th Shuttle mission landed on 18 January 1986, NASA's goal for the year was to ramp up the flight rate. The delay in launching STS-61C had frustrated the hope of exploiting the fact that there were now four orbiters to mount 15 missions in FY86. There was therefore great pressure to make up lost time. The next mission on the manifest, 51L, was to deploy the long-overdue second TDRS satellite. Its launch had been set for 22 January, but when it became clear that Columbia's return would be late, this was postponed to 26 January, and when poor weather was predicted for that date it was put back 24 hours. After several delays, including a crosswind that would have jeopardised a RTLS landing, the window closed and the countdown was recycled for a launch the next day, 28 January.

Although, at 28° latitude, the Kennedy Space Center is almost as far south as it is possible to go within the continental United States, the overnight temperature can plummet in winter. On the night of 27–28 January, the temperature on Pad 39B (the first time that this pad had been used in the Shuttle programme) fell so low that there was a threat to the water pipes on the Fixed Service Structure, and the valves had to be opened to allow water to flow to prevent the pipes freezing. It was so cold that this water froze on the gantry, covering its walkways with sheets of ice and adorning its structure with long icicles, but as the ice was expected to melt at sunrise this was not expected to pose a problem. However, the next morning was unusually chilly and the ice was still very much in evidence when the pad-inspection team was asked to give its approval for the launch. The issue was not the cold, it was that the sonic shockwave would loosen so much ice that it could damage the orbiter's thermal protection tiles. It was decided to hold for two hours to allow the ice to melt, which it duly did. There had been no problems with the Shuttle, so it was launched at 11:38 EST, at which time, even though it was rapidly warming, the pad was 15 °C below that of any previous launch. At Thiokol, the responsible group of engineers, led by Roger Boisjoly, had expressed doubts about the likely resilience of the rubber O-rings that would have to seal the joints of the segments that formed the SRBs, but Joseph Kilminster, the company's vice-president for boosters, was very aware of the imperative to build up the flight rate. Lawrence Mulloy, the manager at the Marshall Space Flight Center who supervised the Thiokol interface, asked Kilminster incredulously: "When do you want me to launch? Next April?" Kilminster overrode his engineers and recommended a launch. In the NASA way, a waiver was drawn up which recertified the SRBs for this new low temperature; it was seen as yet another step in the continuing process of stretching the Shuttle's operating envelope. As the twin pillars of smoky fire lifted Challenger off the pad, the decision to launch was seemingly vindicated. However, although no one was aware of it, the O-ring at the lowest field-joint of the righthand SRB had been so chilled that it had failed to seat properly and had passed a blast of hot gas. The resulting puff of dense black smoke that leaked from the side of the casing was spotted only by the high-speed cameras that filmed each launch close-in for subsequent analysis. A rapid series of smaller puffs followed as the joint flexed in response to the casing absorbing the longitudinal stress of acceleration and a succession of blasts of gas seared further into the O-rings before the joint finally sealed. Flight controllers at the Launch Control Center at the Cape, those in Houston, and Dick Scobee and Mike Smith on board Challenger, were all monitoring their displays, but there was no indication that the vehicle was in distress.

The thrust was throttled back to prevent the stack being overstressed as it went supersonic. After the aerodynamic loads had peaked, the SSMEs were increased to 104 per cent, and the geometry of the solid propellant in the SRBs increased their thrust for the long climb to orbit. The SRBs, which were due to burn out at T + 128 seconds, were barely half-way into their cycle when the leak from the righthand booster reappeared, now as a jet of flame. This blowtorch began to slice through the strut that attached the motor casing to the ET, seared the orbiter's wing, and burned through the insulation at the base of the tank itself. Almost simultaneously at T + 73 seconds, the tank ruptured and spilled its hydrogen into the roiling airflow, the GPC shut down the SSMEs, the strut gave way, and the booster, now attached only at its top end, pivoted and its nose crushed the intertank section, allowing the oxygen to spill out. As the propellants detonated in a massive fireball, the orbiter was ripped apart by the asymmetric aerodynamic forces. The visitors' stand at the Kennedy Space Center held a few stalwarts of the Press, but most of the small gathering were friends of the astronauts and this was the first launch they had witnessed. A few, believing that what they were seeing was a normal staging event, started to cheer, but this petered out as those who recognised the horrible cloud for what it was gasped in shock. The still thrusting SRBs continued to ascend. Those observers familiar with the vernacular muttered "RTLS!" and waited for the orbiter to appear, but all that emerged was a hail of débris that splashed into the sea. After 25 years of launching astronauts on rockets, NASA had finally lost a crew. CNN, the only major network to show the launch live, relayed NASA's video feed. As the controllers stared in disbelief at their telemetry displays, the Agency's PAO, Stephen Nesbitt, explained to his television audience that it was "obviously a major malfunction". He followed a few moments later with: "We have a report from the flight dynamics officer that the vehicle has exploded!" The loss was all the more shocking because there seemed to have been no indication of a problem.

Roger Boisjoly had raised a 'red flag' on the O-ring issue after inspection revealed that, on a launch the previous April, gas had burned completely through a narrow arc of the inner ring and had severely eroded the ring immediately beyond. Thiokol had set up a formal study in October, but the O-ring became just one of many pieces of hardware being considered for upgrade. The data was insufficient to characterise the fault. Given that the problem involved enduring stresses encountered in flight rather than on the horizontal test rig, developing the requisite database required more Shuttles to be launched.

This incremental process has been criticised for unnecessarily placing the Shuttle at risk. However, the argument that NASA ought to have grounded the Shuttle to enable the O-ring issue to be investigated is naïve. Incremental upgrading is standard practice in the aerospace industry. The operation of airliners continues while a 'fix' for a rare fault is developed. When this is ready it is introduced to the production line, and if practicable, retrofitted to existing aircraft. Only a catastrophic failure is permitted to ground an entire class of aircraft, and, even then, to ensure that the aircraft can resume flying as soon as possible, the first step towards a solution is likely to be the imposition of a rule designed to avoid a recurrence of that particular type of failure by constraining the operational envelope. In the case of the Shuttle, upon finding that there was a burn-through problem with its boosters, Thiokol had initiated this formal review process. However, this particular problem became just one more item on a long list of 'issues', and its significance – particularly its sensitivity to low temperature – was not widely recognised. In stretching the envelope to launch at the record low temperature, NASA gambled. It had gambled in the past, and would need to do so in the future if the performance of the Shuttle was to be increased. This time, however, it lost the bet, and Dick Scobee, Mike Smith, Judy Resnik, Ellison Onizuka, Ron McNair, Greg Jarvis and Christa McAuliffe paid the price.

As soon as it had absorbed the shock of the loss of Challenger, NASA's first imperative was to ensure that the flow of Columbia's turnaround continued. It had to be prepared in case it proved possible to proceed with STS-61E in March. To some outside the Agency this decision was met with disbelief, but it was perfectly in keeping with the NASA operating process of not closing off options. It was soon accepted, however, that the Shuttle would have to be grounded while a Presidential Commission investigated what was being called 'the accident'. Chaired by William Rogers, the Secretary of State in the Nixon administration, the Commission included: Neil Armstrong, the Apollo 11 moonwalker; Chuck Yeager, the pilot of the Bell X-1 that had broken the sound barrier in 1947; Shuttle mission specialist Sally Ride; Robert Hotz, a former editor of the leading aerospace magazine *Aviation Week & Space Technology*; and Richard Feynman, a Nobel laureate in physics from Caltech. It was formed on 3 February, and after interviewing 160 individuals, examining 6,000 documents, and instigating 35 panels to chase up specific issues in depth, announced its conclusions on 6 June in the *Report of the Presidential Commission on the Space Shuttle Challenger Accident*. The media focused on the management process that had led to the decision to launch, the organisational changes designed to reduce the 'turf wars' within NASA concerning the Shuttle, and

The crew of STS-51L pose for their official NASA portrait.

When pad technicians opened the water lines on the Fixed Service Structure in order to prevent them from freezing in the overnight chill, icicles formed.

As Challenger lifted off for mission STS-51L, a puff of dense black smoke (arrowed) emerged from the lowest field joint of the righthand Solid Rocket Booster.

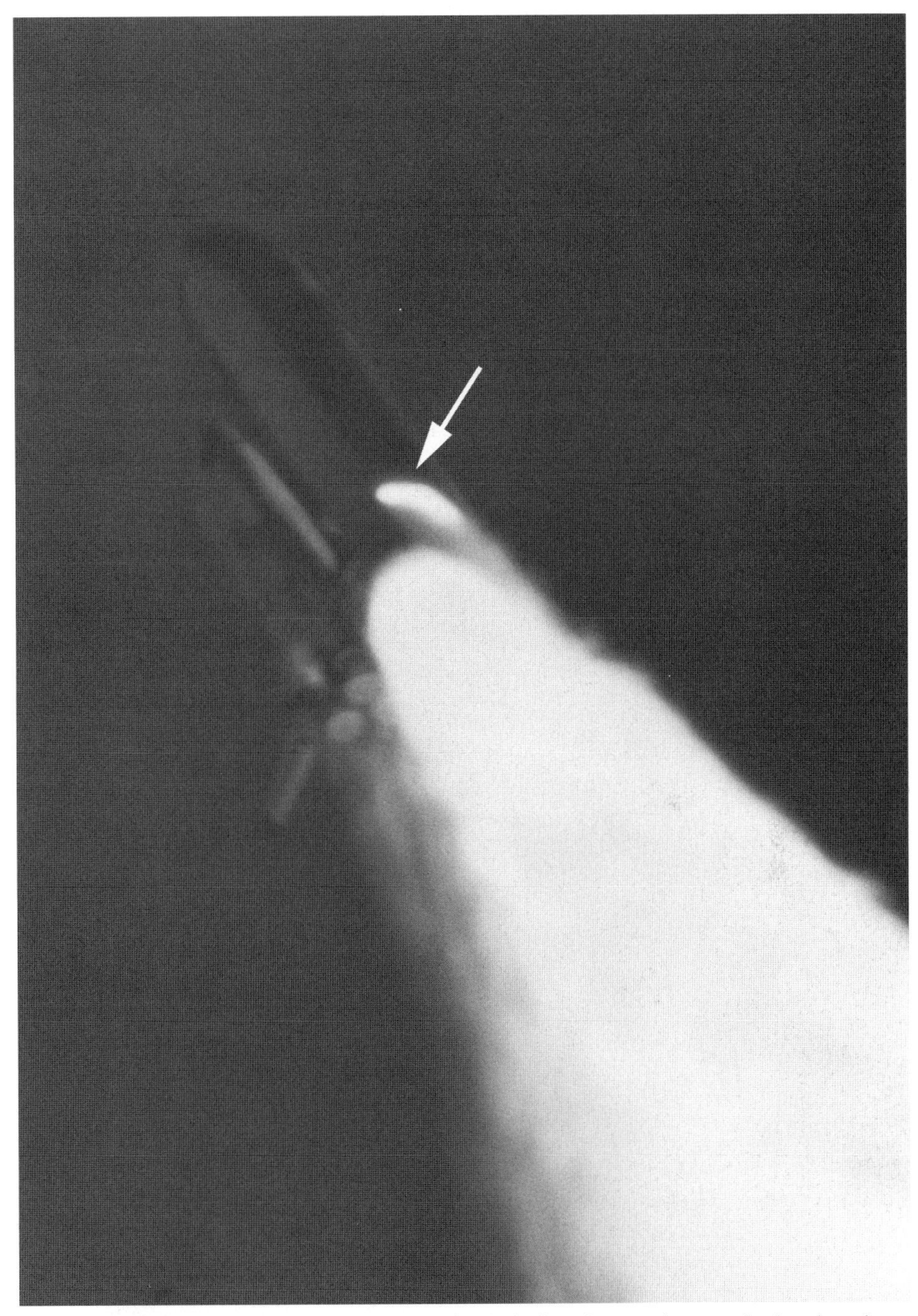

A tracking camera view of STS-51L seconds before it was destroyed, showing the anomalous plume from the lowest field joint of the righthand Solid Rocket Booster.

Fight director Jay Greene reacts to the "major malfunction" that destroyed STS-51L, visible on the screen at the end of the console.

a revision of reporting procedures to ensure that the views of the astronauts and engineers were fully represented in the decision-making process. All of these changes had merit, but they did not address the fundamental issue. The real significance of the loss of Challenger is not that a rocket failed – it was inevitable that one day a component of this very complex system would fail – but that the onset of that failure was not detected in time to initiate the appropriate abort; in this case a contingency-RTLS.

The RTLS abort mode actually addressed a spectrum of situations. If an SSME had to be shut down while the SRBs were firing, the orbiter would have to return to the Kennedy Space Center. However, no action would be taken until the boosters had completed their cycle and been jettisoned. Ten seconds later, on its remaining engines, the orbiter would climb rather steeper than the normal trajectory, level off at an altitude of about 150 kilometres and fly out over the Atlantic for several hundred kilometres for a 'turnabout' manoeuvre in which it would transition from flying on its back with the ET above, to flying tail-first with the ET beneath – a manoeuvre that has never been attempted in practice, but for which all Shuttle pilots train. In this configuration, the craft would slow down and lose altitude. The orbiter would then fly a trajectory calculated to intersect the nominal descent trajectory in order to glide back to the Kennedy Space Center – as if it was returning from orbit – at which time the ET would be jettisoned. This option remained open until T+4 minutes 20 seconds. If an engine failed after this, a Shuttle would attempt either a Transatlantic Abort Landing (TAL) with the intention of landing in either Europe or North Africa, or execute an Abort To Orbit (ATO) and try to achieve a low orbit which the OMS engines could then either abandon in an Abort Once Around (AOA) concluding in a landing in California or augment to continue the mission. The progression of abort options is really meant to deal with SSME failures at various

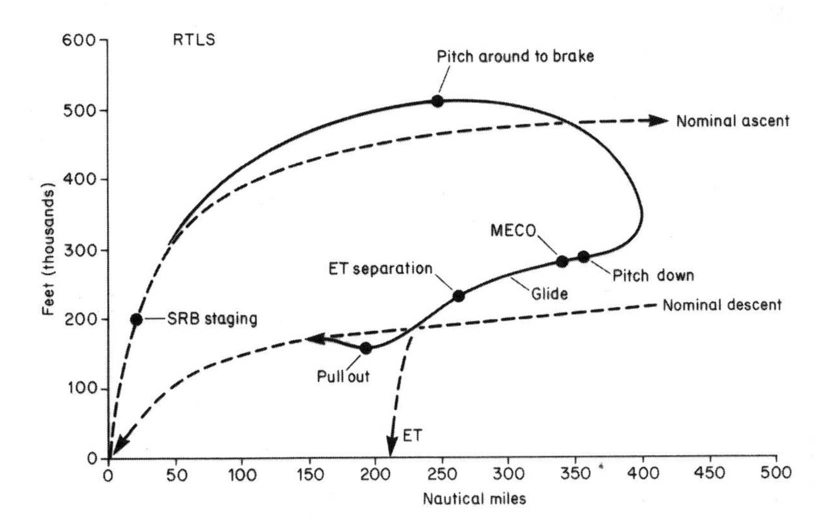

The Return To Launch Site (RTLS) profile.

phases in the ascent, yet the most dangerous phase is the SRB boost. Although the Shuttle's design requirement did not explicitly specify that the orbiter must have a means of escape in the event of an SRB fault, this was not the same as stipulating that there should be *no* chance. In addition to the abort modes outlined above, all of which were expected to have a high likelihood of success, the GPC had been programmed for a 'contingency abort' during the SRB boost. After such an abort, the best that could be hoped for would be a ditching at sea, which would be risky and would probably lead to the break-up of the orbiter. However, a slight chance of survival was better than no chance.

In the circumstances in which 51L found itself, the plan called for the still-firing SRBs to be jettisoned immediately, Challenger to continue its powered flight in order to attain an energy state that would facilitate a turnabout manoeuvre, the ET to be jettisoned, and the orbiter to attempt to save itself. All of the abort plans relied on there being sufficient warning of an impending failure to perform the necessary separation, but the instrumentation monitoring the vehicle's performance did not show any evidence of the burn-through until the SRB's internal pressure began to fall. A RTLS might just possibly have been feasible. Shedding the still-firing SRBs would have been fraught with danger, because the separation might not have occurred cleanly; but by that point Challenger was already in extreme peril. In an organisation like NASA, renowned for its contingency planning, a successful abort, even if the orbiter had to ditch at sea, would have been a vindication of the operating procedures. How much time would have been available for this most difficult of the abort options? The abridged extract from the Commission's Report replays the story in appalling clarity.

Table 3.1 Challenger's countdown to catastrophe

T+ seconds	
0	With all three SSMEs stable at 100 per cent of rated thrust, the two SRBs are ignited, and almost instantly build up to their peak output to lift the stack off the pad.
0.678	A strong puff of smoke spurts from the aft field-joint of the right SRB. There are further puffs at a rate of about one every quarter of a second, which is the frequency of the field-joint's flexure in response to the loads inherent in the onset of flight. The thick dark smoke is suggestive of the O-rings passing hot propellant gas with resultant burning of the joint's grease and insulation in the arc facing the ET.
2.733	The last smoke puff from the right SRB.
19	Smith: "Looks like we've got a lot of wind here today."
20	Scobee: "Yeah."
22	The internal pressure in the SRBs peaked, then began to decrease in order to accommodate increasing aerodynamic forces.
22	Scobee: "It's a little hard to see out my window here."
24	The computer commanded the SSMEs to throttle down to 95 per cent.
28	Smith: "There's 10,000 feet and Mach point five."
35	Scobee: "[Mach] point nine."

Table 3.1 (cont)

T+ seconds	
35.379	The computer commanded the SSMEs to throttle down to 65 per cent.
40	Smith: "There's Mach 1."
41	Scobee: "Going through 19,000 [feet]."
42	The SSMEs were now at 65 per cent.
43	Scobee: "Okay, we're throttled down."
50	Max-Q.
51.860	The computer commanded the SSMEs to throttle up to 104 per cent.
57	Scobee: "Throttling up."
58	Smith: "Throttle up [acknowledged]."
58.788	As the SRBs began to increase thrust, a flame appeared in the same position on the right SRB as smoke had previously been vented.
59	The SSMEs were now at 104 per cent.
59	Scobee: "Roger."
59.262	The flame was now a continuous well-defined plume. The supersonic slipstream played this flame across the base of the ET. As this flame grew, it impinged on the strut holding the SRB to the ET.
60	Smith [remarking on the acceleration]: "Feel that mother go."
60	Uncertain: "Woooohoooo."
60.004	Telemetry now began to show the two SRBs with different internal pressures, that of the right SRB being lower.
62	Smith: "35,000 [feet], going through [Mach] one point five."
62.484	Telemetry indicated that the righthand outer elevon control surface was in motion.
63.924	Telemetry showed a change in this elevon's actuator pressure.
64.660	The flame breached the ET and hydrogen began to vent, and burn in the impinging plume.
64.937	The computer started to gimbal the SSMEs with large pitch variations in order to overcome unwanted vehicle motions.
65	Scobee: "Reading 486 [the meaning of this is disputed] on mine."
65.524	Telemetry showed a pressure change in the actuator for the outboard elevon on the left side.
66	MCC: "Challenger, go at throttle up."
66.764	Telemetry showed a drop in the ET's hydrogen pressure.
67	Smith: "Yep, that's what I've got too."
70	Scobee [answering the ground's call]: "Roger, go at throttle up."
72.204	The strut holding the base of the right SRB to the ET came loose and the booster pivoted around its upper strut, its rear skirt impacting the orbiter's right wing; telemetry indicated divergent yaw rates between the two SRBs.
72.284	Telemetry indicated divergent pitch rates between the two SRBs; the loose SRB continued to thrash about.
72.497	The SSME gimbals were now rolling at 5 degrees per second in an attempt to damp out vehicle motions.
73	Smith: "Oh-oh."
73.124	The bottom cap of the ET failed, dumping pressurised hydrogen into its wake.

Table 3.1 (cont)

T+ seconds	
73.137	The nose of the still-thrashing right SRB struck the ET's intertank, fracturing the base of the oxygen tank, leading almost immediately to an explosive burn of mixed propellant. The asymmetric loading on the vehicle ripped the stack apart and the orbiter disintegrated.
73.143	The SSMEs responded to the loss of propellant and the turbopumps began to overheat.
73.482	The SSMEs initiated a shut down.
73.618	Orbiter telemetry was terminated.
75	PAO: "1 minute, 15 seconds. Velocity 29 hundred feet per second. Altitude now 9 nautical miles. Down range distance 7 nautical miles. Flight controllers here [are] looking very carefully at the situation. Obviously a major malfunction. We have no downlink. We have a report from the Flight Dynamics Officer that the vehicle has exploded!"

Note: Some details of the nominal ascent following the termination of the smoke plume pulsations and prior to the onset of the flame plume have been edited.

In addition to this 'official' timeline, aerospace journalist Tim Furniss wrote in *Flight International* in February 1997 that Ali Abu Taha, who had been given access to the NASA evidence and had carried out an independent analysis, had recovered a photograph which showed a 3.3-metre-long plume of flame emerging from the SRB at about T+20 seconds, but this image has yet to be published. This coincides with the 'elbow' in the SRB's internal pressure profile, so a flame leak at this point may have been caused by the casing relaxing as the pressure began to fall rapidly, in order to ease the stress on the vehicle in the run-up to Max-Q.

The first indication in the telemetry that there was a fault in the right SRB was at T+60 seconds, when its internal pressure began to depart from nominal. Telemetry over the next 13 seconds confirmed and clarified the nature of the developing catastrophe. The issue for the flight controllers was whether the fall in internal pressure was a temporary fluctuation – in which case the focus of attention would move to how well the Shuttle was able to correct its trajectory by swivelling the nominal engines – or whether it represented a divergent trend. The internal pressure could vary within a narrow band and still be acceptable, but at T+63 seconds it departed from this band; it was clearly under-performing, but the reason was not evident, at least not in the telemetry. In principle, a RTLS abort could have been invoked at this point. Putting it into effect would take 3 seconds. The situation was not yet critical, but at T+66 seconds, when the ET was fractured and started to vent hydrogen, the option of a powered-RTLS closed. From this point, Challenger was in dire peril. The SRBs could still have been dumped, but as there was no hydrogen for the SSMEs, the likelihood of the orbiter being able to manoeuvre to a point suitable for an emergency landing was slim indeed. For the next 6 seconds the telemetry continued to track the falling pressures in both the right SRB and the hydrogen tank. Time was running out fast, and when the end came it was essentially instantaneous.

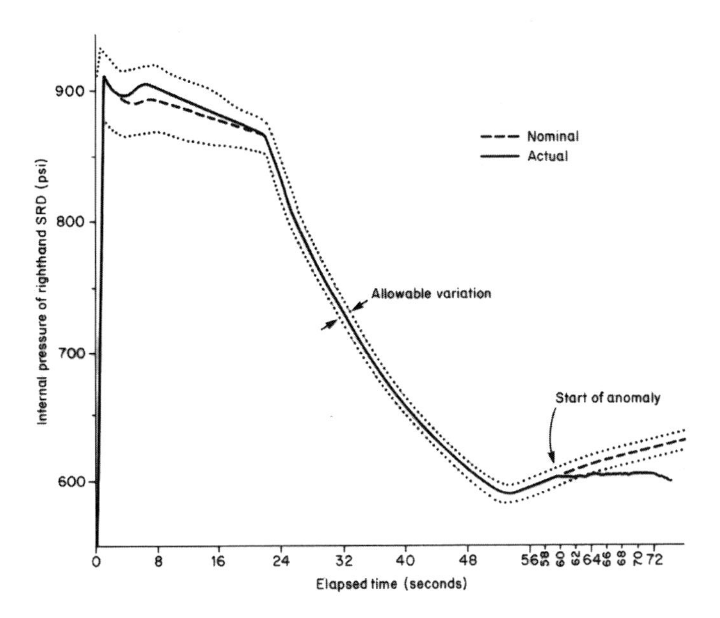

A chart of the internal presure in the righthand Solid Rocket Booster of STS-51L showing the timescale during which the booster's performance degraded.

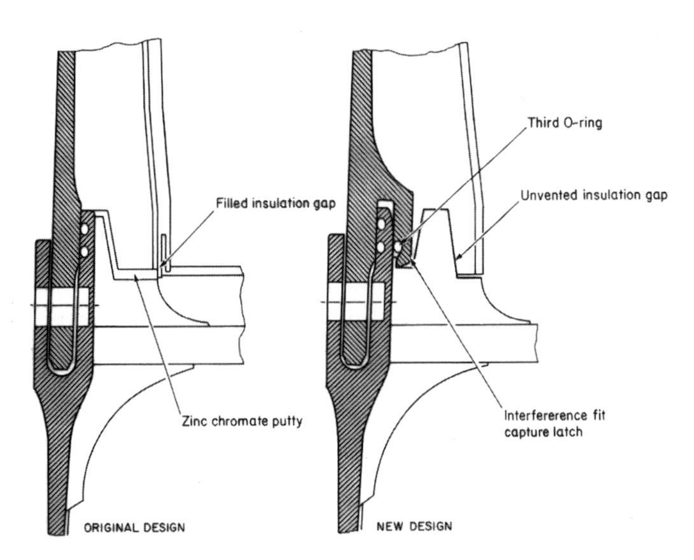

The new design of the field joint of the Solid Rocket Booster had a double clevis and a third O-ring positioned to seal joint flexure in a way that was not possible in the original configuration.

The window for a powered-RTLS abort had lasted only 6 seconds, from the moment that the right SRB began to lose pressure to the time that the associated flame compromised the integrity of the hydrogen tank. This wasn't much, but it was a window of opportunity, so why wasn't a contingency abort ordered? The reason was simple: although the telemetry indicated that there was a fault in the righthand SRB, which grew to include the ET, the telemetry did not reveal the cause. While the flight controllers worked to understand the cause, they were denied the unambiguous indicator that would certainly have prompted them to call for an abort. The terse exchanges between Scobee and Smith in the final seconds have been interpreted as their awareness of the fall-off in SRB pressure. However, even if they had realised what was going on, it is unlikely that they would have initiated an abort without confirmation of this fault from their colleagues on the ground. If a window of only a few seconds seems to be asking too much of the controllers, how long should it have taken to figure out that disaster was looming? The vital information that made the looming catastrophe clear was the imagery from the tracking camera, which showed the plume of flame. However, the view fed to the television networks – the view that was available to the flight director – was a view from which the flame plume was obscured. Might events have turned out very differently if the output from the other camera had been shown live?

GROUNDED

The upshot of the Rogers Commission's report was that NASA must:

- develop a safer SRB field-joint;
- reassess issues of landing the orbiter safely, and pay particular attention to the tyre and brake failures;
- upgrade contingency planning for aborts during the ascent phase;
- stock sufficient spare parts to preclude cannibalising recently returned orbiters in order to enable another one to fly;
- resist the temptation to override established limits by the issuance of waivers in order to increase the flight rate;
- precisely define the authority and responsibility of the Shuttle programme manager; and
- allow astronauts to participate in the management process.

It was additionally decreed that an independent panel should review safety issues, and report directly to the programme manager.

A few weeks later, Congress, aware that the operational safety of the Shuttle had assumed a higher profile, ordered NASA to cancel the Centaur upper stage because it could jeopardise an abort. Even if the orbiter made it to an emergency runway, in a rough landing the Centaur might break free and spill tonnes of cryogenic propellants that would cause a devastating explosion. The cancellation of this powerful rocket threw the planetary programme into chaos.

However, there was also some good news. On 15 August, the White House

The SRBs emerge from the fireball of STS-51L

ordered the construction of a replacement orbiter. It also instructed NASA to cease using the Shuttle to launch commercial satellites, and not to use it to deploy any satellite that could be offloaded onto an expendable rocket. The adoption of this mixed-fleet policy was good news for the rocket makers. When Reagan had declared the Shuttle operational, making it the National Space Transportation System for government payloads, the makers of the rockets had been told that they could sell their services in the commercial sector, but the lower-than-cost fees for satellites booked on the Shuttle had made this an empty gesture. As a result, the lightweight Delta had already ceased production, and the Atlas and Titan were to be scrapped as soon as the Shuttle took over the launching of polar-orbiting satellites. Now, with not only commercial but also government payloads offloaded to expendables, there were rich pickings for the rocket manufacturers – and their only competition would be Arianespace. The production lines were reopened immediately, but until deliveries could be resumed the inventory of expendables would be extremely limited. In fact, there were only five Deltas left, and all the remaining Atlas and Titan launch vehicles were reserved by the government.

Meanwhile, NASA worked on the Shuttle. All but two of the SRBs had been recovered (both of STS-4's had been lost due to identical parachute failures) and evidence of hot gas flowing past the main O-ring had been found in nine casings. Although NASA redesigned the tang-and-clevis joint to reduce flexure, it also minimised the modification in order to be able to use the segments it had in stock. In addition, an electrical heater was installed to protect the two existing O-rings from low temperatures, a third was added, the inside of the joint was sealed with extra putty, and a 'weather strip' was wrapped around the joint to make it waterproof.

To comply with another of the Rogers Commission's recommendations, it was decided to develop a way for the crew to bale out of the orbiter in the subsonic glide after either a launch abort that could not reach a runway or if there were indications during re-entry of a problem with the landing gear. During the ALT and OFT phases of the programme, astronauts had worn full pressure suits and had ridden in ejection seats. Since the ejection seats had been replaced, crews had worn lightweight blue flight suits, with helmets to provide oxygen in the event of a cabin depressurisation. Future crews would wear pressure suits that were high-visibility orange to assist in rescue. The orbiter's side hatch was to be rigged with pyrotechnics to enable it to be blown off. A telescoping pole on the mid-deck ceiling would be extended through the open hatch by the mid-deck crew and the orbiter would be placed on autopilot. Each astronaut, in turn, would sling a harness over the pole and bale out. The pole would ensure that an evacuee passed beneath the leading edge of the wing. Once clear, the astronaut would release the parachute integrated into the suit. Such an evacuation would clearly be a risky process, but it would offer a better chance of crew survival than a ditching at sea.

4

Communications satellites

GEOSTATIONARY RELAY

It was Arthur C. Clarke who first pointed out that as the period of a circular orbit at a height of 36,000 kilometres would be 24 hours, a satellite above the equator would remain fixed in the sky. He published his analysis of this geostationary orbit (GSO) in the *Journal of the British Interplanetary Society* in May 1945. Later that year, in the October issue of *Wireless World*, he pointed out that if three such satellites were positioned at 120-degree intervals, they would be able to relay radio signals between any two points on the planet. Of course, a space-based communications system was beyond the state of the art of the late 1940s, not just in terms of rocketry but also in electronics; and even after the first satellite had been launched, the development of a suitable communications technology still represented a considerable challenge.

Telstar – the first commercial communications satellite – was launched in 1962. It was developed by the American Telephone & Telegraph Company (AT&T) as a private venture, and although it carried the first live TV across the Atlantic, its low orbit restricted its use as a long-distance relay to brief periods. AT&T intended to launch a constellation of Telstars, thereby capturing the entire market, but the US government passed the Communications Satellite Act and ordered that a national consortium be established to develop and operate a system with the capacity to serve the domestic telecommunications industry. The Communications Satellite Corporation (Comsat) was created in February 1963, but rather than pursue the low-orbiting Telstar it decided to use geostationary relays, as these would provide continuity of service. A geostationary satellite represented a considerable technical challenge, but NASA was already developing Syncom as a technology demonstrator. This satellite was positioned above the International Date Line in 1964, just in time to relay live coverage of the Olympic Games in Japan to a fascinated US audience, and the banner 'Live By Satellite' immediately became the defining symbol of that time.

The International Telecommunications Satellite Consortium (Intelsat) was established in August 1964 to deploy a network of geostationary relays to provide telecommunications on a worldwide basis. The consortium approached Hughes, who had built the Syncom satellite, but requested a modified design, and when the Early

Bird was stationed over the Atlantic in April 1965 it became the first commercial geostationary relay. By the mid-1970s Comsat operated a fleet of Comstar satellites and leased transponders to the US domestic market – its main clients being AT&T and General Telephone & Electronics (GTE), which had a lucrative business relaying voice for a diverse community of users. Of the 11 founder members of Intelsat, the largest was Comsat, but as the benefits became manifest, the membership increased ten-fold, with each new member building its own ground stations and contributing funding for satellite development in proportion to its revenues from the business. As a result, the consortium soon controlled the majority of the international telephone traffic and almost all the transoceanic TV-relay business. Owing to the exponential growth in demand, it was financially sensible to ensure the continuous development of new generations of ever more capable satellites.

Having built Syncom, Early Bird and the Comstars, Hughes had established a clear lead in the satellite-construction business, but to maintain its lead it had to exploit every advance in microminiaturised electronics to develop better satellites. In the early 1970s Hughes introduced the 300-kilogram HS-333. This spin-stabilised drum had a conformal array of solar cells and a despun antenna that could relay either a colour TV channel or 1,000 voice channels. The key was the electronics to shape this beam to cover the customer's target area, and only that area. Shaping the beam delivered all the radiated power directly to a specific 'footprint', rather than distributing it over a wider area, enabling neighbouring customers to share frequencies free of interference. As the frequencies in the C-Band were rationed to avoid interference, this ability to double-up was a significant marketing advantage for Hughes.

The Telesat consortium was formed in 1971 by the Canadian government in partnership with the domestic telecommunications industry to develop an integrated national service. It was responsible for voice and data communications across the country, and was the main broadcast service for the widely dispersed communities in the vast northern territories. In deciding to operate its own communications satellites, Telesat rode the crest of the wave made possible by the introduction of the HS-333 configurable-footprint system. Telesat had a requirement for only two satellites, but ordered a third as a spare in case one was lost. The system was called 'Anik', which means 'brother' in Inuit, the language of the Eskimos of northern Canada. Since they were to be replaced by a more advanced variant at some point, the first generation of satellites were marked by the letter 'A': Anik-A1 being launched in November 1972 on a Delta rocket, followed by Anik-A2 in 1973. Having proved themselves, they attracted business, and to satisfy this increased demand the spare was added in 1975 as Anik-A3. Although Telesat hoped that the HS-333 satellites would survive for the advertised seven years, it knew that it would have to start launching replacements by the early 1980s, and began to explore the limits of the technology to determine the requirements for its follow-on system. It signed up the Radio Corporation of America (RCA) to build an experimental satellite to carry both C-Band and Ku-Band transponders. Launched in 1978, Anik-B1 was the first satellite to broadcast TV directly to fee-paying customers. However, as the receiving antenna was a dish two metres in diameter, this service was aimed at hotel complexes. Nevertheless the experiment was so successful that Telesat decided to set up two networks of satellites in parallel: one to continue the C-Band facilities of its then fully operational system, and the other to exploit the increased capacity of the Ku-Band. For the Anik-C series, Telesat ordered the latest satellite produced by Hughes, the HS-376, which was capable of accommodating its Ku-Band requirements. Also, to develop a domestic manufacturing capability, Spar Aerospace bought a licence to build an HS-376 clone as the Anik-D model that was to sustain the C-Band service well into the 1980s.

When the Satellite Business Systems (SBS) consortium was set up in 1975, the majority of the shares were owned by IBM and Comsat. It was established to sell communications services directly to end-user businesses having facilities spread over the continental United States, with Boeing, General Motors and Westinghouse being its initial customers. It ordered a fleet of HS-376s fitted with Ku-Band transponders to provide an all-digital service integrating voice, data, video and e-mail. As before, spare capacity was leased to the telecommunications companies for voice traffic. In addition to members of Intelsat buying their own satellites in order to provide local telecommunications services, it now became possible for individual corporations to buy satellites for private use, instead of buying services indirectly through Comsat's satellites.

The introduction of the Shuttle as the National Space Transportation System was very timely. On the basis of production orders, it looked as if the most popular type would be the Hughes HS-376. At 4.6 metres wide and 18 metres long, the Shuttle's bay was long enough to accommodate five such satellites, and because it had only to deploy the satellites into low orbit – using the same support equipment repeatedly – its turnaround time and operating costs would be minimised. Launching all these satellites was evidently going to be a lucrative operation. Although NASA was given a monopoly in the domestic market, to ensure that the Shuttle would have sufficient work to fly on a weekly basis the agency also set out to attract foreign business. However, the Shuttle had competition in the international market from the European Space Agency, which was developing a range of Ariane rockets to launch commercial satellites. Nevertheless, NASA issued an introductory offer whereby a number of satellites could be launched for a bargain-rate fixed fee during the first five years of Shuttle operations, and even before the Shuttle had made its first test flight several governments had signed up a significant number of satellites. SBS was also seduced, and booked all of its satellites on the Shuttle. However, because it could not afford to wait out the Shuttle's protracted development, the first two satellites were offloaded to Delta rockets, and SBS-1, launched in 1980, became the first HS-376 to enter service. Telesat also booked all of its new satellites on the Shuttle. However, although its HS-333 satellites operated for longer than Hughes had predicted, Telesat had to transfer Anik-D1 to a Delta in order to prevent the degradation of its C-Band system. The Shuttle's entry into service was therefore eagerly awaited by the new service providers of the telecommunications industry.

THE ACES

After four test flights Columbia was declared operational, and when it lifted off for STS-5 in November 1982 it carried two HS-376s mounted vertically in separate cradles towards the rear of its payload bay. Each satellite was a drum coated with solar cells. On station the satellites would be spinning for stability and to smooth out thermal stress, but while sitting in the bay they would be baked on one side and frozen on the other. As soon as the bay doors swung open, clamshell covers were closed over the satellites to protect them from the space environment, and heaters were switched on to keep them warm. The covers were not opened until just before the satellites were deployed. Columbia faced its payload bay forward and tilted its right wing downward. In this orientation, the released satellite would be aligned with its velocity vector, exactly as would have been the case if it had been delivered by an expendable rocket. The final stage of the preparation was to stabilise the satellite by spinning it on a turntable in the cradle. Eight hours into the mission, SBS-3 became the first satellite to be released from the Shuttle. The release, which was videotaped and relayed on the downlink immediately afterwards, showed the spinner being spring-ejected from its cradle, pass the Shuttle's tail, and slowly drift towards the horizon. The National Space Transportation System was finally in the commercial satellite business.

The satellite had a perigee-kick motor affixed to its base. McDonnell Douglas had made its Payload-Assist Module (PAM) compatible with both the Shuttle and its own Delta in order to ease its customers' transition onto the Shuttle. The 2.2-tonne solid rocket could accelerate a 1.25-tonne satellite into geostationary transfer orbit (GTO) and after forty-five minutes, while crossing the equator, the PAM was programmed to fire for about 90 seconds. Because a solid motor tends to accrete a residue of spent propellant in the throat of its nozzle, thereby deflecting its thrust, the rocket would have diverged from its desired trajectory if it had not been put into an axial spin. The spin induced a corkscrew motion ('coning') that wasted energy but did not seriously jeopardise the short burn. As soon as the motor shut down it was jettisoned. Six hours later, on reaching apogee high over the Eastern Pacific, SBS-3 fired its own thrusters to circularise for GSO. If it slowed down marginally, it would descend and increase speed, which would cause it to drift east; if it climbed slightly, it would reduce speed and drift west. Over the next few days it manoeuvred onto its station at 94°W. The HS-376 had been ingeniously designed with a double drum of solar cells, one inside the other. This simple feature meant that the 6.6-metre configuration could be accommodated in a 3.7-metre-tall enclosure, enabling it to double its power supply by extending the outer drum from the base. The second HS-376, Anik-C3, was similarly dispatched the following day and successfully took up its station at 115°W.

The astronauts celebrated their successful double deployment by displaying a placard to the video downlink. This identified their employer as the *Ace Moving Company* and cited the corporate slogan, *We Deliver*.

Table 4.1 Commercial satellites launched by the Shuttle

Satellite	Flight	Manufacturer	Bus	Operator
SBS-3	STS-5	Hughes	HS-376	SBS, USA
Anik-C3	STS-5	Hughes	HS-376	Telesat, Canada
Anik-C2	STS-7	Hughes	HS-376	Telesat, Canada
Palapa-B1	STS-7	Hughes	HS-376	Perumtel, Indonesia
Insat-1B	STS-8	Ford	–	ISRO, India
Westar-6	STS-10	Hughes	HS-376	WU, USA
Palapa-B2	STS-10	Hughes	HS-376	Perumtel, Indonesia
SBS-4	STS-12	Hughes	HS-376	SBS, USA
Telstar-3C	STS-12	Hughes	HS-376	AT&T, USA
Anik-D2	STS-14	Hughes	HS-376	Telesat, Canada
Anik-C1	STS-16	Hughes	HS-376	Telesat, Canada
Morelos-1	STS-18	Hughes	HS-376	SCT, Mexico
Arabsat-1B	STS-18	Aerospatiale	–	ASCO, Arab League
Telstar-3D	STS-18	Hughes	HS-376	AT&T, USA
Aussat-A1	STS-20	Hughes	HS-376	Aussat, Australia
ASC-1	STS-20	RCA	S-3000	ASC, USA
Morelos-2	STS-23	Hughes	HS-376	SCT, Mexico
Aussat-A2	STS-23	Hughes	HS-376	Aussat, Australia
Satcom-K2	STS-23	RCA	S-4000	RCA, USA
Satcom-K1	STS-24	RCA	S-4000	RCA, USA

SATELLITES GALORE

Over the next three years the Shuttle released a total of 20 commercial satellites, all of which were to be boosted into GSO to serve as communications relays. With so much of the geostationary relay business serving the North American continent, the problem of overcrowding in the arc of the equator covering this region grew progressively worse. The satellites had to be distributed several degrees apart, to ensure that they would not interfere with each other. As interference was more apt to occur at lower frequencies, the Ku-Band satellites could be more closely packed, and interleaved with their C-Band predecessors. It was nevertheless often necessary to rearrange several of the existing population to accommodate newcomers.

All three of Telesat's Anik-Cs and the second of its Anik-Ds were deployed by the Shuttle, and all reached their stations and operated successfully for many years. Although two of the Ku-Band relays provided sufficient capacity to carry its load, Telesat exploited NASA's multiple-satellite deal to have the third launched early to serve as an *in-situ* spare. Anik-C1 was placed on station awaiting the command to commence relaying traffic. In light of the scramble for launchers after the loss of Challenger, the early installation of Anik-C1 proved to have been a wise decision. In fact, Anik-C2 was also spare capacity initially, and until May 1984, when Telesat brought it on line, its transponders were leased to evaluate pay-to-view satellite-to-home broadcasting, and in so doing it set the scene for the boom in Direct Broadcast Service (DBS) satellites that followed.

Some 25 per cent of Mexico's population lived in the immediate environs of Mexico City. The remainder had poor communications, and no TV. The Secretariat of Communications and Telecommunications (SCT) decided to remedy this situation using satellites and, following Telesat's lead, it opted for the HS-376. The satellites were named after Jose Maria de Morelos y Pavon, the hero of the revolution, and booked on the Shuttle. The first was launched in June 1985, and even though a severe earthquake in September soaked up the funds intended for the construction of ground stations, it was decided to proceed with the launch of the second satellite in November. Morelos-1 proved its value in coordinating the recovery in the aftermath of the earthquake.

When AT&T became the first private company to design and operate its own communications satellites, it called its system Telstar. When the company bought HS-376s and operated independently of Comsat, it revived the famous brand name. Telstar-301 went up on a Delta in July 1983, but the Shuttle completed the system with two deployments in 1984 and 1985. For the next decade these satellites relayed C-Band communications for the company's domestic customer base.

The American Satellite Company (ASC) – a pioneer in offering fully integrated communications – focused on the various agencies of the US government. It had been formed in 1972 by Fairchild Industries, and was able to design its own satellites. In 1974, in order to build its customer base, it leased transponders on Westar satellites, and later bought a 20 per cent stake in that company. ASC built satellites with C/Ku-Band transponders on an RCA three-axis bus. It booked two on the Shuttle, but as soon as ASC-1 was in service in 1985, the company was sold to the Continental Telephone Corporation (Contel). ASC-2 was to have been launched in 1986, but the loss of Challenger prevented this, and it was not launched until 1991, at which time it rode a Delta. A few months later, Contel merged with GTE, which operated the Spacenet series that used the same RCA bus, and ASC-2 was renamed Spacenet-4.

In 1976, in addition to creating Astro Electronics to sell satellites as a business, RCA set up RCA Americom to beam TV programming to hotels, hospitals, schools and residential complexes possessing 2-metre dishes. The Federal Communications Commission had opened up the market to satellite broadcasting in 1972, and the cable-TV operators eagerly expanded into this new area as it enabled them to reach a vastly enlarged audience free of the expense of laying cables. Satcom-1 was launched in 1975, and by 1982 there were four similar satellites. Home Box Office (HBO) was set up specifically to beam new-release movies to pay-to-view customers using the Satcom satellites. The first generation used C-Band, but because Ku-Band signals could be picked up by a much smaller receiver – further expanding the market to individual households – Ku-Band was used for the follow-on Satcom-K satellites, four of which were booked on the Shuttle. As the Satcom-K was too heavy for the standard perigee-kick motor, it was necessary to employ the uprated PAM-D2. However, following the earlier PAM failures, the insurance premium for the first use of this new motor was so excessive that RCA opted to dispense with insurance – a gamble that paid off! The first two satellites were successfully deployed on flights immediately prior to the loss of Challenger. Satcom-K2 was effectively leased to NBC TV News, and Satcom-K3 was leased to HBO. With no prospect of launching the two final satellites in the near future, RCA decided to sell them. One became a one-off Intelsat TV-relay, and the other eventually appeared as Astra 1B. In 1986, RCA sold Astro Electronics to GE Spacecraft Operations, which built satellites as GE Astro Space. As part of the deal, RCA Americom became GE Americom and continued to run the Satcom satellites until GE Americom launched a new generation of C-Band satellites in the early 1990s.

Comsat sold its holding in Satellite Business Systems in 1984. After launching SBS-3 on the Shuttle, the company had planned to send up another two HS-376s before commencing a new series of the more powerful HS-393, but only managed to launch SBS-4 before the Shuttle was grounded. In 1987, Satellite Business Systems came under the control of the Satellite Transponder Leasing Company. The new HS-393 – the largest commercial communications satellite ever built – had been designed to exploit the Shuttle's voluminous payload bay, with twice the mass and triple the capacity of the HS-376. When SBS launched its first HS-393 in 1990, however, it was on an Ariane rocket. The Shuttle did not encounter one of these satellites until some years later – an occasion on which spacewalking astronauts pulled off yet another triumphant repair operation.

The Shuttle's entry into service was too late to help Western Union, which was in fact the first telecommunications company to use private geostationary satellites to integrate voice, data, video and fax for US customers. The first Westar satellites were HS-333 models. Two were launched in 1974, with a third following in 1979. They were succeeded by a series of HS-376 satellites. Westar-4 and Westar-5 went up on Delta rockets in 1982. Westar-6, which was intended to be the final satellite in the series, was taken by the Shuttle, but the PAM misfired, stranding it in low orbit. The Shuttle had released the satellite perfectly, but the loss of the satellite dented NASA's image. On the other hand, the video of spacewalking astronauts retrieving it a few months later perfectly projected the agency's 'can-do' attitude. In these early years it seemed that for NASA every 'cloud' had a 'silver lining'. The satellite was refurbished and sold to the Asian Satellite Telecommunications Company of Hong Kong who renamed it Asiasat-1, and in April 1990 it became the first US-built satellite to be launched on a Chinese Long March rocket. By this point, however, Western Union had been bought by Hughes, which had replaced the Westars in the busy American arc, on a one-to-one basis, with its own Galaxy satellites, which were also derived from the HS-376 bus.

The overcrowding in the American arc was in marked contrast to the situation in the Far East, where Australia and Indonesia had no problems in finding convenient locations for their satellites. Although both countries were members of Intelsat for international links, neither had operated satellites for local services. As with Canada, the satellites were to overcome geographical adversity, and consortia were formed to undertake the task.

In 1981 the Australian government formed Aussat with Telecom Australia, the domestic telecommunications company. Up to this point, only the five main cities had state-of-the-art but weakly connected local communications, so three HS-376 satellites were booked on the Shuttle. In addition to business communications, the Ku-Band transponders were to relay the national TV networks and the Homestead and Community Broadcasting Service that produced TV for one million people in settlements too remote to be serviced by conventional transmitters. In addition, an L-Band transponder was employed to relay air and maritime traffic. As the footprint covered Australia, a large part of the archipelago to the north, and New Zealand, it represented an enormous advance. The first two satellites were deployed before the Shuttle was grounded, and the final satellite went up on an Ariane rocket in 1987. When the government deregulated the market in 1991 it sold its majority share of Aussat to Optus, which gave its name to the second-generation system. The HS-376 satellites were to have been replaced one-for-one by the new HS-601 models – the first three-axis type built by Hughes – as they became available in the early 1990s, but when the first satellite suffered problems and the second was lost, the Aussats (all of which were still functional) were retained to keep the by-then vital service running.

Indonesia faced an even worse geographical problem. Its 200 million population was fragmented over 13,000 scattered islands, and even substantial communities had little direct contact. A geostationary relay was the only cost-effective way to link up the nation. The Perumtel consortium was created specifically to develop the service. In 1975 it placed a contract with Hughes for an HS-333, and this was launched in 1976 as Palapa ('the fruit of great endeavour'). To follow up, Perumtel ordered the HS-376. As in the case of Telesat – upon which its operations were modelled – the company ordered two satellites and a spare, and booked them all on the Shuttle. Unfortunately, Perumtel was not to have Telesat's luck. Palapa-B1 was delivered to its operating position without incident, but Palapa-B2 became stranded in low orbit when its PAM misfired. As the spare had been purchased for such a contingency, its launch was brought forward but the Shuttle was grounded before this satellite could be deployed, and it was finally launched on a Delta in 1987. The retrieved satellite was refurbished and relaunched on a Delta in 1990, to supplement the network.

India had operated an active space programme for some time. It had developed its own satellite-launching capability and, as a participant in Intercosmos, had sent a cosmonaut to one of the Soviet Union's Salyut space stations. The communications issue facing India was not a scattered population, it was such a large population that it was impracticable to develop a telecommunications infrastructure by conventional means. India had a long-standing interest in acquiring a geostationary capability, but it preferred a multipurpose satellite for a range of applications. In addition to having transponders for C-Band telephone relays and S-Band TV broadcasting, the Indian Space Research Organisation (ISRO) wanted its satellites to provide meteorological and Earth-resources imagery. The contract for this customised platform was given to Ford Aerospace – an American company that specialised in three-axis systems. A fully stabilised bus offered the advantage that it could expose a large array of solar cells mounted on flat panels that could be maintained facing the Sun, and thereby run a more powerful transmitter than was possible on an equivalent spin-stabilised bus. However, a three-axis bus needed additional systems to control the orientation of the bus with sufficient accuracy to maintain the narrow footprint on target and maintain the panels facing the Sun, and if either system failed the spacecraft became useless as a relay. Although the size of a spin-stabilised bus limited its power because only a small fraction of its solar cell array could face the Sun at any given moment, and the fact that its antenna had to be mounted on a despun platform complicated its design, it had the significant advantage that its angular momentum made it inherently stable. Insat-1A was launched on a Delta in April 1982, but was abandoned a few months later after expending its propellant in an attempt to recover from an attitude-control fault. The other satellites in this series had been booked on the Shuttle. Insat-1B left its cradle with all its various projections tightly folded against its boxy bus. As soon as the satellite jettisoned its PAM stage, it used its thrusters to cancel its spin. On assuming its station, at 74°E, a fault developed with one of the solar panels, which took several weeks to flatten out. The astronauts had reported hearing a 'clunk' as the satellite had been spun up in its container, but a replay of the deployment video failed to show any specific signs of damage. Insat-1C had been scheduled for launch in 1986 but was offloaded after the loss of Challenger, and competition for rockets was so intense that it could not be launched until July 1988, when it rode an Ariane. Unfortunately, it lost control after 18 months, and had to be written off. Insat-1D followed on a Delta in mid-1990 and Insat-1B was retired a year later. However, a few months later a power failure disabled half of Insat-1D's transponders and the venerable Insat-1B had to be reactivated until the first of the follow-on series could be launched. For this series, ISRO combined C/S/Ku-Band transponders and, despite having lost two satellites to instability, opted once again for three-axis control. This time, however, it developed its own bus. The first of the new satellites, Insat-2A, was launched in 1992 and went on to function satisfactorily. Of the troubled series, only Insat-1B had been sufficiently robust to deliver the intended service. Almost single-handedly, it had delivered social and educational TV to 75 per cent of India's vast population, and provided hourly cloud surveys spanning the hemisphere from Egypt to the China Sea. In doing so, it confirmed India's need for a geostationary capability.

The Arab League established the Arab Satellite Communications Organisation (ASCO) in 1976. It set the specification for a system to link the whole of North Africa and the Arabian Peninsula using a single C-Band footprint, and ordered a trio of three-axis satellites from Aerospatiale in 1981. Arabsat-1A went up on an Ariane in February 1985, and Arabsat-1B was deployed by the Shuttle a few months later. When these were retired in 1992, the third satellite, bought as a spare, was launched. By this time, demand had outstripped the capacity of a single satellite, and because more advanced satellites were not expected until the mid-1990s, ASCO purchased a 'retired' satellite to increase its capacity. If a satellite exceeds its expected service life and is still operating reliably, it can be sold on the second-hand market where the price is often little more than the insurance premium for the launch of a new satellite. This provides newcomers with a convenient method of establishing an infrastructure before investing in new satellites. Argentina, for example, took over the ageing Anik-C1 and Anik-C2 in 1993 when Telesat (by this time Allouette Communications, as the government had sold its share of Telesat) introduced its C/Ku-Band Anik-E series. Also, when Telesat retired Anik-D2 in 1991, it was leased to GE Americom to fill a gap caused by the loss of Satcom-4. In 1993 ASCO purchased Anik-D2 and operated it as Arabsat-1D. When Telstar-301 was superseded in 1994, AT&T sold it to Telesat, who used it to cover for the ailing Anik-E2, and when this returned to service several months later, sold the Telstar to ASCO, who renamed it Arabsat-1E to back up Anik-D2/Arabsat-1D and further augment Arabsat-1C until the next generation of satellites was ready. A communications satellite is rarely deactivated until it is completely worn out, or until it is no longer cost-effective in terms of the state of the art.

EVER MORE CAPACITY

In a move that was strongly reminiscent of the pioneering spirit of the Syncom, Earth Resources Technology Satellite and Applications Technology Satellite programmes, NASA developed the Advanced Communications Technology Satellite (ACTS) to show the advantages of high-capacity Ka-Band communications in the hope of stimulating the commercial sector. ACTS was sponsored by the Agency's Office of Commercial Programmes and managed by the Lewis Research Center. The Ka-Band offered the prospect of an unprecedentedly high degree of integration in global communications, but its realisation would require the development of new technology. At the heart of NASA's demonstration was a computer processor to sort and route traffic, and a microwave switching matrix to facilitate routeing at much higher data rates than was possible using conventional satellites. In particular, ACTS used steerable pencil-beams that could be aimed at specific localised sites on Earth. As the higher frequency band supported such a high data-rate, it was possible to adopt a 'burst' mode of operations in which a beam would 'hop' between several sites and transmit so much data in each time-slice that it could serve each site as if it had a dedicated link. The processor's task was to exploit this capability. ACTS was to be capable of linking supercomputers and relaying high-definition TV transmissions.

NASA had conceived the project in the early 1980s and had issued the contract to RCA Astro Electronics in 1984. Congress backed the idea because while the move to this higher frequency band was inevitable, domestic service providers would resist rendering their expensive Ku-Band systems prematurely obsolete. The development suffered repeated technical and budgetary crises, however, and the planned launch in 1989 proved impossible. At 2,540 kilograms, ACTS was too large for the PAM-D2 perigee-kick motor, so NASA used the Transfer Orbit Stage (TOS) developed by the Orbital Sciences Corporation. This had successfully acted as the escape stage for the Mars Observer spacecraft in 1992, but its first ride on the Shuttle was on STS-51 in September 1993, with ACTS. The payload was carried in an annular support cradle, and when the command was issued to detonate its pyrotechnics, a malfunction fired both the primary and secondary charges, and although a shower of débris peppered the bay's aft bulkhead there was no serious damage. The TOS successfully boosted ACTS into GTO, which manoeuvred into its slot at 105°W without incident. After initial indifference, its services were in demand 24 hours a day. Critics have argued that there was no requirement for NASA to develop ACTS, because the technology was already imminent. However, the fact remains that even though it was delayed, ACTS was the first on the scene, and its presence did eliminate the technological risk from the subsequent commercialisation.

TRACKING AND DATA RELAY

When the Skylab programme ended, NASA shut down many of the stations in its ground network. To maintain contact with the Shuttle, it planned to build a network of geostationary satellites as a Tracking and Data Relay System (TDRS). Although NASA had planned a simple design that emphasised throughput, the Department of Defense modified the specifications to provide secure communications for its Shuttle missions. NASA had intended to launch them on Atlas–Centaur rockets prior to the Shuttle's first test flight, but in their upgraded form the satellites were too big for the Atlas. As a result, STS-1 was in contact with the flight controllers for at most 20 per cent of each 90-minute orbit, and that time was fragmented. Once the Shuttle became operational, one of its most important early assignments was to deploy this network of satellites.

Table 4.2 Intended TDRS schedule, c. late 1982

Satellite	Launched	Flight	Orbiter
TDRS-A	Jan 1983	STS-6	Challenger
TDRS-B	Jul 1983	STS-8	Challenger
TDRS-C	Jan 1984	STS-12	Discovery
TDRS-D	May 1984	STS-15	Discovery

The TDRS satellites were built by Thompson Ramo and Wooldridge (TRW) for the Space Communications Corporation, a consortium of Contel and Fairchild Industries that was formed specifically to serve NASA's communications requirements. NASA would initially lease the satellites, and take ownership after six years. If the satellites operated well beyond their 10-year design life, NASA would reap the benefit, but if they failed at an early stage it would not be financially exposed. Contel bought out Fairchild's interest in 1985, and was itself taken over by GTE in 1991. The 2.25-tonne three-axis stabilised bus was equipped with a comprehensive suite of S/C/Ku-Band transponders. Each had two redundant primary systems with a 2-metre dish dedicated to the ground link and a steerable 5-metre dish for single-access S/Ku-Band that tracked and relayed signals to and from individual spacecraft. A set of S/C-Band transponders was mounted on the Earth-facing side of the bus for parallel use on a multi-user basis, and electrical power was provided by a pair of solar panels. Each transponder was required only to serve as a 'bent pipe' that sent on a transmission without processing. The system was meant to relay not only voice and telemetry for Shuttle missions, but also data streaming from satellites. Each could simultaneously serve two Ku-Band users at rates as high as 300 megabits per second, and two dozen S/C-Band users at 50 kilobits per second. A ground station was built at White Sands in New Mexico, and the two TDRS relays were designated 'East' (near 45°W) and 'West' (172°W). Arthur C. Clarke's concept had envisaged a third satellite on the far side of the planet, but NASA did not intend to use this relay point because it would require the third satellite to relay via one of the others in order to maintain contact with White Sands. NASA was content with the 85 per cent minimum coverage that a pair of satellites could provide the Shuttle.

The first – TDRS-A – was launched on STS-6 in April 1983. It was too big for the PAM motor that had dispatched the HS-376 commercial satellites on the preceding flight and in any event it was not intended to perform its own circularisation burn. The TDRS was to be deposited in geostationary orbit by the Inertial Upper Stage (IUS) that Boeing had built for the Air Force. This had two motors: one for the GTO-insertion, the other for GSO-circularisation. The satellite had its own propulsion system, but this was designed to enable it to adjust its station, not to perform radical orbital manoeuvres. The payload was carried on an annular cradle at the rear of the bay. The 17-metre stack completely filled the bay. As soon as the bay doors were opened, the orbiter turned its belly to the Sun to shade the satellite, which could not be exposed to the Sun until its environmental system was activated. The first step in bringing the stack to life was to elevate the cradle to 30 degrees. The IUS was heavily instrumented and each stage's redundant control systems were verified. The astronauts used a camera showing the first stage's nozzle to verify that it gimballed properly. The satellite itself was tested while in contact with the control centre, and once the stack had been verified the power umbilical was disconnected and the elevation increased to 60 degrees. At the appointed time, the ring-clamp holding the IUS in place was released, enabling a spring to eject it. This was the first time that the mechanism had been used, and it worked flawlessly. The 18-tonne stack – the heaviest payload yet deployed by the Shuttle – drifted over the top of the orbiter's cabin as it departed. Ten minutes later, by which time Challenger had withdrawn, the IUS activated its flight control system. After a further 30 minutes it performed a series of star sightings to update its inertial platform. As it crossed the equator it ignited its first-stage motor, which delivered 18,500 kilograms of thrust for 150 seconds then shut down. As the vehicle was now in GTO, the first stage motor used its small thrusters to align the stack and set up a slow roll to even out thermal stresses on the 6-hour climb. Several minutes before reaching apogee, the first-stage motor re-orientated the stack for the circularisation burn, then released its payload. The second-stage motor was to have delivered 2,750 kilograms of thrust for 103 seconds but, towards the end, the oil-filled seal of the gimbal deflated, the nozzle slewed off-axis and the offset thrust induced a 30-revolution per minute end-over-end tumble. As soon as the controllers noted this in their telemetry, they commanded that the satellite be released. Although TDRS-A was soon stabilised, it was in a 21,700 by 35,550-kilometre orbit that had little to recommend it as a communications relay. It was not only non-synchronous, because the IUS's burn had not yet cancelled the 28.5-degree inclination of the Shuttle's orbit it was nodding 3 degrees each side of the equator. The IUS had redundant systems to overcome many faults, but as it had no way to recover from the mechanical failure in the manifold, it was immediately withdrawn from service. Satellites that would need the IUS were withdrawn from the Shuttle's manifest pending its recertification.

NASA had set up a carefully interleaved schedule for the deployment of the TDRS network. The second satellite was to have been launched within months to enable STS-9 to fly the first Spacelab mission, as this would require a high-capacity data downlink. Setting up the network would have to await the recertification of the IUS, which might easily take a year. As TDRS-A would be NASA's only asset for the foreseeable future it was vital to manoeuvre the satellite onto its assigned station by using its thrusters. In fact, it took two months to nudge the satellite into GSO, and it did not reach its assigned slot at 41°W until early October, by which time it had consumed half of its propellant – a loss that would seriously reduce its service life. As soon as it was clear that the relay would be on line in the autumn, NASA authorised the Spacelab mission, which would have to manage with a single relay.

When the IUS eventually flew in January 1985 the performance of the first stage was less than expected, but the vehicle was able to overcome the shortfall in velocity and deliver its military satellite as planned. With the IUS reinstated, TDRS-B was assigned to STS-51E and scheduled for March but a few days before launch, a report by engineers who had analysed some timing errors on TDRS-A prompted fears of a design fault, TDRS-B was offloaded for tests and flight 51E was cancelled. Once the satellite had been verified, NASA put it on STS-51L and it was lost with Challenger on 28 January 1986. When Shuttle flights were resumed, the build-up of the TDRS network was given priority – even over the backlog of military satellites. The manifest drawn up in late 1986 envisaged operations being resumed in early 1988. A TDRS was assigned to the first flight, Department of Defense payloads to the next two flights, and another TDRS to the fourth flight, but as the dates slipped, the third and fourth flights were transposed.

The situation had slightly degraded while the Shuttle had been grounded. In late 1986 one of TDRS-A's steerable antennae developed a problem, which denied the satellite half of its Ku-Band capacity. When TDRS-C took up station at 172°W, it encountered difficulty deploying its antennae, but once this was overcome NASA finally had a relay in each primary station – although with one operating at less than full capacity. In March 1989, TDRS-D replaced TDRS-A and the older satellite was relocated to 79°W to serve as an in-orbit spare. Unfortunately, TDRS-C lost part of its Ku-Band capacity in January 1990, and in May it was offset to 174°W to allow TDRS-A to augment it at 171°W. It had been hoped to add TDRS-E in the autumn, but this was delayed to August 1991, at which time it replaced TDRS-C, which was relocated to 62°W as a spare. By this point, NASA had decided to run the satellites in pairs to ensure that neither station would be left uncovered by the total loss of a satellite, and when TDRS-F was launched in January 1993 it was placed at 46°W to supplement TDRS-D at 41°W. As the satellites were proving to be less robust than expected, NASA ordered a replacement for the satellite lost on Challenger, and when TDRS-G was launched in July 1995 it was placed at 171°W to augment TDRS-E at 174°W. Meanwhile, in late 1993, TDRS-A had been relocated to 85°E to assist the Gamma-Ray Observatory which, having lost its tape recorder, was in need of a real-time relay, and since it required only an S-Band link, TDRS-A was fully capable of handling it. A ground station was set up at Tidbinbilla near Canberra in Australia to

TDRS-C is readied for deployment from STS-26.

Table 4.3 TDRS relay network, c. late 1995

Satellite	Launched	Status
TDRS-A	Apr 1983	Retired
TDRS-B	Jan 1986	Lost
TDRS-C	Sep 1988	85°E
TDRS-D	Mar 1989	41°W
TDRS-E	Aug 1991	174°W
TDRS-F	Jan 1993	46°W
TDRS-G	Jul 1995	171°W

receive and relay its data to White Sands via the Intelsat network. When superseded in this rôle by TDRS-C in mid-1995, TDRS-A was finally retired.

It took much longer than expected to build up an effective network, and in a sense the network has yet to be completed, because even although the far-side station was finally occupied, it was 'staffed' with degraded satellites. The primary stations could cover only 85 per cent of the Shuttle's orbit. In addition, because it proved necessary to place two satellites in each station to guard against loss or malfunction, many more satellites than planned had to be launched.

In space, the orbiter swung a small steerable dish over the starboard sill, just aft of the cabin. This was deployed as soon as the payload bay doors were opened, and retracted just before they were finally closed. One benefit of a relay over the Pacific Ocean was that during re-entry, when the orbiter was immersed in ionised plasma, it could still communicate using its omni antenna. It was not of great significance to the crew, but it removed a source of anxiety for the ground controllers during this critical phase of the flight.

MARKET FACTORS

Three conclusions are immediately clear: (1) communications was the commercial satellite business; (2) the HS-376 was the preferred bus; and (3) the Shuttle was the preferred carrier for this type of satellite. This status was a significant achievement for NASA, because half of the HS-376s it released were for non-US clients, and these contracts were won against competition from the European Space Agency's Ariane rockets. In a sense, NASA had succeeded in demonstrating that there was a market for the Shuttle's services. But appearances can be deceptive. NASA's fee for deploying satellites had been calculated using early projections for flight rates, turnaround time and operating overheads. By the end of 1985 it was clear that these expectations were unrealistic. Even in 1985 NASA managed only nine flights. At this rate of flights, and with each flight costing much more than predicted, the fee did not even cover the cost. And it was this artificially low fee that was attracting clients. In fact, the situation was becoming worse. The overly generous deal committed NASA to an unrealistic fee for all satellites throughout the period of the introductory offer, and with each passing year of escalating operational costs the satellites were being ferried into space at an increasing loss. The Shuttle was busy, but it was not a viable business. For the first five years of its operation, NASA was actually trapped; and, worst of all, the US tax payer was actually subsidising foreign satellites.

Intelsat's first geostationary satellite, Early Bird, could relay 240 telephone circuits or a single monochrome TV channel. The state-of-the-art advanced at such a pace that the relay capacity increased dramatically with each new generation of satellites. The communications industry could satisfy this soaring demand only by continuously investing in new technologies. It was a constructive cycle in which the forecasted increase in demand funded the development of the technology, which not only immediately began to repay its investment but also further stimulated demand. It was this phenomenal growth rate that opened up geostationary communications satellites to private operators.

Table 4.4 Generations of Intelsat satellites

Series	First	Voice	Supplier
1/2	1965	240	Hughes
3	1968	1,500	TRW
4	1971	4,000	Hughes
5	1980	12,000	Ford
6	1989	120,000	Hughes

Because the HS-376 dominated the Shuttle's commercial manifest, it is worth examining this satellite further. Not only could its antenna footprint be shaped to the customer's territory, but its bus was sufficiently versatile to accommodate a varied package of transponders in a mix designed to suit the customer. As another crucial factor, the technology had achieved the point of being able to support a 10-year operating life. The life expectancy of the early geostationary communications satellites had been three years, which had imposed an ongoing replacement strategy that only consortia could afford. By doubling this time, the HS-333 had allowed the investment to be recouped, which made the use of this satellite a realistic option for a major telecommunications company seeking to service its clients. The HS-376 enabled individual companies to link their dispersed sites using their own satellite, and its operating life was designed to turn in a healthy profit. However, it was not simply a matter of developing more durable satellites. With technological advances, there was a switch to higher frequencies, and in the latest case, this was the switch from the C-Band to the Ku-Band. As a satellite with a life much longer than 10 years was likely to turn into a liability towards the end through obsolescence, the HS-376 was therefore an effective compromise. In a static market, extending the life of the product would compromise future sales. Telecommunications, however, was not a static market: it had always been – and will likely always be – an expanding market. The longevity of the HS-376 did not serve to limit sales. Quite the contrary:

- its low cost permitted individual telecommunications companies to run their own satellites;
- it stimulated the creation of specialist leasing companies to compete with Comsat in the US domestic market;
- it led to the creation of Comsat-style consortia that provided national services in other countries where there were no integrated communications systems; and
- it later served as the basis for the early Galaxy class, which Hughes set up as the world's largest fleet of privately-owned commercial communications satellites.

As a result, the HS-376 production line expanded with each passing year, which cut the unit cost and thus made it an even more attractive product. It appeared at just the right time to create a critical mass of orders, thus ensuring that it became the *satellite of choice*; it was, quite simply, a product whose time had come.

The pace of technological development in this market is well illustrated by the A-series and D-series of Telesat's Aniks, which had C-Band transponders for either voice or colour TV. By integrating voice, video and data, the telecommunications companies using these satellites were able to support video-conferencing, which was an innovation that benefited customers by enabling them to make more effective use of their time. Businesses buying services on these satellites found themselves faced with an unprecedented degree of interconnectivity.

Communications satellite technology exploited microminiaturisation in electronics. Each generation of solid-state electronics delivered more power at less cost than its predecessor. A satellite operator could pass on a saving to its customers without undermining revenues. When Early Bird entered service in 1965, it cost $9 to make a 3-minute telephone call across the Atlantic. By the time the Shuttle deployed its first satellite in 1982, the cost had fallen to $3. If this had kept pace with inflation, it would have been $30. Over that same time, the traffic increased from 2 million calls to 200 million calls. Thus, while the cost had fallen by an order of magnitude, usage increased by two orders of magnitude. Direct-dialling and the Internet would not be viable without a massively interconnected low-cost global communications system.

Table 4.5 Comparison of Hughes's satellites

First	Type	TV	Voice
1972	HS-333	1	1,000
1982	HS-376	24	20,000

In various forms, the HS-376 is still being manufactured. In 1997, Russia ordered its first US-built communications satellite, which was a customised HS-376. Over the years, more than 50 HS-376 satellites have been placed into geostationary orbit, and even the retired ones are still there. When someone suggested that Early Bird should be briefly reactivated to mark Intelsat's twentieth anniversary, it came back on line. At only 38 kilograms, Early Bird was a truly remarkable device for its day. How many Early Birds could a single Shuttle have deployed? As the payload bay could have accommodated five HS-376 cradles, the Shuttle could have deployed a flotilla of satellites; but it never did so, because the resulting forward centre-of-mass could have jeopardised the orbiter in an emergency landing. The largest number of satellites the Shuttle ever carried was three, which it did on four occasions, and in each case only two were HS-376s. During Columbia's refit in 1985 it was equipped with instrumentation to measure the orbiter's hypersonic flight characteristics. Once this was understood, the payload was to be increased. The Department of Defense intended to exploit this increase in capability by deploying its constellation of 24 NAVSTAR satellites in batches of four.

Under the Shuttle-only policy, the prospects for the rocket-builders had been bleak, but NASA rode the wave of commercialisation in triumph because, although it was not running profitably, it was at least able to demonstrate that the Shuttle had successfully met one of its primary technical objectives. Unfortunately, the agency's opportunity to profit from commercial satellites ended in a cloud of destruction in the chilly Florida sky on 28 January 1986.

During the years that the Shuttle was active in the commercial satellite business, eight Ariane rockets deployed 13 communications satellites, one rocket lost two others, and one dispatched the Giotto probe to Halley's comet. It is noteworthy that Ariane attracted only one HS-376: Brazil's first domestic relay satellite.

Despite the intention to phase out the Atlas, it, too, was very busy during these years. In the period when the Shuttle was actively deploying commercial satellites, 17 out of 18 Atlas rockets were successful. However, their payloads were mostly for the Department of Defense and the National Oceanic and Atmospheric Administration, and only Intelsat assigned communications satellites to the Atlas. The 13 Titan III launches each carried a single Department of Defense satellite, one of which was lost. The Delta had long been the rocket-of-choice for the small satellites sponsored by the many government agencies, but it had also attracted the earliest Anik, Westar, Satcom and SBS satellites, and over the early period of the Shuttle's activity, 6 of the 12 launches carried commercial communications satellites. Overall, therefore – even discounting the Palapa and Westar satellites that were left in low orbit by faulty kick-motors – the Shuttle was able to carry as many commercial satellites as the Delta and Ariane rockets combined, and considering that it did this in the context of a wider programme, it really performed well.

The Shuttle's grounding led to a complete reversal of the policy of operating the single National Space Transportation System. Immediately after receiving the report of the inquiry into the loss of Challenger, the Reagan administration reverted to a *mixed-fleet* policy, the Department of Defense was authorised to buy more heavy launchers and to order improvement programmes to update the rockets, and the manufacturers eagerly restarted their production lines.

Table 4.6 Satellite launches and rocket losses while the Shuttle was operational

Vehicle	Launches		Satellites		
type	Total	Lost	Total[1]	Lost	Commercial[2]
Shuttle	25	1	33	1	20
Ariane	10	1	16	2	13
Delta	12	0	13	0	6
Atlas	18	1	18	1	5
Titan	13	1	< Classified >	< Classified >	0

Notes:
1 Satellites of all types, but in the case of the Shuttle not free-flyers.
2 Commercial communications satellites.

If Challenger had not been lost, and if the Shuttle had continued deploying commercial satellites, NASA would not have been able to lift its unprofitable fee until the five-year term expired in 1988. It is debatable how much business it would have attracted if it had charged customers an economic rate. In the event, while the Shuttle was grounded, the Reagan administration ordered NASA to withdraw from the commercial satellite business, and in doing so it threw the US market wide open. Free of the Shuttle-only policy, McDonnell Douglas with its low-end Delta, General Dynamics with its medium-lift Atlas, and Martin Marietta with its Titans were all eager to exploit what were clearly highly profitable opportunities. In previous years, the rocket manufacturers had sold their goods to the government, whose agencies had worked with either NASA or the Air Force to prepare and launch their payloads. Now that NASA had left the arena, they could offer a 'full service' encompassing payload integration and launch – a service for which a fee could be charged. It was not only the commercial satellite business that was up for grabs; all the government satellites that were to have been carried by the Shuttle, including the 24 NAVSTARs, were also in need of rockets. In the past, each rocket had been tailored to a given payload mass and during these lean years competition had been squeezed out, so in this new world of rocket supremacy there seemed to be a convenient niche for each of the survivors. However, because it would require reversing the process of decay that had accompanied the phasing out of expendable rockets, it would take time for production to match this sudden rise in demand, and in the meantime all the remaining Atlas and Titan III rockets were reserved by the government.

The Delta was an effective rival for Ariane but, in contrast to the Shuttle, it would have to be run economically. Unfortunately, production had terminated when the Shuttle entered service, and there were only five left. Two of these rockets went to the Strategic Defense Initiative Organisation, two went to the National Oceanic and Atmospheric Administration (one of which was actually lost), and the last was used to compensate Indonesia for the embarrassing stranding of its Shuttle-deployed Palapa satellite.

There were simply not enough rockets to meet the sudden demand, and only

Table 4.7 Satellite launches and rocket losses while the Shuttle was grounded

Vehicle type	Launches		Satellites		
	Total	Lost	Total[1]	Lost	Commercial[2]
Ariane	9	1	17	1	13
Delta	5	1	5	1	1
Atlas	8	1	8	1	0
Titan	5	1	< classified >	< classified >	0

Notes:
1 Satellites of all types.
2 Commercial communications satellites.

Arianespace was in a position to increase its operations. It immediately won three commercial satellites withdrawn from the Shuttle: Aussat-A3, Insat-1C and SBS-5. The 9 Ariane launches during the 32 months that the Shuttle was grounded sent 16 satellites successfully on their way and wrote off another. In effect, therefore, during this time Arianespace had the commercial business all to itself.

One final point deserves to be stated explicitly: the Shuttle is a rocket; no rocket is infallible; and even the most reliable rockets will fail occasionally. The Atlas and the Titan III were considered to be well-proven systems yet, in the short time that the Shuttle was grounded, one of eight Atlas rockets failed, as did one of five Titan III rockets. In the light of these loss rates, it is clear that during its brief period in the commercial satellite business, the Shuttle was a remarkably productive transportation system.

5

Department of Defense involvement

'PACT WITH THE DEVIL'

In 1970, when NASA decided the type of Shuttle it would need, it opted for a fully reusable configuration of two piloted, winged vehicles bolted together for a vertical launch. In the upper atmosphere, the big carrier would release its small companion, which would continue on to orbit. The carrier would fly back and land on a runway at the launch site, just as the orbiter would on conclusion of its mission.

NASA's main requirement was that orbiter's payload bay should be 4.6 metres wide to enable it to accommodate the modular component of the space station that it hoped eventually to build. It opted for a bay 12 metres long with a payload capacity of 12 tonnes.

Because this innovative configuration was so different from the enormous and powerful Saturn V rocket built for Apollo, development would be expensive. In fact, if NASA was to be the Shuttle's only user, as it had been for the mighty Saturn, the cost would be prohibitive. NASA therefore set out to broaden the Shuttle's utility, and thereby build up the political support required to secure the necessary funding. The first step in this process was to refer to the Shuttle as the Space Transportation System to show that it was a *utility*, and not an end in itself. The second step was to carry out a cost–benefit analysis to indicate that if the Shuttle flew sufficiently often (on a weekly basis, in fact), the cost-per-kilogram of payload would fall below that of expendable rockets. This meant that these rockets could be scrapped, that the Shuttle could become the *National* Space Transportation System, and that the high cost of its development could be offset against the savings that would accrue over a 25-year period of operation. The most crucial step, however, was NASA's seeking a Department of Defense involvement.

In effect, the military's space programme was run by the Air Force, which had a fleet of reliable launchers derived from the Atlas and Titan missiles, so it was not very receptive to NASA's overtures. In fact, Robert Seamans, then Secretary of the Air Force, told Congress that the Shuttle was "not essential" for national security. The only way that James Fletcher, the newly appointed NASA administrator, could secure Seamans's support was to let the Air Force specify the Shuttle's operational requirements. In return, however, to ensure that the Shuttle became an essential military requirement, Fletcher demanded that the Air Force phase out its rockets and adopt the Shuttle-only policy. Rather reluctantly, the Air Force agreed, but in order to ensure the continuation of the scheduling flexibility that it had enjoyed in using its own rockets, it insisted on the right to requisition Shuttles in the interests of national security.

The most powerful rocket in the Air Force's inventory at that time was the Titan III. The Air Force insisted that the Shuttle provide twice the Titan's payload mass and three times its payload volume. Specifically, the Air Force demanded that the Shuttle's payload capacity be increased to 30 tonnes (half that into polar orbit), and that the payload bay be lengthened to 18 metres. Scaling up the orbiter for these requirements ruled out a fly-back carrier. The orbiter would fire its own engines for lift off and be augmented by a pair of massive strap-on solid rockets. However, as it would be impractical to accommodate the vast amount of propellant required within the body of the orbiter, it was decided to fit an external tank. Although a partially reusable system such as this would be cheaper to develop, it would be rather more expensive to operate. The orbiter's shape also had to be revised. NASA had planned to launch from and return to the Kennedy Space Center in Florida. It had envisaged an aircraft-like vehicle that would re-enter the atmosphere at a high angle-of-attack in order to minimise thermal stresses by using its belly as a brake, and only lowering its nose once it had slowed down. It was to have stubby straight wings for its subsonic approach and landing. The Air Force, however, wanted the flexibility to land at any convenient airbase. It demanded that the orbiter be able to lower its nose early, in order to diverge from its straight-in path. This hypersonic soaring capability dictated a delta wing for a better lift-to-drag ratio, and it also substantially increased thermal stress. In effect, the Air Force saw the Shuttle as a means of reviving the rôle of the X-20, a project it had been forced to abandon a decade previously, in which a winged spaceplane was to be launched on a rocket, make a surprise reconnaissance pass over a target on the other side of the world, and land 90 minutes later. If it was to land at the launch site, it would require a cross-range capability of 2,500 kilometres because the Earth would have rotated 22 degrees in that time. In fact, the Air Force was not merely specifying a Shuttle that would supersede its rockets; it was also expanding its own operational capability. On the other hand, NASA saw an opportunity for a trade. The Air Force would need to launch the Shuttle into polar orbit. This was not feasible from the Kennedy Space Center because to do so would involve making the initial phase of the ascent over land, and raining débris on an angry population in the event of a catastrophic failure. The highest inclination attainable from Florida was 57 degrees. At Vandenberg Air Force Base in California it was possible to launch south out over the Pacific. Given that the Air Force would have to build a Shuttle complex at Vandenberg, NASA said that it would assign one of its orbiters to the Air Force in return for gaining access to polar orbit, thereby expanding *its* operational capability. Both sides therefore gained from the deal.

This process of horse-trading led to the Shuttle's configuration being dominated by the requirements of the various government agencies within the Department of Defense.

AN INAUSPICIOUS START

Many of the satellites launched by the Air Force were not only heavy and bulky; they were also sent up to geostationary orbit (GSO). Having agreed to use the Shuttle, the Air Force set the requirements for the two-stage manoeuvring unit that was to fly its satellites to their final destinations after the Shuttle had released them in low orbit. Boeing won the contract for this Inertial Upper Stage (IUS). In case the Shuttle's development was delayed, the Air Force required the IUS to be Titan-compatible. In the event, the IUS suffered from political infighting and did not become available until the Shuttle was already flying. The first IUS actually rode a Titan into orbit in October 1982, and delivered a pair of military communications satellites to GSO. NASA had assigned the IUS to STS-6, STS-8, STS-12 and STS-15 to deliver its TDRS communications satellites, with a view to flying Spacelab on STS-9 as soon as the first two relays were operating. In December 1982, the Air Force requisitioned a flight in 1983, and was assigned STS-10. Challenger's maiden mission was to have been in January 1983, but this was postponed due to a problem with its engines. Alarmed by this three-month delay, the Air Force placed a late order for a dozen expendable rockets to guarantee that it would have a stock of launch vehicles with which to dispatch any satellites that NASA could not carry at short notice.

Six hours after launch on 4 April as STS-6, Challenger deployed the TDRS/IUS stack without incident. The IUS's first stage's solid-propellant rocket motor burned successfully to enter geostationary transfer orbit (GTO), and on reaching apogee the second stage's motor ignited for the circularisation burn. But towards the end of this burn, the actuator for the thrust-vectoring system failed and forced the gimballed nozzle off axis, sending the vehicle tumbling. Ironically, the IUS had been fitted with redundant avionics to make it the most reliable transfer-stage ever built, but it could not recover from a mechanical fault of this nature. As soon as the flight controllers at the Air Force's operations facility in Sunnyvale, California, realised what had happened they commanded the upper stage to release the satellite, which was eventually able to reach its operating station by firing its own thrusters. All the Shuttle payloads that required the IUS were postponed pending the resolution of the problem.

Table 5.1 Military Shuttle assignments, c. mid-1982

Flight	Date	Orbiter
STS-10	Nov 1983	Challenger
STS-13	Mar 1984	Challenger
STS-22	Nov 1984	Discovery
STS-25	Feb 1985	Discovery
STS-27	Apr 1985	Columbia
STS-30	Jul 1985	Columbia

RECONNAISSANCE

When the IUS flew again, it was on STS-51C in January 1985, and this time it was the first stage that caused concern. Although the solid motor fell short of its nominal thrust, the inertial flight control system noted this and fired its thrusters to make up the shortfall in velocity of 16 metres per second. The circularisation burn placed the payload on its assigned station. Eavesdropping 'ferret' satellites were introduced in the 1960s. To be effective, the capabilities of such satellites rely on secrecy, but it is likely that the 300-kilogram Rhyolite that made its appearance in 1973 was sent up to listen in on telemetry from weapons systems, and the rôle of the 1.2-tonne Chalet that joined it in 1978 was to monitor voice links. This new satellite's rôle was meant to be a secret, but the *Washington Post* reported that it was called Magnum, and had been ordered by the National Security Agency when it lost its listening post in Iran following the overthrow of the Shah in 1979. Its rôle was to listen to the entire radio spectrum emanating from the Warsaw Pact and relay its 'take' to Washington. Since the IUS/Magnum stack was 17.5 tonnes, the second one could not be offloaded to an existing rocket in the aftermath of the loss of Challenger, so it had to wait until the Shuttle resumed flying, and was deployed by STS-33 in November 1989.

Another classified reconnaissance satellite was exposed in the press long before it could be launched. In mid-1987, with the Shuttle grounded, *Aviation Week & Space Technology* noted that an imaging radar satellite with 1-metre resolution would not only be able to track armoured vehicles in the field, but would also be able to classify them and thereby provide an order of magnitude improvement in battlefield reconnaissance capability. No such radar had yet been demonstrated, however. The US Navy's SeaSat, which had operated for a few months in 1978, had a radar with a resolution of approximately 25 metres, as did the Soviet ocean-surveillance satellites, but a radar operating at a shorter microwave frequency would be necessary to track a vehicle. In addition to monitoring armoured forces, such a radar could easily identify mobile ICBM launchers, and find targets for follow-up inspection using imaging satellites, thereby policing the treaty limiting the deployment of these missiles. This high-resolution radar would also enable the Defense Mapping Agency to survey the perimeter air defences to produce safe routes for terrain-following cruise missiles. The multifaceted benefit to be derived from possessing such a system – not least that it could see through cloud – guaranteed that it would be built as soon as it was technologically feasible. The first such satellite, named Lacrosse, was launched by STS-27 in December 1988. Its importance was indicated by the fact that it was assigned to the first military flight after the Shuttle re-entered service. As radar reconnaissance was not feasible from geostationary orbit, this satellite had no need of the IUS. However, for its ground track to cover most of the Soviet Union it had to be placed in a high-inclination orbit. Atlantis was launched at 57 degrees, which was the highest inclination it could use without having to fly a dog-leg over the Atlantic. The satellite was lifted out of the bay by the RMS and checked out. The 23-metre solar panels proved difficult to deploy, but once everything was verified the satellite was released. Once the orbiter had cleared the scene, the satellite boosted itself up to almost 680 kilometres. If Challenger had not been lost, this satellite would probably have been launched in 1987, but because it had been designed to exploit the payload bay, it had to await the Shuttle's return to service.

At the end of the Second World War, the United States had felt invulnerable to attack. It alone had the atomic bomb, and it believed it had a decade's lead on the Soviet Union. It came as a shock when the Soviets detonated their own bomb in 1949. President Truman immediately ordered development of the more powerful thermonuclear fusion bomb. The Strategic Air Command's mission was to retaliate if the United States was attacked. During the 1950s, it progressively modernised its fleet of strategic bombers to be ready to carry out the policy of massive retaliation, but it had little knowledge of its targets. At first, it issued pre-war commercial maps to its pilots, and while these were sufficient to find a city, military targets were not marked. It needed photographic reconnaissance to compile a meaningful target list. The super-high-flying subsonic U-2 aircraft was built in the mid-1950s specifically for this purpose, but it proved vulnerable to a newly deployed surface-to-air missile. What was required was a camera in orbit! As soon as the Air Force had converted its Thor missile into a satellite launch vehicle it developed a system for returning a film canister to Earth. Several cameras, collectively called Keyhole (KH), were introduced in the 1960s. The KH-4 Corona had a resolution of several metres from an altitude of 200 kilometres, and was used to find interesting targets for the KH-7 and KH-8 Gambit that skimmed the upper atmosphere at perigee to take detailed pictures with a reported resolution of 15-cm. As both recorded imagery on film and returned this to Earth for processing, their individual missions were of short duration – typically two weeks. The Corona was superseded in the 1970s by the KH-9 Hexagon, which to some extent combined these rôles and had several film-return canisters to increase its useful life. The operating procedure was transformed in 1976 by the introduction of a CCD-based imaging system whose imagery could be radioed to Earth to provide near real-time intelligence, and since there was no need to make film drops, a satellite could be used almost indefinitely. The KH-9 model, which was appropriately known as Big Bird, rode a Titan IIID. By the time the KH-11 Kennan had grown to need a Titan 34D, it was a 20-metre-long cylinder weighing in at over 13 tonnes. In effect, each of these later satellites was "a Hubble Space Telescope aimed at the Earth rather than into space". Although supported by the Air Force, these satellites were operated by the National Reconnaissance Office for the multifaceted US intelligence community.

It was widely speculated that an even bigger reconnaissance satellite had been designed for deployment by the Shuttle, and this was referred to as the 'KH-12' in expectation. The fact that the classified satellite deployed by STS-28 in August 1989 was dropped off in a 57-degree orbit and did not require an IUS strongly suggested that it was the long-awaited newcomer. However, all speculation as to its nature was greeted by official silence. After Columbia had moved clear, the satellite's orbit was raised to 460 kilometres and then it seemed to disappear, prompting the suggestion that it had manoeuvred into a highly eccentric high-inclination orbit of the type utilised by the Satellite Data System (SDS) that made their appearance in 1976 (the same year as the KH-11) and spent most of their time high above northern mid-latitudes acting as real-time relays for the imagery produced by KH-11s in low orbit, and indeed this theory was later confirmed. A second such satellite was released by STS-38 in November 1990 and a third by STS-53 in December 1992.

The satellite that was released by STS-36 in February 1990 caused considerable speculation because Atlantis flew an inefficient dog-leg over the Atlantic to achieve a 62-degree inclination. The Stabilised Payload Deployment System, which was being used for the first time, deployed the 17-tonne satellite by rolling it over the port sill. When a tracking radar observed four distinct objects in its orbit three days later, the Press jumped to the conclusion that the satellite had blown up, and criticised NASA for another "wasted" Shuttle flight. But six months later, it emerged that the satellite was in an 800-kilometre circular orbit at the increased inclination of 65 degrees – an orbit that gave excellent coverage of the Soviet Union. Evidently the débris observed by the radar consisted of the various covers jettisoned during its check-out sequence prior to manoeuvring to its operating orbit. It was later acknowledged to have both digital imaging and signals-intelligence receivers for the Central Intelligence Agency and National Security Agency. It is thought to have been an experimental 'stealthy' reconnaissance satellite, and to have been a disappointment.

COMMUNICATIONS

The Department of Defense had been very impressed by NASA's Syncom, the first relay satellite to be placed in geostationary orbit. As it was eager to integrate its communications into a single global system, it ordered the development of a series of satellites that would provide secure communications. This required a significant advancement of the technological state-of-the-art because each satellite was to provide 1,300 channels, with transponders operating in the X-Band. And whereas Syncom had been a mere 65 kilograms, this new satellite was to be a 600-kilogram monster that would require the Titan IIIC, at that time the most powerful rocket in the inventory. As development was expected to be protracted, a compromise was ordered as a stop-gap. A single-channel relay would be small enough for a Titan IIIC to launch them in batches of eight. A 24-satellite constellation was to be established in low orbit, which, although not providing a global uninterrupted network, would be a major step towards integrated communications. The operational network was to be the Defense Satellite Communications System (DSCS). Deployment of the Phase I stop-gap was completed in mid 1968. Although the initial satellite had a design life of two years, most of the system was still operational when the first of the Phase II geostationary relays was introduced in 1971. Developed by Thompson Ramo and Wooldridge (TRW), it was a 3-metre diameter spin-stabilised drum with a despun antenna, and was powered by peripheral solar cells. Each satellite was expected to last five years. The Phase III offered only a 50 per cent increase in capacity, but its electronics were better hardened and its service was considerably more secure. Built by GE Astro Space, it was a three-axis stabilised bus powered by large solar panels. The first was launched in 1982, but because the Titan 34D was unable to carry two of the 850-kilogram satellites, this inaugural flight of the IUS delivered a Phase II and a Phase III.

Shortly after the Shuttle's first test flight in 1981, the Air Force announced that building up this new DSCS network would be a high priority for the Shuttle, but the grounding of the IUS in 1982 forced it to continue to use the Phase II satellites. After the successful IUS flight in early 1985, the Air Force booked a Shuttle to build up its degraded DSCS network. When STS-51J flew this mission in October 1985, it climbed to 500 kilometres – slightly higher than usual – to enable the IUS to deliver two Phase III satellites. Another pair was scheduled for November 1986 to complete the network, but in the aftermath of the loss of Challenger, as part of its withdrawal from the Shuttle, the Air Force offloaded them. However, a Titan 34D had failed in late 1985 and a second was lost in April 1986, both of which carried reconnaissance satellites. After successive failures, the Titan 34D was grounded. It did not fly again until February 1987, and because reconnaissance took priority over communications the next DSCSs were not sent up until September 1989, on the last such rocket. The Shuttle's rôle in this programme, although curtailed by the loss of Challenger, at least had the advantage of being able to dispatch the satellites in pairs. As a result of the scramble for rockets after the Shuttle's grounding, and the mass offloading of Air Force payloads, the network was not completed until 1993, by which time the early ones were in need of replacement.

In 1978, the US Navy decided to design the next generation of Syncom satellites to exploit the Shuttle's capacious payload bay, and awarded the contract to Hughes. The resulting HS-381 took advantage of the manner in which NASA's cost–benefit analysis had deemed a squat satellite to be 'cheaper' than a long thin one that did not occupy the full width, because such a satellite occupied more of the payload bay's length. In contrast to the HS-376 that Hughes had designed for the commercial sector as a 2.8-metre-tall drum of just over 2 metres diameter in order to suit the shroud of a rocket such as the Delta, the HS-381, which was only marginally taller, was over 4 metres wide. Whereas the HS-376 had been mounted on a perigee-kick motor and set upright in the bay, the HS-381 – which incorporated its own motor – was carried on its side and deployed 'frisbee-style'. Although it weighed 7 tonnes, half of this mass was propellant for the trip up to GSO. As a result, this series was not delayed by the grounding of the IUS. The network required four satellites, but a fifth was built as a spare. As the satellites were to remain the property of the manufacturer and be leased to the Navy, they were called Leasats.

Between August 1984 and August 1985, Discovery deployed four Leasats, the first on its maiden trip – a flight that was memorable for being the first to release three satellites. The spring-loaded cradle not only set the satellite spinning at 2 revolutions per minute for stability, it also triggered the switch that activated the sequencer. The satellite's first act was to raise its omni antenna to enable Hughes to check it out prior to it setting off for GSO. With two Leasats safely on station, it came as a shock when the third rolled out of the bay and failed to activate. It seemed as if the switch had jammed. In an attempted repair, the astronauts cut and shaped the stiff plastic cover from a flight plan into a 'fly swatter', taped this to the end of the RMS, and then dragged it across the switch in an attempt to flip it, but it had no effect and they were obliged to abandon the satellite. Once Hughes realised what had gone awry, it devised a way to kick-start the satellite. After Discovery had deployed the next satellite, Leasat-4, it rendezvoused with Leasat-3 and two spacewalkers installed the bypass that brought the derelict satellite back to life. Although Leasat-3 was successfully manoeuvred up to its assigned station, the network remained incomplete because Leasat-4 fell silent soon after reaching its operating station. As the terms of the lease did not oblige the Navy to pay for a satellite until it was on station and fully operational, the time Leasat-3 spent stranded in low orbit, and the loss of Leasat-4, were borne by Hughes. The spare satellite, which had been booked on STS-61C, was withdrawn for inspection with the intention of reassigning it once it had been verified, but the Shuttle manifest for 1986 was dominated by missions to observe Halley's comet, to launch the Galileo and Ulysses interplanetary probes, and to deploy the Hubble Space Telescope. The Challenger accident rendered this schedule irrelevant. It was not possible to offload Leasat-5 because it was too wide for a rocket, and had to wait. However, the fact that it was not launched until almost 18 months after the Shuttle had resumed flying showed its relatively low priority.

The Shuttle not only took Leasat in its stride; it demonstrated its versatility by retrieving and activating a satellite that otherwise would have been written off, doing so at minimum cost since the repair was added to a previously planned flight. If the OMV-tug had been built, it may have been feasible to retrieve Leasat-4 from GSO to enable the Shuttle to return it to Earth for repair and relaunch, and if these tasks had been integrated into established missions, this could also have been achieved on the fly.

EARLY WARNING

Although the first intercontinental-range ballistic missile (ICBM) had been tested in secret by the Soviet Union, its power was dramatically demonstrated on 4 October 1957 when it was used to launch Sputnik, the world's first artificial satellite. From the point of view of the Department of Defense, the threat it faced was transformed overnight. After losing its monopoly over the atomic bomb the United States had created a sophisticated air defence system integrating perimeter radars and missile-armed interceptors, and in the Strategic Air Command it had created the world's most powerful strike force. It had been confident of its ability to detect Soviet bombers and shoot them down before they could do damage, and meanwhile launch an overwhelming counterstrike. It had erected a fence to protect itself, but the ICBM gave the Soviets a means of hopping over this barrier. Furthermore, it could not be intercepted. Even worse, its great speed cut the time between mounting an attack and having it delivered from half a day to half an hour, which exposed the United States to a devastating surprise attack. This threat created a new strategic mission for the Department of Defense. It was crucial that it be able to provide the President with the greatest possible warning that an attack was underway. The first step was to build powerful radar systems to scan the approach routes in order to detect missiles as they rose over the horizon. However, to gain the maximum warning time, it would be necessary to spot the missiles as they were launched, and this could be done only from space.

In early 1960, the Air Force started to send up Missile Defense Alarm System satellites to test sensors for detecting missiles as they rose out of the atmosphere. Detecting a plume of hot exhaust gases proved to be straightforward, but as signal-processing technology was crude it was difficult to eliminate the false signals from cloud reflections, forest fires and oil fields burning off gas. As it was also necessary to experiment with operational strategies, some satellites were put into low polar orbit, others into high polar orbit and, later, into geostationary orbit. It was 1970 before a mature system – known as the Integrated Missile Early Warning System – could be made operational. Five satellites had been sent up to geostationary orbit by 1975. This TRW design had a telescope projecting from the top of a drum-shaped body fitted with a cruciform of fold-out solar panels augmenting the conformal cells, the telescope tube being inclined to the satellite's rotational axis so that the narrow field of view of its 2,000-pixel focal-plane-array infrared detector swept a wide arc across the Earth's surface. The first Phase II was launched in 1976, under the banner of the Defense Support Program. It could detect a missile in the first few minutes of flight, pinpoint the site, and, as it tracked the missile, report its azimuth. It was not as if this capability was difficult to test – at that time the Soviets were launching rockets with satellites every few days.

As the Integrated Missile Early Warning System satellites grew, the Titan IIIC gave way to the more powerful Titan 34D. In the early 1980s the Air Force set out to take full advantage of the Shuttle's cavernous payload bay for its Phase III model, and produced a 10-metre-long giant that weighed in at 2,360 kilograms and required an IUS for the ride up to GSO. The problem with the IUS totally disrupted the Air Force's schedule. Unable to launch a Phase III, the Air Force built a hybrid satellite in which the new 6,000-pixel detector and its associated signal processor were retrofitted into a left-over Phase II. It was launched in December 1984 as a stop-gap. As a result, when the IUS resumed flying in 1985, the Phase III was a lower priority and, unfortunately, the Shuttle was grounded by the loss of Challenger before one could be launched. The plan to send up another hybrid was frustrated by the grounding of the Titan 34D in 1986, and this did not appear until November 1987. The importance of orbiting the first Phase III is evident from the fact that it was assigned to the inaugural launch of the Titan IV. Although delayed by six months of teething troubles, the new rocket performed flawlessly when it finally got off the ground in June 1989. After the second Defense Support Program satellite followed in 1990, it was announced that STS-44 would carry the third. Named 'Liberty' by the astronauts, it was released without incident in November 1991. In 1994 the Air Force considered launching another one on the Shuttle, but this was not done, and when this satellite was launched in early 1997 it was on the first of the upgraded Titan IV-B rockets.

Superficially, therefore, the Shuttle could be said to have played an insignificant part in the Integrated Missile Early Warning System's Defense Support Program. This view, however, fails to acknowledge the fact that this satellite, like Magnum and Lacrosse, had been designed with the Shuttle in mind. Having been quick to exploit the Shuttle's payload bay capacity, the Air Force was then at a disadvantage when the Shuttle was grounded. As none of its remaining rockets could lift its new satellites, it had to await the development of the Titan IV, which was intended to deliver the performance originally set for the Shuttle. There can be little doubt that the Air Force would have preferred to develop the Titan IV a decade earlier, and, in retrospect, this would have been wise, but providing rockets to complement the Shuttle had been contrary to the entrenched policy at that time. If the IUS had not suffered teething troubles, then the Shuttle would have been able to carry many more of these Air Force satellites before it was grounded. With more of its high-priority reconnaissance, communications and early-warning satellites already in space, the crisis that faced the Department of Defense following Challenger's loss would not have been so serious.

NAVIGATION

When the US Navy placed ballistic missiles in submarines, it had to devise a more accurate means of navigation. The great operational advantage of a submarine was that it could remain out of sight, yet it had to be capable of navigating with sufficient accuracy to enable it to target its missiles. The Navy adapted an inertial navigation system developed by the Air Force for one of its missiles, but because such devices suffer from drift, they have to be periodically updated. This merely reimposed the original problem, but with a much more demanding requirement for accuracy. If the submarine needed the inertial system to find its approximate position, how would it find its actual position with sufficient accuracy to cancel the inertial unit's drift?

It was possible to navigate offshore by triangulating signals from several Loran radio beacons, but this was not practicable for a submarine submerged in mid-ocean. However, a system of satellites would allow radio triangulation. All the submarine needed to do was raise its radio mast. The Navy deployed the Transit system in the 1960s, and over successive generations this system served through to the late 1970s. Its utilisation was essentially limited to stationary submarines, however, because it required bearings from several satellites over a period of about an hour to establish an accurate fix. Certainly it was no use to an aircraft in flight, but the Air Force wanted to find a way of navigating in real time with a very high degree of accuracy. Furthermore, whatever system was developed, it had to be usable by a compact and lightweight receiver, because there was no room for an electronics cabinet on a fighter aircraft, a missile or a smart bomb. The design it selected was the NAVSTAR Global Positioning System (GPS). Several sets of satellites in 20,000-kilometre-high orbits inclined at 55 degrees – with the planes of the various sets distributed in longitude – would provide an accurate fix, but the trick in simplifying the receiver was to design the complexity of the system into the satellites. The first six satellites were sent up between 1978 and 1980 to test the feasibility of the system, and the results were excellent. Transit could provide a fix within 200 metres, and it was strictly two-dimensional. When in operation, the GPS network was to have sufficient satellites to fix the receiver's position to within 15 metres, to do so on a continuously updated basis, and to supply altitude. Although it was intended for military use, it was clear from the beginning that the system would have a multitude of civilian applications. A parallel triangulation algorithm therefore was added to the system – one that did not operate in an encrypted mode and offered 100-metre accuracy. Also, true to the requirement that the receiver be simple, it was made the size of a pocket book and could be bought for a few hundred dollars. The operational GPS system was to have six sets of four satellites, and establishing the network was to be a major contract for the Shuttle. By ferrying satellites up in batches of four, each flight would be able to populate one plane of the network. Even though they were not destined for GSO, at 1,600 kilograms each satellite would require the PAM-D2 kick-motor. Deployment was to start in 1987, and the network was to be operational within two years. After the loss of Challenger it was hoped that deployment would slip by no more than a year, but in January 1987 the Air Force offloaded the programme from the Shuttle and placed an order for the new Delta II, even although this would be able to install them only one at a time. The first satellite went up in February 1989, and the last in March 1994. If the Shuttle had not been grounded, and deployment had proceeded as originally planned, the GPS system would have become available five years earlier.

STAR WARS

STS-4 in June 1982 carried the first secret payload for the Department of Defense. It comprised a pair of instruments on a pallet in the payload bay. The telescope of the Ultraviolet Horizon Scanner measured the far-ultraviolet characteristics of the horizon to provide basic data for the development of a system that was to give early warning of ballistic missiles by detecting them as they came across the horizon. This task was complicated by the fact that ballistic missiles from the Soviet Union would approach over the north pole, and be masked by emissions stimulated by the Earth's magnetic field. The Strategic Defense Initiative Organisation (SDIO), established by Ronald Reagan in 1983 to develop technologies by which ballistic missiles could be detected and intercepted, was to devote considerable effort to solving this problem. In addition, the Cryogenic InfraRed Radiance Instrument in Space (CIRRIS) was to look down at the Earth to record the spectral signature of the exhaust plumes of air-breathing engines. This data was to calibrate another package, called Teal Ruby, that was then in an advanced stage of development and was to be tested on the Shuttle before being deployed operationally as a satellite. Teal Ruby was to detect and track aircraft – particularly cruise missiles – in order to provide early warning of an attack. The advantage of tracking an aircraft passively from space, in contrast to tracking it by radar, was that the target was not made aware that its presence had been noted. A constellation of such satellites in polar orbit would be able to monitor the areas in which Soviet cruise-missile submarines would be obliged to operate in order to strike the continental United States. Furthermore, even if the launch of a missile was not directly observed, the exhaust trail was expected to remain detectable long enough to identify the point of origin, so that a hunter-killer submarine could be sent in to track down and dispose of the launch platform. The telescope utilised an infrared detector that was cryogenically chilled by liquid helium. Although sophisticated processing was required to distinguish an exhaust plume from the thermal background, this task was simplified by the uniformity of the open ocean background. Unfortunately, the trial was frustrated when the cover of the telescope failed to open.

The sensor technology needed for Teal Ruby involved a significant advancement of the state of the art. It used a 3-metre focal length telescope and a 250,000-pixel detector. Dividing the image into small cells provided high spatial resolution, and by measuring different spectral bands it was possible to identify the nature of any given source. From the intended 600-kilometre operating altitude, the telescope provided a 2-degree field of view and the detector had a surface resolution of about 100 metres. The use of a staring array detector was crucial because it continuously sensed the entire field of view, whereas a scanning detector, by its nature, would have sensed points within the field in sequence. It was not the time that it took to sense a specific field of view that was important, it was the time the detector could devote to each cell within that area. In the time it took a scanner to sample the entire field of view, it would have sensed each cell for only a fraction of that time. With a staring array, each cell would have been sampled for that entire period. Since the sources were so faint, relative to the background, it was this ability to integrate that made detection possible. In 1977, in a climate of concern that its air defence system was vulnerable to the latest Soviet cruise missiles, the Air Force set out to test whether this space-based tracking system would work in practice. With the Shuttle expected to start flying in 1979, it was hoped to have Teal Ruby ready for a trial in 1980, and it was given the payload designation DoD80-1. In the event, the Shuttle's protracted development did not hold up Teal Ruby, as the cryogenically cooled sensors proved more difficult to make than expected, and it was not actually ready to fly until 1983. By this time, it had been decided to fly Teal Ruby from Vandenberg so that it could be realistically tested in a polar orbit. Unfortunately, construction of Space Launch Complex 6 (SLC 6) had been postponed when it became evident that the Shuttle would not be available for military service as early as originally expected. Although it was hoped to finish SLC 6 in late 1985, this slipped to early 1986. Nevertheless, as soon as everything was finalised, Teal Ruby was assigned to the first flight from SLC 6. As part of its deal with NASA, the Air Force was to have virtually exclusive use of Discovery, which was to make flight 62A in July 1986. The Teal Ruby test was regarded as so important that Air Force Under Secretary Pete Aldridge was to accompany it as an observer. It was to be deployed and left for a year, during which time its abilities would be assessed, then be retrieved. In the aftermath of the loss of Challenger, the Air Force lost interest in the Shuttle, halted work on SLC 6 and, in a series of steps, mothballed and then decommissioned the facility.

When the Shuttle resumed flying in 1988, Teal Ruby was assigned to a classified Department of Defense flight at that time scheduled for early 1990, but six months were lost in 1990 by hydrogen leaks which grounded two of the orbiters, and it was slipped to early 1991. However, during this hiatus the payload was changed. It had been planned to fly Teal Ruby as a technology demonstrator prior to deploying the system operationally, but by this point the sensor that had seemed so innovative when it had been proposed in the mid-1970s had been rendered obsolete. As a result, when the Department of Defense flight finally lifted off in April 1991, as STS-39, it carried CIRRIS to obtain spectral signature data on airglow and auroral emissions in the 2- to 25-mm wavelength range to identify wavelengths that would facilitate tracking ballistic missiles in the coasting phase of their path over the pole, free of ionospheric emissions. This task was assisted by the occurrence of a major auroral display early in the flight. In addition, the SPAS free-flyer that had flown twice on early missions had been purchased from Germany and fitted with a suite of sensors for the Infrared Background Signature Survey (IBSS). After the SPAS had been deployed, Discovery moved 10 kilometres ahead of it. The free-flyer carried a multispectral suite of imaging sensors and an imaging spectrograph sensitive in the infrared. It observed the orbiter execute a series of tight manoeuvres and measured the spectral characteristics of the thruster plumes as the OMS engines were fired one at a time to fly in circles. Xenon, neon, carbon dioxide and nitrous oxide were sprayed in turn from canisters in the bay, while the free-flyer observed their interaction with the Earth's magnetic field to establish whether exhaust gases made a characteristic glow. The Chemical Release Observation experiment involved ejecting canisters that released a selection of rocket propellants once 200 kilometres away, so that the effect could be studied by the free-flyer, the orbiter and airborne sensors. The aim of these experiments was to determine whether it would be feasible to spot a ballistic missile's bus as it manoeuvred to dispatch its warheads to their individual targets. This could be done either directly, by observing the bus's thruster plumes, or indirectly, by observing the glow from the exhaust products reacting with the Earth's magnetic field. Another objective was to observe a simulated leakage of propellant from a damaged missile. It had been deemed unacceptable for an anti-missile system to rely on a terminal-phase interception, and if the missiles were to be intercepted in the post-boost cruise, such testing was crucial to the SDIO's goal of developing the technologies to track them while they were above the atmosphere.

In February 1990 a Delta rocket launched the Laser Atmospheric Compensation Experiment (LACE) satellite into a 550-kilometre-high circular orbit for the SDIO. Its primary rôle, as its name suggested, was to serve as a target for ground-based laser beams. Once on station, it extended a cruciform of sensors. When a laser was aimed at the target, the sensors measured the extent to which the beam was disturbed by its passage through the atmosphere. In this way the efficiency of a corrective-optics system designed to stabilise the beam could be evaluated. The laser was fired by the Air Force's Maui Optical Station (AMOS) in Hawaii. The ability to direct a laser onto a small target in space was another objective of the Star Wars scheme. In June 1985 AMOS fired a laser at Discovery to test its ability to put the beam on a target in low orbit. For this high-precision tracking experiment, the crew placed a 20-centimetre-wide corner-cube reflector in the porthole of the orbiter's hatch. LACE also carried the UltraViolet Plume Instrument (UVPI). This was a CCD imager designed to detect a rocket's exhaust plume. As airglow and auroral emissions in the 2,300- to 3,450-Ångstrom wavelength range were familiar, it was hoped that UVPI would be able to detect its targets against this background. This followed earlier experiments in which, in February 1988, a Delta rocket launched a spacecraft that ejected a series of packages and then tracked them by their exhaust while they manoeuvred nearby, and in 1989 a satellite successfully tracked sounding rockets by their exhaust plumes. In contrast to these technology trials, UVPI was a long-term engineering evaluation, and every opportunity was taken to test it with a well-defined target. It had attempted to spot thruster firings on STS-36 and STS-39, but had failed; and the plan to do so on STS-44 was curtailed after an inertial measurement unit failed; but was successful on later missions. AMOS routinely observed Shuttles to secure spectral signatures of thruster firings and water dumps, and tried to track them by their interaction with the atomic oxygen in the ionosphere – the so-called Shuttle glow phenomenon.

Prior to the loss of Challenger, Atlantis had been scheduled for a May 1987 flight with a Spacelab devoted to SDIO experiments. This Starlab comprised a full-length pressurised module and a pallet. A laser in the main module was to have fired through an axial porthole in the aft bulkhead and been redirected by a mirror on the pallet to evaluate its ability to 'hit' a variety of test articles carrying laser detectors. When the laser had been calibrated using LACE, small Starlet satellites were to have been released by the Shuttle, and Starbird sounding rockets fired from the ground to serve as targets. In the immediate aftermath of the accident, NASA had hoped to be able to resume flying in early 1987, and Starlab was reassigned to June 1989. Once it became apparent that the Shuttle would not fly until 1988, Starlab slipped to March 1990, and when the impracticality of attempting to build up to a monthly flight rate was finally recognised after the Shuttle had resumed service, crucial payloads were brought forward in the schedule and others were either offloaded or permitted to slide. The Starlab was pushed back, first to September 1991, then to January 1992, and finally deleted. But by this time, the international situation that had defined the SDIO's mission had been transformed. In 1991 the Soviet Union – the entity that Ronald Reagan had dubbed the 'Evil Empire' – collapsed and was superseded by the Commonwealth of Independent States, of which Russia was the largest member, and Russia and America were partners in a 'New World Order'. For America, the new threat was from 'rogue states' equipped with nuclear-tipped missiles. In 1993 the SDIO was renamed the Ballistic Missile Defense Organisation and refocused for Theater rather than Strategic Defense. Terminal-phase interception is viable in such an attack – although only just – so the plan to develop an orbital battle-station was dropped. The space-based sensors, however, were retained in order to provide early warning, and the Shuttle continued to play a rôle in testing. By February 1995, when the SPARTAN-204 free-flyer carried the Far-UltraViolet Imaging Spectrograph for the Naval Research Laboratory, the state of international relations were such that it was a Russian cosmonaut, Vladimir Titov, who conducted the test. Operating the RMS, he hoisted the free-flyer from the bay and aimed its instrument towards the orbiter's tail to monitor the Shuttle glow, the propellant that was leaking from one of the aft thrusters, and the plumes from thruster firings. It was then placed back in the bay. A few days later, he deployed it to carry out observations of the ultraviolet sky in order to characterise the background against which ballistic missile sensors would have to work.

SPACEFLIGHT ENGINEERS

In the mid-1960s the Air Force recruited pilots for its Manned Orbiting Laboratory. They were to ride a Gemini spacecraft mated to a pressurised module that would be launched by the powerful Titan IIIM. Their primary task was to conduct military reconnaissance with the KH-10 camera. After a tour of duty lasting about a month, they would pack up their film, vacate the module and return to Earth. As the project was preparing for its first test flight in 1969, it was cancelled by the incoming Nixon administration because the automated form of the camera, the KH-9, promised to be able to achieve the same results at much reduced cost and greater operational flexibility. With the plan to run its own man-in-space programme in jeopardy, the Air Force told its astronauts that they could either return to their old jobs or transfer to NASA. The youngest members of the team – Dick Truly, Karol Bobko, Bob Crippen, Gordon Fullerton, Hank Hartsfield, Bob Overmyer and Don Peterson – opted for NASA despite the fact that Deke Slayton, who was responsible for crew assignments, said that they had no chance of flying Apollo or Skylab missions. Over the years they made themselves invaluable behind the scenes, and when the Apollo astronauts retired in droves, these 'cuckoos' ruled the roost and claimed the Shuttle as their own. The days of plucking a capsule out of the ocean were over. They would land their spaceplane on a runway. Only a few of NASA's own astronauts stayed on, the most notable being Fred Haise, Vance Brand, Joe Engle, Jack Lousma and, of course, John Young, who went on to command the Shuttle's inaugural flight.

In 1979 the Air Force recruited a new class of 'spaceflight engineer' to ride the Shuttle and supervise classified payloads. In effect, they were to be the military equivalent of what NASA dubbed payload specialists, who flew for institutional collaborators and commercial customers to oversee specific payloads. They would receive complete briefings on how to check out, deploy and, if necessary, spacewalk to tend to secret payloads, so that it would not be necessary to supply such details to NASA. Each secret payload was to shepherded by at least one member of the Air Force's elite cadre. The first flight opportunity was to be in 1983, but the fault with the IUS led to a postponement. In the interim, the Air Force assigned David Vidrine to the SolarMax repair mission to observe this ambitious spacewalk, but when it heard that there would be live video of the spacewalk it decided that *in-situ* observation would have "no value" and he was withdrawn from the crew. It was January 1985 before Gary Payton became the first spaceflight engineer to supervise the deployment of a classified payload. He was followed by William Pailes a few months later.

Once the Shuttle was operating on a more or less routine basis, NASA set out to broaden the public's sense of participation by instigating Teacher-In-Space and Journalist-In-Space. But first, Senator Jake Garn and Congressman Bill Nelson – the chairmen of the Congressional committees controlling NASA's budget – were given flights. Air Force Under Secretary Pete Aldridge ousted one of the two spaceflight engineers assigned to oversee the deployment of Teal Ruby, so that he could fly the first Shuttle from Vandenberg, but his effort was foiled when Christa McAuliffe, the winner of the Teacher-In-Space competition, perished on Challenger on that fateful day in January 1986. On being promoted to Air Force Secretary, Aldridge initiated development of the Titan IV, scrapped SLC 6, and wound down the Air Force's commitment to the Shuttle as quickly as he could. The spaceflight engineers were disbanded in August 1988, just one month before the Shuttle resumed flying, and the dream of 'blue suits' in space faded once again.

MILITARY-MAN-IN-SPACE

Even although the Air Force disbanded the spaceflight engineers, this did not mean that there was no place on the Shuttle for a military officer. Paul Scully-Power was a Navy oceanographer who flew on Challenger in 1984 to study oceanic phenomena. He was astonished by the subtle texture of the sea, and by the fact that this had not been visible on the film he had studied. The human eye was much more adept than film at distinguishing subtle variation in hue. In fact, he discovered a system of giant eddies in the Mediterranean Sea that had not been known to seafarers. When STS-31 climbed to 600 kilometres in order to deploy the Hubble Space Telescope, its crew observed that the scope for visual study of oceanic phenomena was improved rather than degraded, because larger-scale structures were more easily discerned. To convey some idea of the scale of the view from that altitude, Story Musgrave pointed out that, on turning around, he found the entire Australian continent, with all the subtle shades of the outback, staring him in the face.

Despite its reliance on imaging satellites, in November 1991 the Department of Defense flew Tom Hennen on Atlantis to assess the potential of the human eye as a reconnaissance sensor. He was well trained for this assignment because he had spent years scrutinising reconnaissance imagery as an Army photo-interpreter. By going into orbit, he would not only be able to test his highly trained vision, but would also be able to report whether a camera missed anything important. For the Terra Scout project he had the Direct-View Optical System, which was a pair of binoculars on a stabilised mount that compensated for the Shuttle's motion once it had been set on a target. He had been assigned a variety of 'test sites' typical of those monitored by reconnaissance satellites, and was to pursue a schedule of observations, noting what could be seen at each site, and he was astonished by the detail that could be resolved. The blandly designated M88-1 project was more ambitious, and turned Atlantis into a sophisticated reconnaissance platform. Hennen used a telescopic CCD camera to take high-resolution imagery of ships in port and at sea, airfield operations, artillery firing, and manoeuvring armour. He had an image-processing system to analyse each site in real time. Mario Runco reported to *in-situ* teams by a secure UHF radio called Night Mist. In return, the supervisor could ask Hennen to confirm specifics, but Hennen had to be wary because some sites were misleading. The pressure imposed by the interactive nature of this process was a vital part of the experiment. In 1963, during his day in space on the final Mercury flight, Gordon Cooper had spotted trains in open country, but his observations had been dismissed as wishful thinking. On Cooper's Gemini flight two years later, Pete Conrad had been present to provide corroboration. With his imaging system, Hennen had demonstrated that he could not only distinguish different types of transport aircraft, but that he could also produce militarily useful data in real time.

The Navy's main interest in the Shuttle was in taking pictures of the ocean; but to be of any value, it was necessary to know the latitude and longitude. How does an astronaut locate an object in open ocean? An automated method of calculating the geographical location of the centre of an image was required. The first attempt, the Latitude and Longitude Locator (L3), measured the angle between the centre of the frame and the horizon, and utilised a knowledge of the orbiter's path to calculate the location of an image. Trials conducted on STS-28 and STS-32 showed this system to be accurate to within 12 kilometres, but, because it was awkward to use, the Naval Research Laboratory built the more sophisticated Hercules system using a 35-mm Electronic Still Camera and a ring-laser gyroscope. On its first trial, on STS-53, it proved able to fix a site to within 4 kilometres. On STS-56 its digital imagery was relayed to Earth in real time. The initial version required its inertial reference to be initialised by taking star sighting, but an upgraded version could use GPS to fix sites to within 2 kilometres. On a stabilised mount, Hercules could be set up to wait for a specific site to come into view, and then to snap a series of pictures. It was an improvement over the L3, and because the horizon did not need to be in the frame, it was suitable for nadir-viewing imagery.

For the Navy's Military Applications of Ship Tracks, the astronauts took high-resolution images on infrared-sensitive film with the objective of determining how a ship affects its immediate environment. Most oceanic life inhabits the uppermost 200 metres. Due to a chemical reaction between two enzymes, some species emit light – bioluminescence. At night, it was hoped to observe the bioluminescence from fauna feeding on plankton stirred up by a ship's high-speed passage, and, during the day, from the wispy cloud that is 'seeded' by the increased evaporation of dimethyl sulphide phytoplankton waste in the wake of a ship. Starting with STS-65 in 1994, this became a routine experiment. A long-term goal was to develop a way to track a submerged submarine from space. In open ocean, a submarine running at 30 knots in shallow water might reveal itself by the bioluminescence in its wake. When a shallow sea of clear water is viewed at a low angle of lighting, it becomes semi-transparent and it is possible to see the shape of its floor, and in such conditions it was possible to *see* a submerged submarine.

The Cloud Logic to Optimise Use of Defence Systems project made its debut on STS-41D in August 1984. It involved no more than the use of a 35-mm camera with an infrared filter to take high-resolution pictures of clouds. It was later pursued on STS-39, STS-45 and STS-53. In the procedure, meteorologists were to specify the cloud structures they wished to have recorded (they were particularly interested in those associated with severe weather and wispy cirrus) and the crew were to focus on the target the moment that it appeared over the horizon, and track it, snapping pictures at regular intervals using the motorised camera to record it over as wide a range of viewing angles as possible, until it could no longer be seen. The goal was to quantify apparent cloud cover as a function of the angle at which clouds of different types are viewed, to provide data to calibrate the sensors used by the Defense Meteorological Satellite Program.

THE RETURN OF THE TITANS

The Shuttle never really achieved the operational requirements set by the Air Force. The orbiter's payload bay was of the specified size, but its volumetric capacity was undermined by the fact that the stack was somewhat heavier than expected – an issue on which NASA was progressively making progress. It had been possible to test the orbiter's subsonic handling in a wind tunnel and by releasing it from the back of the Boeing SCA, but because there was no way to verify that it would be able to withstand the stresses associated with hypersonic soaring until it was returning from orbit, Columbia had been heavily instrumented to provide this information. The later orbiters came off the production line successively lighter. Other measures addressed the rest of the stack: for example, the payload capacity was increased by 275 kilograms simply by not painting the ET.

Even although the Shuttle had been declared operational following its fourth test flight, it was still undergoing development. At first it was criticised for not meeting its specifications, and was then criticised for undergoing continuous refinement. The progressive uprating of the SSME thrust was not taking 'unnecessary' risks; it was a process of 'stretching the envelope' to work towards optimum performance. With a research aircraft like the X-15 rocket plane, which was an end in itself, this could be done inconspicuously, but the National Space Transportation System had to be *used* while its performance was being refined. Similarly, although the orbiter's re-entry cross-range was initially restricted, a series of trials were progressively exploring its true potential.

On the other hand, several of the Air Force's requirements turned out to be short-lived. The cancellation of a new reconnaissance camera in 1980 not only eliminated the single-orbit mission, but also the 30-day endurance requirement that was to have revived the Manned Orbiting Laboratory. As introduced, the orbiter was restricted to about 10 days in space, but this was just a consequence of the size of the cryogenic tanks it carried. A wafer of stores was later installed in the payload bay to facilitate flights of extended duration for the Spacelab programme. One by one, therefore, NASA was addressing the requirements laid out by the Air Force so long ago. But it was not just NASA that was behind schedule. If the Air Force had *really* wanted to fly in polar orbit, it would have had its Vandenberg facility ready and waiting. It was not until 1985 that it signed an agreement with Chile to upgrade the airport on tiny Easter Island, 4,000 kilometres off the coast, to take an orbiter making an emergency descent after it was beyond the point at which it could return to Vandenberg. As part of this deal, NASA was to fly a Chilean national on the Shuttle as an observer, but this never occurred.

How did the Shuttle perform from the point of view of the Department of Defense? Its late introduction undoubtedly held up the novel payloads that had been built to exploit it, but after it was declared operational it was the IUS that imposed further delay. All the satellites were successfully deployed. One that failed to start up was repaired, and one that failed in geostationary orbit was beyond the Shuttle's reach, but neither of these failures was in any way attributable to the Shuttle.

When Challenger was lost and the fleet was grounded, the Air Force had just

An IUS with a pair of DSCS-III satellites in STS-51J's payload bay.

begun to increase its use of the Shuttle. Initially, NASA had expected to resume operations within a year. Pete Aldridge had secured a stockpile of a dozen Atlas and Titan III rockets as a "contingency", so the Air Force believed that it would be able to cover this gap, but when the first of the Titans exploded in April, the US found itself in the unprecedented position of being unable to launch heavy satellites. On receipt of the Rogers Commission's report in June, the Reagan administration not only abandoned the Shuttle-only policy and authorised the Air Force to order more rockets, but also initiated improvement programmes. More significantly, however, it resumed development of the powerful Titan IV to ensure that national security payloads would not rely on the Shuttle. The newly liberated Air Force effectively withdrew from the Shuttle, first by mothballing and later writing off its Vandenberg facility – a decision that had the side-effect of denying NASA access to polar orbit. As a result of this divergence, the Air Force now has, in the shape of the Titan IV-B, a rocket possessing a shroud even more voluminous than the Shuttle's payload bay, and is all the more useful because it *can* attain polar orbit. In the end – as Robert Seamans had testified to Congress – Nixon's legacy, the Shuttle, was not essential for national security.

6

A multiplicity of missions

STRIPPING THE MANIFEST

At the time of the Challenger accident, NASA had been endeavouring to ramp up the launch rate to 'service' the long manifest of payloads for which the Shuttle – as the National Space Transportation System – was solely responsible. While the fleet was grounded, some payloads (such as the one to study Halley's comet in the summer of 1986) had been rendered obsolete. Prohibiting the Shuttle from launching commercial satellites and offloading government satellites to expendable rockets dramatically cut down the manifest, but it resulted in a frantic scramble for the few remaining rockets.

NASA initially hoped to be able to complete its modifications within a year and resume flying in early 1987, but in the autumn of 1986 it announced that the Shuttle would stay grounded until mid-1988. The manifest published in July 1987 reflected both NASA's priorities and its obligations to the Air Force, which, notwithstanding its disenchantment with the Shuttle, had exercised its right to requisition flights in the interest of national security – until the Titan IV became available, the Air Force had no other way to launch its out-sized satellites. In resuming operations, NASA decided to revert to the sequential numbering scheme, even though it was likely that events would lead to missions being flown out of order.

NASA's priority was to build up of the TDRS communications network. After two classified missions had been flown, and before it sent Magellan to Venus, the agency hoped to have sufficient time to complete this network. Magellan assumed priority over the Hubble Space Telescope (HST) because the launch window for the interplanetary mission was defined by the dynamics of the Solar System. The assignment of the HST to Columbia was unfortunate; as the heaviest of the fleet, it was the least able to lift the 11-tonne telescope to the record 600-kilometre altitude required to minimise orbital decay. By early 1988, it was clear that preparations to launch Discovery were running about two months late, and the manifest was revised accordingly.

Table 6.1 Immediate post-Challenger manifest

Flight	Date	Orbiter	Objective
61E	Mar 1986	Columbia	ASTRO for Halley's comet
61F	May 1986	Challenger	Ulysses deployment
61G	May 1986	Atlantis	Galileo deployment
61H	Jun 1986	Columbia	Communications satellites
62A	Jul 1986	Discovery	Classified
61M	Jul 1986	Challenger	TDRS satellite
61J	Aug 1986	Atlantis	Hubble Space Telescope
61K	Sep 1986	Columbia	Earth studies Spacelab
61I	Sep 1986	Challenger	Retrieve LDEF
62B	Sep 1986	Discovery	Classified
61L	Oct 1986	Atlantis	Astrophysics Spacelab

Table 6.2 Shuttle manifest, c. mid-1987

Flight	Date	Orbiter	Objective
STS-26	Jun 1988	Discovery	TDRS-C
STS-27	Sep 1988	Atlantis	Classified
STS-28	Nov 1988	Columbia	Classified
STS-29	Feb 1989	Discovery	TDRS-D
STS-30	Apr 1989	Atlantis	Magellan
STS-31	Jun 1989	Columbia	HST

Table 6.3 Shuttle manifest, c. early 1988

Flight	Date	Orbiter	Objective
STS-26	Aug 1988	Discovery	TDRS-C
STS-27	Oct 1988	Atlantis	Classified
STS-29	Jan 1989	Discovery	TDRS-D
STS-28	Mar 1989	Columbia	Classified
STS-30	Apr 1989	Atlantis	Magellan
STS-31	Jun 1989	Discovery	HST
STS-32	Jul 1989	Columbia	Leasat-5; retrieve LDEF
STS-33	Aug 1989	Atlantis	Classified
STS-34	Oct 1989	Discovery	Galileo

Although the recycling of Discovery, Atlantis and Columbia was disturbed, the timescale was compressed to preserve the April and June dates because the October window for the Galileo spacecraft was fixed. A fortunate effect of swapping STS-28 and STS-29 was the reassignment of the HST to Discovery, which was better suited to this rôle.

RETURN TO FLIGHT

By the time Discovery finally returned to space on 29 September 1988, the manifest had been rearranged to work within the constraints imposed by the fixed windows of the interplanetary spacecraft. The TDRS relays had retained their priority, but the HST had lost out to the Air Force, slipping eight months to early 1990.

The deployment of TDRS-C was completed on the fifth orbit, as was usual for an IUS payload. The IUS worked perfectly. The TDRS satellite experienced difficulty deploying its twin primary antennae, but this was soon rectified and the new satellite supplemented TDRS-A, which had lost half of its Ku-Band capacity in late 1986. With two of these relay satellites now in service – one above the Atlantic and the other above the Pacific – the network was finally able to provide the service that it was to have had within six months of the Shuttle being declared operational, providing continuous contact for 86 per cent of each orbit; the dead zone being low over the Indian Ocean. Having achieved their primary mission, the crew settled down to a programme of microgravity experiments on the mid-deck. The nosewheel steering had been certified, but as the tests had been in the absence of a crosswind, the Shuttle Landing Facility (SLF) in Florida was still rated as an emergency-only recovery site. Immediately upon touching down on the dry lake at Edwards, Discovery used differential braking to move off the centreline and back again to test the revised carbon–beryllium brakes that were later to be replaced by new carbon–carbon brakes. All in all, this mission was an excellent resumption of flight operations.

The launch of STS-27 was slipped a day due to cloud that would have inhibited a RTLS, but it left on time the next day and deployed its satellite which, despite the secrecy, was "reliably reported" to be the first Lacrosse radar-imaging satellite for all-weather reconnaissance. In contrast to previous Department of Defense satellites, which had ridden IUS stages to geostationary orbit, this one boosted itself up to an altitude of about 700 kilometres.

Table 6.4 Shuttle manifest, c. autumn 1988

Flight	Date	Orbiter	Objective
STS-26	Sep 1988	Discovery	TDRS-C
STS-27	Nov 1988	Atlantis	Classified
STS-29	Feb 1989	Discovery	TDRS-D
STS-30	Apr 1989	Atlantis	Magellan
STS-28	Jul 1989	Columbia	Classified
STS-33	Aug 1989	Discovery	Classified
STS-34	Oct 1989	Atlantis	Galileo
STS-32	Nov 1989	Columbia	Leasat-5; retrieve LDEF
STS-36	Dec 1989	Discovery	Classified
STS-31	Feb 1990	Atlantis	HST

With two missions flown, NASA was confident that it would be able to fly seven more in 1989. However, a routine post-flight inspection of STS-27's SSMEs revealed that one of the three liquid oxygen turbopumps had a cracked bearing-race. It was therefore decided to exchange the pumps on STS-29, which had already been rolled out to the pad, delaying it by almost a month and marking a poor start to the new year's schedule. After the deployment of TDRS-D, the only problem was the failure of the Station Heatpipe Advanced Radiator Experiment, but this was in the bay and there was nothing the crew could do to fix it.

The dispatch of the Magellan spacecraft was the first 'fixed point' in the year's schedule. The window opened on 28 April, but the optimal date was 5 May as this would result in the best orbit around Venus, and thereafter the energy of the transfer would deteriorate again until 22 May, when the window would close. If Magellan missed its window, it would have to wait 18 months. Rather than wait for the ideal date, and risk a delay, NASA scheduled STS-30 for the first day of the window, but the clock was stopped at T–31 seconds because of an electrical fault in a pump in one of the SSMEs. The first available date was 4 May so, in the end, Magellan was given the best possible start. The IUS stack was checked by Mark Lee and released six hours into the mission. The Shuttle lingered until Magellan's solar panels were deployed, and then withdrew. Half an hour later, the two-stage IUS accelerated the spacecraft away with such precision that the option of a trajectory correction using its own thrusters was deleted. Despite having dispatched Magellan in the first week of its window, NASA postponed one of the classified missions until after Galileo was also safely on its way, thereby further eroding the schedule. The effort to keep STS-28's payload secret was more successful than the case of the 'open secret' of the Lacrosse; the only confirmed fact was that it did not use an IUS. Columbia still had the SILTS sensors in place, so as it made its way back to Edwards it gathered further data on its re-entry characteristics.

The Galileo spacecraft, the flagship of NASA's planetary programme, was years late. In the aftermath of Challenger's loss it had had to be redesigned and, ironically, would now have to pursue a five-year roundabout route to Jupiter that began with a minimum-energy transfer to Venus. The window opened on 12 October and lasted until 21 November. As with Magellan, the launch was set for the first available date,

Table 6.5 Shuttle manifest, c. mid-1989

Flight	Date	Orbiter	Objective
STS-28	Aug 1989	Columbia	Classified
STS-34	Oct 1989	Atlantis	Galileo
STS-33	Nov 1989	Discovery	Classified
STS-32	Dec 1989	Columbia	Leasat-5; retrieve LDEF
STS-36	Feb 1990	Atlantis	Classified
STS-31	Mar 1990	Discovery	HST
STS-35	Apr 1990	Columbia	ASTRO-1
STS-37	Jun 1990	Discovery	CGRO

but the countdown was scrubbed when a fault showed up in one of the SSMEs long before the crew were due to board Atlantis. This forced a five-day postponement. After a hold for weather, the countdown was abandoned. Everything went perfectly the next day, however, and over the next few hours Shannon Lucid verified that the IUS and its payload were healthy. Ellen Baker and Franklin Chang-Diaz then suited up and entered the airlock to be ready to address any issue that might impede the deployment, but the IUS was dispatched without incident. Atlantis was recalled two orbits early, to sneak into Edwards ahead of approaching inclement weather. This involved modifying the computer that used roll-reversals to bleed off kinetic energy during the hypersonic phase. In fact, most of the descent from the Mach 25 entry interface down to the Mach 13 point was spent in a left-bank designed to swing the ground track 1,000 kilometres westward. Unfortunately, Atlantis was not fitted with the SEADS and SUMS sensors to document this first high-cross-range re-entry. On this occasion, the nosewheel steering was subjected to a high-speed test immediately after touchdown.

With the fixed points in the schedule satisfied, NASA could switch to building up the flight rate. First in the queue was STS-33, the Department of Defense flight. Whatever the nature of its classified payload, orbital mechanics imposed the first night-time launch since operations had resumed. Another aspect of the orbiter's flexibility during the descent phase was demonstrated 35 minutes after the de-orbit burn, when a change in winds prompted the southerly approach to the dry lake to be switched to a north-easterly landing on the concrete runway. Accommodating the interplanetary spacecraft had resulted in only five of the seven missions planned for 1989 being launched, but hopes were running high for 1990, and the manifest was flexible through to October, when the Ulysses spacecraft was to be dispatched to Jupiter.

All the Shuttle flights since operations had resumed had been dispatched from Pad 39B. Columbia was to have been the first to use Pad 39A in December, but the work on the modifications to the pad was running late and STS-32 was postponed until early in the new year. In its own way, this mission was time-critical. It was to dispatch the final Leasat communications relay for the Navy and then retrieve the LDEF satellite, which had been in orbit since 1984. This materials-exposure satellite had not been meant to spend such a long time in space. Its orbit was decaying, with re-entry predicted for March. NASA wanted the LDEF back, and if STS-32 proved difficult to get off, it could not be permitted to slip too far down the schedule. This situation was aggravated by the fact that Atlantis was due to fly soon as STS-36. In the event, although Columbia was delayed a day due to bad weather, it was launched on 9 January, the Leasat was deployed, and the LDEF was retrieved as planned. As a bonus for the programme of mid-deck science, the descent was put back a day by fog on the dry lake. To provide the longest possible launch window, Columbia had carried more propellant than usual so that, in the event that it launched late in the window, it would be able to effect the plane change required to rendezvous with the LDEF. The launch had been perfect, however, and Dan Brandenstein had made a smooth rendezvous, so Columbia had 2 tonnes of propellant in reserve. In order to dump this excess, the de-orbit burn was performed with the vehicle's axis offset 50

The Long Duration Exposure Facility which was deployed by STS-41C in 1984 and retrieved by STS-32 in 1990.

degrees from the velocity vector. The data from the aerodynamic sensors was very welcome as the 10-tonne cargo not only made this the heaviest orbiter yet to return, its centre of mass was the farthest forward. All of the flights thus far since Return-To-Flight had been brief – four or five days. At 10 days, STS-32 was to be the first full-duration flight. In fact, the wave-off made it the longest to date. Despite having been delayed (although through no fault of the Shuttle itself) this otherwise textbook flight fortunately had not disrupted the new year's schedule.

After a few days lost to weather and faulty ground systems, STS-36 launched on 28 February. It was the first Shuttle to make a dog-leg as it flew out across the Atlantic to attain an inclination higher than the 57-degree limit that is imposed by the shape of the Eastern Seaboard; the 62-degree orbit was dictated by the mission of its classified satellite.

STS-31 was next in line. Its payload, the Hubble Space Telescope, was the first in a series of big-budget, big-science satellites – the Great Observatories – which it was hoped would revolutionise astronomy. In this case, the satellite was named after Edwin P. Hubble, the Mount Wilson astronomer who discovered that the Universe is expanding. Organising the Shuttle manifest was a tricky task. The HST could not be launched until the TDRS satellites were operational, because it would rely upon their relays. The first two of these satellites were to have been launched in 1983, but a fault in the IUS stage carrying the first prompted the postponement of the second, and, consequently, of the HST. Recertifying the IUS took longer than expected. The first opportunity to dispatch the second TDRS was on STS-51L, in January 1986. That summer's manifest was reserved for observations of Halley's comet and for launching the Galileo and Ulysses spacecraft, and the HST got the next available slot thereafter, but the loss of Challenger, which had been carrying that TDRS satellite, grounded the fleet. Despite lobbying from the astronomical community for the HST to be launched as soon as the TDRS satellites were available and the Magellan and Galileo spacecraft had been dispatched, the Department of Defense had precedence. After the HST was assigned to December 1989, it was realised that the orbit of the LDEF was decaying more rapidly than expected because recent solar activity had inflated the Earth's upper atmosphere. Unless its retrieval was brought forward, this satellite and the samples it carried would be lost, so the HST slipped into early 1990.

Table 6.6 Shuttle manifest, c. early 1990

Flight	Date	Orbiter	Objective
STS-36	Feb 1990	Atlantis	Classified
STS-31	Apr 1990	Discovery	HST
STS-35	May 1990	Columbia	ASTRO-1; BBXRT
STS-38	Jul 1990	Atlantis	Classified
STS-40	Aug 1990	Columbia	SLS-1
STS-41	Oct 1990	Discovery	Ulysses
STS-37	Nov 1990	Columbia	CGRO
STS-42	Dec 1990	Atlantis	IML-1

The 1986 HST deployment was to have been commanded by John Young, but because he had retired soon after the Challenger accident Loren Shriver inherited both the mission and its crew. The launch was set for 18 April, but, for once, NASA found itself ahead of schedule and brought the mission forward to 10 April. On that date, however, a fault developed in one of the APUs and, despite the T–31 seconds hold, the count had to be scrubbed. The auxiliary power unit was readily replaced, but the resulting delay meant that the satellite's battery had to be recharged, which took a week. The countdown on 24 April proceeded without incident. To ensure that the satellite's orbit did not suffer undue decay, it was to be released at an altitude of 600 kilometres, the highest to which a Shuttle had yet ventured. To achieve this, Discovery flew a steeper than usual climb through the atmosphere so that, at MECO, this 'direct ascent' placed it in a 50 by 600 kilometres orbit which was circularised by the OMS-2 burn at the first apogee.

The next day, the HST was deployed in a tricky operation that involved Steven Hawley, operating the RMS, first lifting the enormous bulk of the 12-tonne satellite out of the bay, then rotating the 13-metre-long stepped-cylinder to allow him to see the indicators on its display panel from his work station on the aft flight deck. As the telescope was hoisted up out of the bay, it disconnected itself from the orbiter's power supply. The deployment process was now running against the clock. The spacecraft's battery was good for just eight hours. If the solar panels had not been deployed by that time, Bruce McCandless and Kathryn Sullivan would have to make a spacewalk to plug in an umbilical. While the Goddard Space Flight Center methodically checked out the satellite, Discovery faced its belly towards the Sun to shade it until its thermal regulation system was brought on line. Finally, it was time to deploy the appendages. The first of the two 12-metre-long solar panels unrolled perfectly, but it required several attempts to complete the deployment of the second panel because the system that monitored the tension in the control wire overrode the motor. However, when the astronauts reported that they could see no fault, this safety feature was itself overridden and the panel rolled all the way out. While the problem with the panel was being studied, the clock had been ticking, and Discovery was now reoriented to enable the solar panels to recharge the battery. Once the two TDRS antennae had been deployed, the satellite was released as planned. Discovery stood by 60 kilometres away, and its crew performed mid-deck experiments for three days while Goddard continued to work through the HST's start-up procedure. It had once been intended simply to

Table 6.7 Shuttle manifest, c. June 1990

Flight	Date	Orbiter	Objective
STS-38	Jul 1990	Atlantis	Classified
STS-35	Aug 1990	Columbia	ASTRO-1; BBXRT
STS-41	Oct 1990	Discovery	Ulysses
STS-37	Nov 1990	Atlantis	CGRO
STS-40	Dec 1990	Columbia	SLS-1
STS-42	Jan 1991	Discovery	IML-1

deploy the spacecraft 'clean' and leave it to deploy its appendages, but the difficulty encountered in deploying a solar panel had demonstrated the wisdom of not releasing the spacecraft until its power supply was secure. As Goddard neared the point of unlatching the telescope's aperture, Discovery prepared to return, just in case the astronauts would be called upon to open it manually, but, once it was verified that the flap was open, the Shuttle was cleared to withdraw. Before the Space Telescope Science Institute that had been established in Baltimore, Maryland, could take operational control of the telescope, three months would be spent calibrating its instruments. Discovery's return from its unprecedented altitude involved the longest de-orbit burn, and the longest coast down to the entry interface of the programme to date. During its roll-out on the concrete at Edwards, it tested the new carbon–carbon brakes.

THE LONG HOT SUMMER

What was to have been a bumper year for the astronomical community turned out to be an unmitigated disaster. Columbia was to have flown the ultraviolet telescopes of the ASTRO-1 Spacelab, but the countdown on 30 May had to be abandoned at T–6 hours, when a hydrogen leak was noted while the ET was being loaded. When this was traced to the 17-inch disconnect valve in the aft compartment of the orbiter for the pipe that fed hydrogen from the ET, it became necessary to roll the stack back to the VAB. NASA decided to bring forward Atlantis's classified STS-38 mission to July, fly the STS-35 ASTRO mission in mid-August and postpone Columbia's next mission (as STS-40) to the end of the year in order to allow time for the turnaround. The fixed point amid this reshuffle was the dispatch of the Ulysses interplanetary spacecraft in October.

Meanwhile, on 20 May the 2.4-metre-diameter mirror in the HST was found to be suffering from spherical aberration and could not focus properly. A commission of inquiry under the chairmanship of JPL director Lew Allen reported in November that the laser interferometer that had been used to test the primary mirror during its manufacture had been incorrectly configured during its installation in 1981, with the result that the mirror had been extremely precisely polished, but to the wrong figure! To save money, no integrated testing had been undertaken, and so there had been no opportunity to discover the spherical aberration. It was the same kind of integration problem as caused so many headaches during the development of the Shuttle itself. Eliminating preflight testing was a risky business. Whatever the additional cost of an integrated test of the telescope's optics might have been, it could hardly compare to the cost of overcoming the flaw now that the telescope was in space.

And worse was to come. At the end of June, a hydrogen leak was discovered on Atlantis in the vicinity of its disconnect valve, and it suddenly began to seem that there might be a generic fault in this crucial component. This appalling prospect forced NASA to ground all three of its orbiters, which threatened the chances of Discovery meeting the window for Ulysses. Fortunately, Columbia's problem was found to be a seal in the flapper valve, whereas Atlantis's was due to a leaky flange at the ET end of the pipe. It was *not* a generic flaw. The flange problem would take time to fix, but as Columbia's problem was easier to repair it was reinstated at the top of the queue in the hope that it would be possible to launch it ahead of Discovery. This hope was frustrated, however, when a fault was found in one of the SSMEs, and so Discovery, with its fixed window, became the focus of attention. The window opened on 5 October. When STS-41 was launched from Pad 39B on 6 October, it drew this five-month hiatus to a close. It was a smooth mission. As with Galileo, Ulysses was checked out and dispatched within six hours of achieving orbit. The Jovian slingshot was to twist the inclination of Ulysses's orbit perpendicular to the ecliptic and deflect it back into the inner Solar System to gain a new viewpoint of the Sun's magnetic poles. Because Ulysses did not have a motor of its own for a substantial mid-course correction, it relied on the IUS, which had to deliver an unprecedented degree of accuracy. The IUS established a slow roll to even out irregularities in the burning of its solid rocket motor, fired its first stage for 148 seconds, shut this down and jettisoned it, coasted for 125 seconds, fired its second stage for 108 seconds and then released its payload. Although Ulysses was now travelling at 11 kilometres per second, this was only just sufficient to escape the Earth. To put it on the fast track to Jupiter, four solid charges distributed around the waist of the specially modified PAM motor fired to spin it up to 70 revolutions per minute for enhanced stability during the motor's 88-second burn, and then the PAM released two small masses on 12-metre-long wires and promptly reduced the spin rate to 7 revolutions per minute by the conservation of angular momentum to prepare to release its 370-kilogram payload. Having been accelerated to 15.4 kilometres per second, Ulysses became the fastest probe sent into deep space. The dispatch of this final Shuttle-deployed interplanetary probe restored flexibility to the manifest.

On the way back to Earth, Discovery demonstrated a procedure that would be required in the event of a RTLS abort. Apart from when returning with a Spacelab or the LDEF, the centre of mass had been far to the rear. At launch, however, with a full bay, the centre of mass was generally much further forward – a fact that could jeopardise an already tricky emergency landing. The only way to slip the centre of mass towards the rear would be to dump propellant from the forward RCS tanks. The objective of Discovery's test was to demonstrate that the RCS thrusters could safely be fired deep in the atmosphere. The PTIs involved making yaw manoeuvres

Table 6.8 Shuttle manifest, c. October 1990

Flight	Date	Orbiter	Objective
STS-38	Nov 1990	Atlantis	Classified
STS-35	Dec 1990	Columbia	ASTRO-1; BBXRT
STS-39	Mar 1991	Discovery	Classified
STS-37	Apr 1991	Atlantis	CGRO
STS-40	May 1991	Columbia	SLS-1
STS-43	Jul 1991	Discovery	TDRS-E
STS-42	Aug 1991	Columbia	IML-1

with the forward thrusters as the orbiter flew through Mach 13, Mach 6 and then Mach 4. This test was not just a step towards increasing confidence in the RTLS procedure, but by showing that an orbiter could adjust its centre of mass, it also increased manifesting flexibility. The centre-of-mass restrictions would also apply to a TAL abort, but in that case there would be more time to reconfigure the vehicle. In addition to demonstrating the new carbon–carbon brakes, Discovery exercised the nosewheel steering and thereby recertified the orbiter for a crosswind landing.

The manifest could change significantly upon a scrubbed launch attempt. When an orbiter slipped down the manifest, all of its downstream assignments also slipped unless the orbiter/payloads pairings were reassigned to maintain mission order. As a result, some missions could slip as much as six months each time. Nevertheless, with the hydrogen leaks resolved, NASA was confident that it would be able to make up for lost time.

CATCH-UP

Rumour has it that shortly before the long-delayed STS-38 was finally launched, its classified payload was swapped to support Operation Desert Shield. The countdown for 9 November had to be abandoned at an early stage when a problem was found with the payload. Although this was soon rectified, by missing its slot on the Eastern Test Range schedule the Shuttle was obliged to slip in behind the Air Force's first Titan IV with a classified satellite. As it took two days to reconfigure the Test Range for a different type of launcher, STS-38 followed on 15 November. After its secret satellite had been deployed, Atlantis prepared to return to Edwards in the hope that the prevailing poor weather would clear up, but four minutes before the de-orbit burn was due, the descent was slipped 24 hours. The weather the next day was no better, and as White Sands was not available either, NASA recalled it to Florida rather than wait another day. Thus, despite the fact that it still had the older carbon–beryllium brakes, Atlantis became the first orbiter to return to its launch site since Discovery had blown a tyre in 1985.

The long-delayed STS-35 was scheduled on 2 December. ASTRO was the first Spacelab to be devoted to a single subject – in this case astronomy. The Instrument Pointing System (IPS) mounted on a pair of pallets at the front of the bay had three ultraviolet telescopes that were controlled jointly by the astronauts and astronomers at the new Payload Operations Control Center (POCC) at the Marshall Space Flight Center, call sign 'Huntsville'. In 1984 two astronomers, Sam Durrance and Ron Parise, had been named as the payload specialists on ASTRO-1, which was to fly in March 1986 to combine its ultraviolet studies with observations of Halley's comet by a Large Format Camera aft of the IPS pallet. In October 1986, ASTRO-2 should have deployed the SPARTAN free-flyer, and the third mission, in 1987, was to have released the CRRES magnetospheric research satellite, but the loss of Challenger rendered this plan obsolete. When the Shuttle resumed flying, ASTRO-1 was set for mid-1990. The comet had long gone, but there was now a supernova in the Large Magellanic Cloud and the rear of the bay carried BBXRT, an X-ray telescope that was remotely controlled by the Goddard Space Flight Center. In the event, the flight slipped six months because of hydrogen leaks. It was not a smooth mission. No sooner had observations begun than the primary instrument controller failed; as did its backup. As time was the most critical commodity, the crew hastily improvised an unrehearsed manual procedure to enable observations to resume. Although, with practice, they proved adept at slewing the IPS onto a target and stepping through the telescope procedure, they were not as fast as the automatic system. It had been hoped to examine 230 individual objects by picking off a few on each orbit, but the list had to be pared to the top priority items. It is worth noting, however, that if ASTRO had been deployed as an independent satellite it would have been a write-off. Fortunately, ASTRO was the first mission since Return-To-Flight to carry payload specialists. The crew split into two shifts to keep the telescopes running 24 hours a day, and when the results were examined it was found that 70 per cent of the objectives had been achieved. Unfortunately, the programme had to be curtailed when it was decided to land a day early to enable the orbiter to slip into Edwards ahead of bad weather. In order to pave the way for the Extended Duration Orbiter (EDO) system that would significantly increase the output from a Spacelab flight, STS-35 evaluated a trash compactor and a modified toilet (one that could be unloaded in space). One aspect of this mission drew comment in the debriefing: a nine-day flight using a pallet-only Spacelab configuration, with a crew of seven working on a 24-hour basis was, as Mike Lounge put it, "pretty stressful", as the off-shift crew members had to try to sleep on the mid-deck while the others worked 'upstairs' on the flight deck. In terms of its scientific yield, the mission had been highly successful, but only because the crew, ably assisted by support teams on the ground, overcame severe equipment malfunctions. The inescapable conclusion was that humans were a crucial part of the system.

By mounting STS-41, STS-38 and STS-35 at monthly intervals, it finally began to seem that it would indeed be possible to increase the flight rate beyond the record of nine in 1985. There was then every expectation that the new year's schedule, which did not appear to be too demanding, would be met. In retrospect, it is evident that in the immediate Return-To-Flight period, missions were precisely defined and so straightforward that they mirrored the military's philosophy of launch, complete the primary mission as soon as possible and then land; there was none of the vibrant spontaneity of the pre-Challenger days, and, very noticeably, no spacewalks.

Table 6.9 Shuttle manifest, c. early 1991

Flight	Date	Orbiter	Objective
STS-39	Feb 1991	Discovery	Miscellaneous DoD payloads
STS-37	Apr 1991	Atlantis	CGRO
STS-40	May 1991	Columbia	SLS-1
STS-43	Jul 1991	Discovery	TDRS-E
STS-44	Aug 1991	Atlantis	DSP
STS-48	Nov 1991	Discovery	UARS
STS-42	Dec 1991	Atlantis	IML-1

The ASTRO 1 telescopes on the Instrument Pointing System flown on the STS-35 mission.

Soon after Discovery was rolled out to Pad 39A for STS-39, cracks were found in a subassembly on the ET and it was returned to the VAB. Meanwhile, Atlantis had been prepared on Pad 39B with the Compton Gamma-Ray Observatory – the second of the Great Observatories – in its bay. This was named in honour of A.H. Compton, the Nobel Prize-winning physicist who made pioneering studies of high-energy physics in general, and of cosmic rays in particular. The count proceeded exceptionally smoothly and STS-37 made a direct ascent into a 450-kilometre circular orbit exactly on time. At 16 tonnes, this was the most sophisticated gamma-ray package ever to be put in space. It was a 3.2- by 6.3-metre aluminium tray some 0.6 metres deep, with propulsion and communications on its base, a solar panel on each side and a variety of instruments set around the periphery and on top. With its various appendages stowed, it was almost 8 metres long and 4 metres wide. Linda Godwin, operating the RMS, hoisted the satellite out of the bay and ran the pre-release checks. The solar panels unfolded as planned but when the command was sent to unstow the solitary TDRS antenna, the 3-metre articulated boom remained tight against the underside of the satellite. Although Godwin moved the arm back and forth in an effort to shake the boom free of its clamp, it remained stuck, and the only option was to send astronauts out to free it. Two astronauts on each crew are trained to make a 'contingency EVA' to overcome any of a variety of problems that could jeopardise the orbiter, such as closing a stuck payload bay door. The first contingency was on STS-51D, when David Griggs and Jeff Hoffman placed the 'fly swatter' on the RMS to try to activate the dormant Leasat. More recently, Bruce McCandless and Kathryn Sullivan were depressurising the airlock to go out and deploy the HST's stuck solar panel when the Goddard Space Flight Center finally coaxed it to unroll. As STS-37 was the first mission since Return-To-Flight to be *assigned* a spacewalk, Jerry Ross and Jay Apt promptly suited up and traversed the length of the RMS to inspect the satellite. Ross verified that the clamp had released the boom but its hinge seemed to have frozen, so he gave the other end of the boom a hefty shove and it slowly rotated to its deployed position. As this action had taken only 20 minutes, they decided to remain out and make an early start on their own external programme, which was to test a number of translation aids that were under consideration for use on the space station. These tests were continued the following day, when they made their scheduled EVA. Atlantis landed on 11 April.

NASA had hoped to launch Discovery for STS-39, which had been rolled back out to Pad 39A on 1 April, on 23 April – this being the shortest interval that the rule book in use at that time allowed – but this attempt had to be abandoned at an early stage after an oxygen pump in one of the SSMEs showed signs of distress. However, it was launched without difficulty on 28 April. Although STS-39 was a Department of Defense mission, it was the first such flight not to be completely classified; a few aspects of the mission remained secret, but most of the payload was announced. The scientific value of the observations that were to be made was explained, but it was evident that the sensors were really being evaluated to support military applications. In terms of orbiter manoeuvres, this was one of the most demanding missions to date; the plumes of the orbiter's engines were studied both by a free-flyer released by the orbiter and by an SDI satellite. The most mysterious item was the small satellite

The Compton Gamma-Ray Observatory, shortly after its release by the Robotic Manipulator System.

that was ejected from a militarised GAS can dubbed the Multipurpose Experiment Canister (MPEC). As the weather at Edwards was unacceptable, Discovery was ordered down to the Kennedy Space Center. When it arrived, there was a slight crosswind; however, as this orbiter had carbon–carbon brakes no problems were expected and, indeed, it was viewed as an opportunity to test a crosswind landing. In the event, the starboard gear greased the concrete before the left, and one of the tyres was shredded as differential braking was used to line up on the centreline prior to lowering the nose. Clearly, differential braking could not be used at high speed. A form of stabilisation system would have to be developed to maintain the orbiter's line after the main gear touched down. The obvious solution was a drag chute, so work was started on its development. Despite the problem in this case, the status of the SLF was upgraded to be 'first equal' with Edwards.

Next in line was the long-delayed Spacelab for Life Sciences (SLS-1) assigned to Columbia as STS-40. The count on 1 June had to be scrubbed at T–23 minutes due to a fault in one of the three IMUs, but it was completed four days later. For this, the first Spacelab to be devoted to life sciences, three of the astronauts were medical specialists. Their objective was to make the most systematic study so far of the first phase of the body's adaptation to microgravity. The primary programme comprised 18 experiments to monitor the response of the heart, lungs, kidneys, and endocrine glands, and the degradation of muscle and bone. All of these experiments formed an interrelated study, and the in-flight data was compared to post-flight tests of the re-adaptation to gravity. Columbia remained in space for the maximum nine days in order to provide the longest possible data set. Although this brief study could not possibly rival the year-long missions of cosmonauts on the Mir space station, the fact that the 10-tonne Spacelab carried a comprehensive suite of apparatus meant that SLS-1 established the 'gold standard' for data on the first week of adaptation to the space environment. After landing at Edwards, Columbia was returned to Rockwell for its second refit, part of which would include the installation of the EDO facility.

The rest of the year's manifest was to be flown by Atlantis and Discovery. NASA was undoubtedly buoyed up by the fact that, despite minor problems, the manifest published at the start of the year was running on time. Misfortune struck again on 23 July, when the launch of STS-43 had to be halted due to a problem with the ET, and again, the next day, when one of the SSMEs needed attention, but both of these problems were noted at an early stage. Although the count on 1 August ran to the hold at T–9 minutes, the weather at the SLF was unacceptable for a RTLS and the window expired before conditions improved. Everything went smoothly the next day, and the primary objective, the release of a new TDRS satellite, was achieved without incident.

In contrast to earlier straightforward satellite-deployment missions, in which the Shuttle had landed after four or five days, STS-43 was to stay up for the maximum time permitted by Atlantis's consumables in order to obtain biomedical data for comparison with that from SLS-1. This required John Blaha's crew to maintain round-the-clock operations without a Spacelab module. The ASTRO mission, which had involved a palletised Spacelab, had demonstrated the difficulty of trying to run split shifts in a crew confined to the orbiter's cabin; the activities of those at work inhibited those supposedly sleeping, with the result that everyone was exhausted in several days. The situation had been exacerbated on STS-35 by the presence of two payload specialists, but, having no passengers, Blaha's crew fared rather better. This aspect of STS-43 showed that operating the Shuttle required a careful balance of conflicting requirements. The astronomers had been eager to operate the telescopes on a 24-hour basis to maximise the data from the limited time available. They had also wanted to fly the telescopes' designers to ensure that the best opportunity was taken of the situations that arose. These desires, however, had conflicted with long-term efficiency. The medics had wanted the best possible output from SLS, and the presence of the laboratory in the bay had made round-the-clock work sustainable. A satellite-deployment mission could not carry a laboratory as well, so, although such a flight could be extended and a shift cycle employed, it could not sustain the high-intensity science that was possible on a dedicated mission with a laboratory full of specialised apparatus. Making this balance would be more difficult once the EDO entered service, because 16 days was a long time for seven or eight people to be confined to the orbiter's cabin.

Atlantis had been scheduled to land at the Kennedy Space Center, but upon reviewing Discovery's difficulty in landing in a crosswind, STS-43 was authorised to use the SLF only if the crosswinds were negligible, which, in the event, they were. Edwards was to remain the prime landing site until the orbiters had been fitted with tyres of a harder compound. This was yet another demonstration that, even after in excess of 40 flights, the Shuttle was still in a process of incremental development. In a very real sense, each mission was a test flight that 'stretched the envelope' in some way, and, as a result, the Shuttle was becoming progressively more effective as an operational system. The manifest had unfolded essentially as planned, but flying STS-39 after STS-37 meant that Discovery's turnaround could not be guaranteed for STS-43, so that mission's TDRS had been switched to Atlantis, which had received a head start by having flown earlier. One result of this exchange was that STS-44 and STS-48 had also to be exchanged. This marginally affected the year's schedule, but as this was running only a few weeks late, the 'flow' at this flight rate was evidently sustainable. The Upper Atmosphere Research Satellite (UARS) – the first element of the Mission To Planet Earth (MTPE) programme (or Earth Science Enterprise, as it was later renamed) – was the largest environmental monitor to date. It carried such a suite of sensors that it required two days to check it out prior to deployment, but the satellite was released on schedule. As with its predecessor, STS-48 was to land on the SLF if conditions were perfect, but they were not, and it was diverted to Edwards.

The launch of Atlantis as STS-44 on 24 November was delayed only by the need to perform work on the Mobile Launch Platform. Although it was a Department of Defense flight, the Defense Support Program satellite – which the crew nicknamed 'Liberty' – was not classified and the video of its deployment on an IUS was shown on NASA TV. In contrast to previous DoD missions, this was to be a full-length flight in order (1) to allow an SDI satellite to observe the plumes of the orbiter's thrusters, (2) to secure biomedical data and (3) to enable Tom Hennen, an expert in interpreting reconnaissance imagery, to test the ability of the human eye to resolve detail on the surface of the Earth. However, owing to a fault in one of the IMUs, it had to come back three days early, and unacceptable weather in Florida resulted in a diversion to Edwards.

Table 6.10 Shuttle manifest, c. late 1991

Flight	Date	Orbiter	Objective
STS-42	Jan 1992	Discovery	IML-1
STS-45	Mar 1992	Atlantis	ATLAS-1
STS-49	May 1992	Endeavour	Intelsat repair
STS-50	Jun 1992	Columbia	USML-1
STS-46	Aug 1992	Atlantis	TSS-1; deploy EURECA
STS-47	Sep 1992	Endeavour	Spacelab J1
STS-52	Nov 1992	Columbia	USMP-1; LAGEOS-2
STS-53	Dec 1992	Discovery	Miscellaneous DoD payloads

The final mission for 1991 was postponed until after the costly holiday season, and dispatched on 22 January 1992, after an hour's hold for weather. As the first International Microgravity Laboratory (IML), this Spacelab was outfitted for life sciences and materials-processing. Ulf Merbold, one of the two payload specialists, had flown on the first Spacelab, and his presence reflected the fact that Germany had built much of the hardware. Roberta Bondar's inclusion resulted from NASA's long-term collaboration with Canada. Although a laboratory was carried in the bay, and a two-shift cycle was invoked, only a seven-day flight had been assigned. However, by conserving the orbiter's consumables it was possible to extend the flight by a day to enhance the scientific yield, which was already high because the experiments had proceeded refreshingly flawlessly.

STS-45's count on 23 March was scrubbed during ET tanking, when a sensor reported a concentration of gaseous hydrogen in the orbiter's engine bay. When an inspection failed to locate a leak, the launch was rescheduled for the following day, at which time Atlantis departed on time. In effect, ATLAS was a second strand of the MTPE programme. Its rôle was to complement the UARS. Like ASTRO, this was a full-length Spacelab mission in which the bay was full of pallets, and as the crew also included two payload specialists, the cabin was crowded. The crew's spirits were raised, however, because the instruments were able to collect a vast amount of data on the state of the Earth's atmosphere and the way in which it is stimulated by the energy received from the Sun. Several of the experiments were flown because on their initial outing on earlier Spacelab missions their operation had been frustrated; and unfortunately one of them (the SEPAC electron-beam) was foiled again, this time by a blown fuse, but not before it had returned a substantial amount of data. It was intended to fly ATLAS on a regular basis in order to characterise the solar–terrestrial relationship at different points in the 11-year sunspot cycle.

ENDEAVOUR'S DEBUT

The most significant item on the 1992 manifest was the debut of Endeavour, the orbiter that had been built to replace Challenger. The count for STS-49 was held for

34 minutes to allow weather at the TAL site to clear, but was otherwise uneventful, and the new orbiter launched on schedule on 7 May. Its task was to rescue Intelsat 603 which had been stranded in low orbit by a fault in the upper stage of its Titan launcher. There was a certain irony in this rescue, because this type of satellite had been designed to exploit the Shuttle's payload bay, but, after the loss of Challenger, NASA had been told to cease its commercial satellite deployments.

The plan was for spacewalking astronauts to catch the satellite, mate it with a new kick-motor, and release it to enable it to resume its journey to geostationary orbit, where it would serve as a communications relay for at least a decade. Special tools had been made to handle the satellite, and training in the hydrotank had suggested that the operation should be able to be completed in two spacewalks. However, capturing the satellite proved to be more difficult than expected, with the result that it was still drifting alongside the orbiter when the time expired. In opting to continue with the Intelsat rescue using impromptu procedures, NASA set several spacewalk records, including duration and the number of people involved, but the difficulty encountered in capturing the large satellite was a cause for concern. The secondary objective of testing procedures for assembling structures had to be truncated and squeezed into the final spacewalk. As Endeavour landed, a drag chute was deployed from the base of its tail. Once the chute had slowed the vehicle to 100 kilometres per hour, it was jettisoned to preclude it snagging on the SSMEs. On this occasion, the chute had been deployed immediately after the nosegear touched down, with the aim of testing the deployment mechanism. In service, it was to be deployed immediately after the main gear touched down in order to slow the vehicle and maintain the centreline, and be released after the nosegear made contact, so that the brakes need not be applied until later in the roll-out for wheel-stop. Finally, it seemed that the Shuttle was capable of landing on a narrow strip in a crosswind.

Next was Columbia as STS-50. It carried a Spacelab of microgravity experiments supplied by US establishments, including the various NASA-sponsored CCDS, and, by analogy with IML, was designated USML. It spent most of its time in gravity-gradient mode, which locked the vehicle stable without needing thrusters to be fired, to provide the materials-processing experiments with the best possible microgravity environment. The life sciences work was greatly enhanced by the fact that this was the first use of the EDO. Although the regenerative unit (the RCRS) failed, there were sufficient lithium hydroxide canisters for the backup system to prevent a build-up of carbon dioxide. Two payload specialists were included in the crew, two shifts were operated and, for the first time, NASA assigned each member of the crew a half-day off during which he could do whatever he wished – even if it was more work! After two weeks, Columbia returned to the Kennedy Space Center because Edwards was 'socked-in'.

Flight operations were now proceeding smoothly. Not only was the manifest unfolding as planned, but NASA was able to dispatch STS-46 a few weeks ahead of schedule. This was not to be a straightforward flight. Radio interference delayed the deployment of the European Space Agency's free-flyer EURECA by 24 hours. This carried experiments requiring an extended period of high-quality microgravity, and was to be left in space and retrieved by another Shuttle a year later. The Tethered Satellite System, which had been jointly funded by the US and Italy, was to have

STS-52 climbs into the Florida sky on a pillar of flame and smoke.

unreeled a test satellite to a distance of 20 kilometres, but it repeatedly jammed after just a few hundred metres and had to be reeled in. As a tethered payload offered significant scientific results, it was virtually assured of a re-flight once the problem in the deployment system was rectified. On its return to space, Endeavour had a Spacelab in its bay. This had been jointly funded by the US and Japan, and involved a large number of materials-processing, technology and life sciences experiments. The launch was on time and, in contrast to its predecessor, caused no excitement, and Endeavour managed to touch down at the Kennedy Space Center between two periods of unacceptable weather. It was followed by STS-52. Located at the rear of the bay was the LAGEOS-2 satellite that had been supplied by Italy. This boosted itself to a high orbit in order to reflect lasers, not for SDI weapons testing but for geophysical research, such as measuring the rate of continental drift and of uplift in volcanoes. Apart from a few GAS cans, the only other item in the cavernous bay was a strikingly small pallet carrying three microgravity experiments, but the status of this payload was acknowledged by its designation – the US Microgravity Package (USMP). This Columbia mission carried such a lightweight payload that it was criticised for not being worth the cost of the launch. The launch of STS-53 – the final mission of the year – had been planned for dawn, but it was delayed for 90 minutes to allow the overnight ice to melt, in order to preclude it shaking loose and striking the orbiter's tiles. Although the temperature had dropped almost as low as that which Challenger had endured the night before its final launch, this was not seen as posing a problem for the redesigned SRBs because the O-rings were now electrically heated. On this mission, the primary payload was a classified satellite, blandly designated 'DoD-1', that was released a few hours after launch. However, other aspects of the flight were conducted openly, and many of these experiments were continuations of earlier projects. The crew had to remain in the orbiter for two hours after landing at Edwards while the recovery team dealt with a leaking RCS thruster in the nose.

For once, the *entire* year's manifest had been flown as planned. By this point, NASA knew that it would never be able to fly a dozen missions a year, and while it had sometimes managed to mount missions within weeks of each other, it was clear that a supreme effort would be required to *sustain* a monthly flight rate. The agency therefore settled down to an annual cycle of six to eight flights, which experience had shown *could* be sustained.

Table 6.11 Shuttle manifest, c. late 1992

Flight	Date	Orbiter	Objective
STS-54	Jan 1993	Endeavour	TDRS-F
STS-55	Feb 1993	Columbia	Spacelab D2
STS-56	Mar 1993	Discovery	ATLAS-2; SPARTAN-201-1
STS-57	Apr 1993	Endeavour	Spacehab 1; EURECA retrieval
STS-51	Jul 1993	Discovery	ACTS; ORFEUS–SPAS-1
STS-58	Aug 1993	Columbia	SLS-2
STS-60	Nov 1993	Discovery	Spacehab 2; WSF-1
STS-61	Dec 1993	Endeavour	HST-1 service

The count for STS-54 was uneventful, and apart from a short hold to enable high-level winds to be assessed, Endeavour was once again launched on schedule. The deployment of a TDRS satellite was now a matter of routine, but in this case, during the coast up to the top of the geostationary transfer orbit, the IUS's primary computer failed and the backup had to step in to deliver the satellite. This flight marked a significant change in NASA's attitude towards spacewalking. The capture of the Intelsat satellite had proved to be much more difficult than the hydrotank training had suggested. Of course, masterful improvisation had overcome these problems, but it had taken significantly longer than the time assigned, and the price had been the curtailment of another task. NASA's immediate concern was that the training for the servicing of the Hubble Space Telescope, which would be far more demanding and more time-critical, might be inadequate. In the longer term, there was some concern that it might be unwise to design a space station that relied on spacewalking for its assembly, and it was therefore decided to add 'generic' tasks to already planned missions in order to build up experience in EVA, enable procedures to be developed, and evaluate the fidelity of underwater training. Another test was performed to reduce the technical risk in operating the station. Standard operating rules required a Shuttle flight to be cut short if a fuel cell had to be shut down, but as an orbiter was to power down after docking with the station it had been decided to verify that a fuel cell could be restarted in space. On the last day of the flight (when it would not matter if the fuel cell could not be brought back on line) one of the three cells was shut down and successfully restarted. The descent was delayed for an orbit to allow the Sun to burn off the early morning mist that was obscuring the SLF at the Kennedy Space Center.

While the newest orbiter was proving to be an exceptionally lucky ship, gremlins were once again at work on Columbia – the oldest member of the fleet – and its February launch as STS-55 had to be postponed to service the high-pressure oxygen pumps of all three SSMEs. The launch was initially set for 21 March, but as the Shuttle had to compete with other rockets for the tracking facilities of the Eastern Test Range it had to be put back a day when a Delta II was delayed. The count was then aborted at T–3 seconds, when an incorrectly set valve inhibited the start of one of the SSMEs. It was the first post-ignition abort since Return-To-Flight. The rule called for engines that had undergone a prestart to be serviced, and this imposed yet another delay. Attention was therefore switched to Discovery for STS-56. This had already slipped several weeks from its planned late-March date. The countdown on 6 April was scrubbed at T–11 seconds when it seemed that a valve in one of the SSMEs had not cycled correctly. Frustratingly, it was found to be a faulty sensor; the valve had operated normally. The launch two days later was perfect, however. On this occasion, a reduced suite of ATLAS instruments had been squeezed onto a single pallet, leaving room at the rear of Discovery's bay for a free-flyer. The 100-class SPARTAN could only carry 140 kilograms of payload, while its successor could carry 500 kilograms. The first one (designated SPARTAN-203) was fitted with two ultraviolet spectrographs to investigate Halley's comet in 1986, but was lost with Challenger. The 200-class SPARTAN-201-1 carried on STS-56 observed the Sun for two days. (The 201 bus would later make a series of outings with an assortment of payloads, none of them related to astronomy.) This second ATLAS flight, following precisely one year after the first, provided a corroborative check on the state of the atmosphere immediately after a northern winter. Meanwhile, Columbia had been serviced sitting on the pad, and NASA wanted to launch it as soon as possible after Discovery's return in order to minimise the disruption to the overall schedule. In the event, Discovery was waved off for an extra day to wait for better weather at the Kennedy Space Center, and did not land until 17 April. Nevertheless, NASA set out to launch Columbia on the first available date, on 24 April. This valiant effort was foiled at T–6 hours when one of the IMUs failed to start properly. Two days later, the count was flawless, and Columbia launched with the second German-sponsored Spacelab. This represented a turning point for the Spacelab programme, because the German Space Research Agency (DLR) had decided that this would be its final 'national' mission. NASA's inability to reduce the cost of flying the Shuttle to anything approaching that of the original hype had made mounting a Spacelab mission so expensive that, from now on, Germany would fly its experiments in the context of the international Spacelabs. Although the D2 Spacelab was on Columbia, the EDO wafer was not carried. However, because the orbiter spent most of its time in the gravity gradient, the rate at which it consumed resources was low enough to allow an extra day for the crew – which included two German payload specialists – to carry out the 88 experiments of the multidisciplinary programme. Much of the laboratory apparatus was being evaluated for use on the Columbus module that the European Space Agency planned to add to NASA's space station.

The manifest was now running significantly late. The factor dictating the launch date of Endeavour's STS-57 mission was the orbit of EURECA, which was to be retrieved. It would have been possible to launch in mid-May, but this would have meant lifting off and landing in darkness, so, accepting the resultant slippage to the schedule, NASA delayed it to 3 June to guarantee a daylight landing. However, when a damaged spring was found in an SSME that was being serviced, it was decided, as a precaution, to replace that part on Endeavour's engines because if the spring fractured in flight an explosion would occur. On 20 June the window expired before rain falling over the SLF lifted sufficiently for a RTLS abort, but conditions were ideal the following day and the countdown proceeded smoothly. This was a landmark mission for NASA because it carried the first 'mid-deck augmentation' module. This facility had been built as a private venture – with NASA's encouragement – by Spacehab Incorporated to carry microgravity experiments on a commercial basis. It provided space for the type of lockers used on the orbiter's mid-deck and its carriage in the front of the bay doubled the usable volume of the orbiter's cabin. For this inaugural flight, NASA had booked half of the module's lockers for experiments by its CCDS; the other experiments were on behalf of commercial customers. The retrieval of the EURECA free-flyer became a race against time. To supply power to the satellite in the short interval between retracting its solar panels and depositing it on its cradle (after which it would be able to draw power from the orbiter) a socket had been installed in the end-effector of the RMS. This socket had, however, been set in reverse, and the satellite was forced to live off its battery. This would not have been a problem if the satellite had been stowed immediately, but its whip antennae refused to retract. The rules prohibited the stowing of EURECA until its antennae had been retracted, but as time was running out it was decided to lower the satellite carefully onto its cradle. Another generic spacewalk had been assigned to this mission, and before they began testing the tools that were to be used in working on the HST, David Low and Jeff Wisoff retracted the antennae to ensure that they could not thrash about during the landing and damage the GAS Bridge Assembly that straddled the bay immediately in front of EURECA's cradle.

With the manifest slipping ever further behind, NASA's priority was to mount the HST servicing mission as planned in December 1993. Although the telescope was returning amazing views, its flawed mirror was still a supreme embarrassment. The successful refurbishment of the telescope was expected to ease the space station's budget through a hostile Congress. If the astronauts damaged the telescope, NASA would probably be forced to abandon its plans for a space station. Given the delays, it was hoped to fly two of the three missions in the queue ahead of the HST service, but STS-60 was actually slipped to the new year.

The launch of STS-51 Discovery had been set for two weeks after Endeavour's return, but the count on 17 July was foiled when a fault was spotted in the system which armed the explosive bolts that would have to fire at T=0 to release the stack. On 24 July, the count was scrubbed at T–19 seconds due to a faulty power unit in one of the SRBs. NASA then decided to wait until after the Perseid meteor shower (and any possible attendant hazard from meteoroid impacts) which peaked on 11 August. The next day produced a post-ignition abort at T–3 seconds, which required that the engines be refurbished. The schedule was being ripped to shreds. Discovery did not finally launch until 12 September. The main objective was the release of the Advanced Communications Technology Satellite (ACTS), which had been built by NASA's Lewis Research Center so as to demonstrate the utility of the high-capacity Ka-Band. This 2.5-tonne satellite was to be lofted to geostationary orbit by the new Transfer Orbit Stage (TOS), which was being used for the first time on the Shuttle. When the command was sent to detonate the pyrotechnics to eject the stage from its cradle, a wiring fault fired both the primary and the secondary systems and while this did not affect the payload, some of the resulting débris punched through the aft wall of the bay; fortunately, the APUs in the compartment beyond were not damaged. The next task was to deploy the ORFEUS–SPAS free-flyer. Germany had sold the original SPAS to the Department of Defense to fund the development of a new heavyweight version. Of its 3.6-tonne mass, fully 50 per cent was payload; and in its new guise it was capable of sustaining a week of independent activities. A 50-gigabit tape unit was fitted to store observations during this extended time, while an S-Band system relayed telemetry via the Shuttle. Two were built. The Orbiting, Retrievable Far- and Extreme-Ultraviolet Spectrometer (ORFEUS) was for ultraviolet astronomy. While it was performing its programme, the Shuttle's crew ran mid-deck experiments and made another spacewalk to test procedures for the HST service. In the final phase of the rendezvous with ORFEUS–SPAS, a laser rangefinder (the TCS) was evaluated as part of the preparations for the recently agreed Shuttle–Mir programme. After bad weather prompted a day's extension,

Endeavour made the first ever night landing at the SLF on 22 September – an experience that indicated the requirement for improved runway lighting.

The second life sciences Spacelab in Columbia's bay for STS-58 was to exploit the EDO to significantly extend the database on human adaptation. The countdown on 14 October was scrubbed at T–31 seconds when a problem was discovered with the ground equipment of the Eastern Test Range, and a fault in the orbiter's S-Band communications link ended the count at T–9 minutes the next day. On 18 October, however, the problem was an aircraft that flew into the restricted airspace over the Cape – it was hardly credible that a private sightseeing aircraft could cause a Shuttle to miss its window, but as soon as the plane had been shooed away Columbia was dispatched. The long-term objective of the biomedical programme was to develop countermeasures against the effects of weightlessness, particularly the bone mass loss that appeared not to stabilise and was slow to recover following return to Earth. Cellular activity is slowed in the absence of gravity, and the terrestrial equilibrium in the relative rates of cell production and destruction is upset in space; the result is a set of imbalances which affect body chemistry – a process that is readily studied in rodents. Previously, rats had been dissected on their return to Earth, but by that time their organs had begun to readapt to gravity, so on STS-58 the rats were killed and dissected in space to preserve the state of their organs in their space-adapted state. Martin Fettman, a veterinarian, flew as a payload specialist specifically to attend to this delicate task. A new experiment, dubbed PILOT, was a programme on a laptop computer to enable the flight crew to refine their flying skills by simulating re-entry, approach and landing. Since in normal circumstances astronauts regularly flew T-38 jets, two weeks of not-flying represented an aberration. This experiment not only maintained their 'proficiency', it also produced quantitative data on how flying skills degraded as a result of being in space. As this was the longest mission to date, Columbia was to return to Edwards to exploit the flexibility of the multiplicity of runways on the dry lake, but in the event it landed on the concrete strip.

FIXING HUBBLE: SPACEWALKERS ON TRIAL

With STS-58 safely back on the ground, attention switched to the attempt to service the Hubble Space Telescope, which was expected to be the most demanding mission of the programme to date. The original operational concept was that the HST would be placed in orbit by a Shuttle and periodically returned to Earth for refurbishment. On reflection, however, it was concluded that this would be impracticable because it would require two Shuttle flights – one to retrieve it and another to replace it in orbit – and the astronomical community would probably be denied the use of its telescope for up to a year on each occasion. To enable the instruments to exploit the inevitable advancements in the state of the art of detector technology over the HST's projected 15-year operating life, the satellite was designed to be serviceable in orbit. This strategy represented a considerable risk, because it required spacewalking astronauts to exchange the bulky and very delicate instruments without damaging the telescope. But NASA had high hopes for its spacewalkers. Routine servicing in orbit was one thing, but overcoming the flawed main mirror promised to be an unprecedented spacewalking task. The light path to *each* of the five instruments needed to be corrected. One of the instruments comprised a broad wedge-shaped package set just behind the mirror; the others were further back in telephone-booth-sized units clustered around the axis. JPL had planned to replace its camera on the first service call, so it installed corrective optics directly into the new camera. As it was deemed impractical to replace all the instruments, it was decided to sacrifice one to enable a 'corrective optics' unit to be installed to restore the vision of the remaining three *without modification*. The job facing the astronauts was daunting. The 300-kilogram COSTAR and JPL camera packages were to be ferried up on a pair of Spacelab pallets and inserted by hand. If the 2.2-metre-tall axial replacement unit was to mate with its guide rails, it would have to be positioned to an accuracy of 1 millimetre. After extensive training in the WETF, the astronauts exuded confidence, but others feared that they might wreck the telescope. In early 1993 Congress considered prohibiting the agency from attempting the repair operation, but the astronauts were eager to try. Although NASA relished the prospect of success, it feared that failure would provide the ammunition for its Congressional opponents to kill the space station, whose assembly would rely on spacewalkers. To ensure that it covered all of its bases, NASA asked Joe Shea at MIT to undertake an independent review of the plan. In May, Shea's panel recommended that NASA be allowed to proceed. In fact, in November 1992 NASA had lost the option of simply retaining the HST in orbit with its degraded optics. The HST had six gyroscopes for attitude control, of which one failed in December 1990, and a second failed in June 1991. Now a third had ceased operating. If a fourth failed, the telescope would have to be shut down. In fact, there was some concern that it might not survive until the Shuttle arrived in December. Even as the MIT panel was considering the triple-EVA plan, one of the three Fine Guidance Sensors started to misbehave, as did one of the electronic drive controllers for the solar arrays – giving two problems that impaired the telescope's utility. With each passing month, therefore, the scale of the repair was increasing. The number of spacewalks increased to five on the optimistic plan, and seven on the pessimistic plan. NASA tentatively assigned a follow-up mission in late 1994 to complete the repair and to rectify any faults caused during the first attempt. Story Musgrave, Jeff Hoffman, Tom Akers and Kathryn Thornton spent *hundreds of hours* immersed in the WETF rehearsing each specific task, followed by integrated simulations in which the tasks were performed in sequence to make sure they did not conflict. To minimise the risk of being taken by surprise, the tools were tested as secondary activities on two Shuttle flights in the summer.

Although the first count on 1 December had to be scrubbed because the weather would have inhibited a RTLS, there were no issues the following day and the launch was on time, which was fortunate as NASA was eager to have the mission finished before Christmas in order to avoid the cost of encroaching on the holiday break. After a two-day chase, Endeavour took up station directly beneath the HST so as to use the differential gravity as a brake as it closed in, both to preclude a collision in the event of a problem and to reduce the degree to which the thruster plumes would contaminate the telescope's optics. Claude Nicollier then used the RMS to grasp the telescope and mount it on a rotating tilt-table (known as the Flight Support System) at the rear of the bay. The plan called for one EVA each day for the next five days, with each excursion maximally exploiting the life-support system of the EMU. Pairs of astronauts were to go out on alternate days in order to ease the load and provide a day's rest between excursions. It was to be a carefully phased programme of work requiring the delicate and precise manipulation of bulky objects. Seven hours was a long time to sustain concentration and a misjudged action could all to easily ruin the telescope. Also any unforeseen problem could cause an EVA to fall behind schedule. If the work went well, but slower than planned, the astronauts were to focus on the primary tasks and leave other activities to the follow-on mission. In fact, in more than 35 hours of external work, every item on the list was accomplished more or less on the expected timeline. On alternate days Musgrave and Hoffman, then Thornton and Akers, progressively repaired the telescope. On the first excursion, a new set of gyroscopes was installed. The following day, the solar panels were removed. One was so badly distorted that it would not roll up and Thornton cast it adrift, but the other was returned for examination. Improved panels of the same configuration were then mounted on the telescope, but not deployed. The third day saw JPL's camera replaced and the photometer replaced by COSTAR. Finally, a failed power supply on one of the spectrometers was replaced, the faulty solar array drive controller was replaced, and the new solar arrays were unrolled. As the fault in the Fine Guidance Sensor had been overcome, the task of replacing this had been deleted from the programme. Almost everything had gone to plan, the only concern being a door that was difficult to close on the axial-instrument compartment. All the special tools had worked. The bulky instruments had come out on their rails exactly as expected, and the replacements had slipped in without difficulty. The HST, whose configuration was dominated by its need to be serviceable in space, had proved to be exceptionally well designed, and the improvement in resolution delivered by the corrective optics was all the more appreciated for having been denied for so long. Barbara Mikulski, chair of the Senate Committee that set NASA's budget, gave the HST the official seal of approval. "The trouble with Hubble is over," she pronounced. The 1994 contingency mission was quietly dropped from the manifest. Fixing the HST was a magnificent demonstration of the ability of astronauts to work in space. With this success, and the agreement to work with Russia to assemble an *International* Space Station (ISS), NASA's long-term future seemed assured.

Table 6.12 Shuttle manifest, c. late 1993

Flight	Date	Orbiter	Objective
STS-60	Jan 1994	Discovery	Spacehab 2; WSF-1
STS-62	Mar 1994	Columbia	USMP-2, OAST-2
STS-59	Apr 1994	Endeavour	SRL-1
STS-63	Jun 1994	Discovery	Spacehab 3; SPARTAN-201-2
STS-65	Jul 1994	Columbia	IML-2
STS-66	Sep 1994	Endeavour	ATLAS-3; CRISTA–SPAS-1
STS-64	Sep 1994	Discovery	LITE; SPARTAN-204
STS-68	Dec 1994	Atlantis	SRL-2
STS-67	Dec 1994	Columbia	ASTRO-2

The new year began with the postponed STS-60. The launch was set for 27 January, but was delayed to enable a leak in one of the aft RCS thrusters to be investigated, and it was finally dispatched on 3 February. Discovery had a Spacehab and a novel free-flyer in its bay, but the media's interest was the fact that cosmonaut Sergei Krikalev was on board. His inclusion in the crew was the first result of the agreement signed by President George Bush and President Boris Yeltsin on 17 June 1992. NASA had booked half of the lockers on Spacehab for experiments supplied by its CCDS. The Wake Shield Facility (WSF) had been privately developed to demonstrate the commercial potential for 'growing' semiconductors in space. This free-flyer was to have been released for two days, but technical problems prevented its deployment, which was frustrating for Ron Sega, a member of the development team who had become an astronaut in order to supervise the test. Conditions at the Kennedy Space Center were marginal, but after a one-orbit wave-off permission was given for the de-orbit burn. However, the weather closed in, and when the orbiter arrived it found 80 per cent cloud. Nevertheless, upon emerging from the final turn of the heading alignment cylinder, mission commander Charlie Bolden reported that he could see the runway through a small clearing, and made a perfect landing.

STS-62 Columbia, which was put back a day because poor weather was forecast, was launched on 4 March. In addition to the USMP pallet, it carried a Hitchhiker Bridge with an assortment of apparatus on trial for use on future spacecraft. As previously, the lightweight payload attracted criticism. This time, the microgravity experiments were able to exploit the presence of the EDO wafer, and to provide the best possible conditions the orbiter adopted gravity-gradient stability for most of that time. In effect, this flight served as a trial run for the kind of research that was to be undertaken on the ISS.

Although delayed a day due to weather, Endeavour's launch on 9 April marked another step in the unfolding planned manifest. STS-59's bay was dominated by the Shuttle Radar Laboratory (SRL), which combined the L/C-Band radars used on two previous occasions with an X-Band radar developed jointly by Germany and Italy. At 10 tonnes, the new SIR-C configuration was the largest instrument ever designed to work in the bay. Between them, the radars scanned a swath centred on the ground track for over 100 hours, and mapped 25 per cent of the Earth's land surface with a linear resolution of better than 25 metres – and in ideal conditions 10 metres. This extraordinarily productive mission required the orbiter to make a record 400 thruster firings to adjust its orbit. Towards the end, the ground track was refined to attempt stereo imaging. The microwave radars proved to be particularly adept at penetrating sand in arid regions to detect the topography of the underlying bedrock, and they revealed several previously unsuspected ancient river beds in terrain that is now desert. The flight was extended by a day to await better weather in Florida, but when this failed to materialise Columbia was ordered to land at Edwards.

It had become clear that, apart from the lockers booked by NASA, the Spacehab was so under-subscribed that its flight on STS-63 could not be financially justified. This mission was therefore postponed while the company sought customers. To the distress of all concerned in the venture, the market for commercial microgravity applications had not developed as rapidly as hoped. There was significant industrial participation in the semi-academic CCDS, but this market was directly serviced by NASA's block bookings. There were too many such payloads for the mid-deck, but insufficient for a Spacehab flight every six months. As Columbia's turnaround after STS-62 could not be accelerated, STS-65 could not be advanced. It was launched on time on 8 July for a record 15-day flight of the second IML Spacelab to assess a new level of interactivity between the apparatus on the orbiter and the scientists who had built it. This 'telescience' method of working was to be used on the ISS. On landing, Columbia was returned to Rockwell for a refit. Meanwhile, NASA had revised some orbiter/mission assignments: ATLAS-3 on STS-66 was reassigned from Endeavour to Atlantis, which was just out of refit, and Endeavour became STS-68 with SRL-2 and advanced to 18 August, but its launch was aborted at T–2 seconds because one of the oxygen turbopumps overheated. Rather than wait for these fired engines to be refurbished, the SSMEs on Atlantis were commandeered, and while they were being installed in Endeavour, Discovery was dispatched as STS-64 on 9 September. This was a busy mission. In addition to deploying the SPARTAN-201 free-flyer for the second time, it included a spacewalk to test a new manoeuvring unit and the testing of a new instrument to investigate the atmosphere. The Lidar-In-space Technology Experiment (LITE), which was carried on a pallet in the bay, was a laser with a bore-sighted telescope. The pulsed laser illuminated a column of the atmosphere and a matched detector in the telescope noted the way in which the light was scattered. By slewing the laser back and forth across the ground track, it was possible to make a fine-resolution survey of the three-dimensional distribution of particulate matter (aerosols) in the atmosphere. Although lidar is standard equipment on meteorological research aircraft, this was the first time such an instrument had been flown in space. A faulty tape unit interfered with the trial, but 50 hours of data was collected, much of it in concert with aircraft whose data was to enable the space sensor to be calibrated. Once refined, such an instrument will be able to be flown on an automated satellite. To assess the extent to which the orbiter's thrusters would impinge upon Mir in the forthcoming rendezvous, the RMS retrieved the 10-metre SPIFEX boom and used the sensors at its end to sample the plumes directly. The visual highlight of Discovery's mission, however, was the spacewalk by Mark Lee and Carl Meade to test the SAFER backpack-augmentation thruster pack affixed to the base of the life-support backpack – developed for spacewalkers working outside the ISS. The unit's nitrogen-gas jets would enable an astronaut with a loose tether to return to the safety of the station's structure. Discovery was ready to come home, but Florida weather forced a day's extension and diversion to Edwards. Ten days later, on 30 September, Endeavour was launched for the second flight of the SRL. This slippage meant that it would not be able to be turned around in time for the January STS-67 mission that it had inherited from Columbia as part of the re-manifesting that occurred earlier in the year, and this ASTRO re-flight was therefore postponed to March 1995, leaving a timely gap in which to reschedule the long-delayed STS-63. The immediate priority, however, was to fly STS-66. As it had taken longer than expected to rustle up new SSMEs for Atlantis, its launch was slipped a week, to 3 November. On this mission to monitor the atmosphere just prior to the onset of the northern winter, it had the ATLAS pallet and a new version of the

Table 6.13 Shuttle manifest, c. late 1994

Flight	Date	Orbiter	Objective
STS-63	Jan 1995	Discovery	Spacehab 3; SPARTAN-204
STS-67	Mar 1995	Endeavour	ASTRO-2
STS-71	May 1995	Atlantis	SMM-1
STS-70	Jun 1995	Discovery	TDRS-G
STS-69	Jul 1995	Endeavour	WSF-2; SPARTAN-201-3
STS-73	Sep 1995	Columbia	USML-2
STS-74	Oct 1995	Atlantis	SMM-2
STS-72	Nov 1995	Endeavour	SPARTAN-206; SFU retrieval

upgraded SPAS free-flyer – in this case the Cryogenic Infrared Spectrometer and Telescope for the Atmosphere (CRISTA). On its return, Atlantis became the fourth flight of the year to be diverted to California owing to inclement weather in Florida.

The multifaceted science programme was now unfolding nicely and since orbital operations had become routine the media took little interest. Nevertheless, political events now spawned an entirely new programme that would repeatedly raise the visibility of the Shuttle to the lead item on the nightly news shows: America was not the only space-faring nation.

VISITING RIGHTS

For STS-63, Discovery was still to carry both the Spacehab and SPARTAN-204 payloads, but it had acquired an extra crewman and an important new objective: cosmonaut Vladimir Titov was to supervise a rendezvous with the Mir space station. The greater the interval between the time that Discovery achieved orbit and the time Mir's orbital plane intersected that position, the more difficult would be the rendezvous. The window was particularly narrow owing to the Russian requirement that the final part of the rendezvous should take place within range of the Russian communications network. In the event that the Shuttle could not achieve the desired initial orbit, the rendezvous with Mir would be cancelled because, its historic nature notwithstanding, it was only the secondary objective and its achievement could not use up fuel required for the deployment and retrieval of the SPARTAN free-flyer. In fact, after a day's delay to replace a faulty IMU, Discovery lifted off at the optimum moment on 3 February. It was flown by Jim Wetherbee and Eileen Collins, the first female Shuttle pilot. Despite concern over leaking RCS thrusters which might contaminate the instruments on Mir's surface, the rendezvous three days later was flawless and the orbiter drew to a halt just 10 metres from the docking port at the end of Mir's Kristall module. For 10 minutes the two 100-tonne vehicles maintained formation, then Discovery withdrew 100 metres for a fly-around to photograph the station prior to departing to conduct its primary mission.

Lacking any commercial payloads, Spacehab Incorporated had renegotiated its contract such that NASA would utilise all of the lockers on the next two modules instead of half of the lockers over the next four flights. One of the experiments on this flight was a robot called Charlotte, which had been developed to tend to other experiments. During a spacewalk to assess procedures for the ISS, Bernard Harris and Michael Foale remained motionless in the orbiter's shadow in order to test how cold they became, and when Foale reported that it felt as if his fingers were in an icebox the test was curtailed; clearly, the gloves of the spacesuit would need some modification.

The repeatedly delayed STS-67 finally lifted off without incident on 2 March for the ASTRO-2 mission. In early 1990, before ASTRO even made its first flight, its re-flight in late 1992 had been cancelled, but after the first results were analysed it was reinstated, although with a low priority. On the other hand, as the EDO system had since been introduced, this second flight was to be twice as long as the first. Indeed, by being extended by a day, Endeavour set a new Shuttle endurance record. It was something of an ordeal for the crew (which once again included Sam Durrance and Ron Parise) operating round the clock in the confines of the cabin. Fortunately, it went extremely well, the only issue being a fault in one of the telescopes that ruined several of its images. Overall, the mission was deemed 98 per cent successful. No secondary payload could be carried because the EDO wafer was located in the rear of the bay. At this point, the ASTRO project was curtailed, in part due to cost, but primarily because NASA had to sacrifice half a dozen Spacelab flight opportunities to free launches for the Shuttle–Mir missions that had been introduced to pursue the joint programme with the Russians.

On 14 March, while the ASTRO-2 mission was underway, astronaut Norman Thagard was launched on a Soyuz rocket from the Baikonur Cosmodrome in Kazakhstan and, two days later, boarded the Mir space station. In the agreement that was signed with Russia in 1992, an astronaut would be allowed to pay a visit to Mir in return for a cosmonaut flying on a Shuttle mission. With the subsequent decision to merge the two national programmes, it had been decided to conduct the next crew handover as part of the Atlantis docking mission. Although this STS-71 flight dominated the summer schedule, the date could not be set until the Spektr module had docked with the Mir complex, and the launch of that module, which was to deliver most of Thagard's scientific apparatus, had been repeatedly delayed. As STS-71 slipped, so did the next mission, STS-70, but in April, when it became clear that Mir would not be ready to receive Atlantis until early July, NASA decided to launch Discovery on 8 June, as originally scheduled, but only for a truncated five-day mission. If STS-70 suffered a substantial delay, it would be postponed to August to enable STS-69 to fly in late July, after Atlantis. On 2 June, STS-70 had to be returned to the VAB to enable the insulation on its ET to be refurbished, since woodpeckers had made holes in the foam while trying to make nests. The mission that would symbolically end the Cold War was thus assigned to NASA's 100th crew. The count on 23 June was abandoned early on due to a nearby storm. The next day it was scrubbed when the hold at T–9 minutes, which was extended awaiting an improvement in the weather, encroached on the 10-minute window. However, on 27 June Atlantis lifted off precisely on time. It was a significant occasion – for once, a

Shuttle had had somewhere to go. Two days later, commander Robert 'Hoot' Gibson eased his vehicle into the Kristall module's docking ring. When the hatches were opened an hour or so later, he and Vladimir Dezhurov, the Mir commander, shook hands to formalise the union. In the control centre at Kaliningrad, Dan Goldin, the NASA administrator, and Yuri Koptev, the chief of the Russian Space Agency, did likewise. When the crews congregated in Mir's base block, it became evident that the station's designers had never expected that it might one day host 10 people. For Bonnie Dunbar this was a bitter-sweet mission. As Thagard's backup, Dunbar was initially to have swapped places with Thagard to serve with Anatoli Solovyov and Nikolai Budarin – the new residents who had flown up on Atlantis as passengers – but Mir's Soyuz escape capsule was limited to three people and the European Space Agency had booked a tour of duty on board the complex for Thomas Reiter. In addition to the historic ceremony, there was scientific data to be gleaned from subjecting the departing Mir crew to a full battery of biomedical tests with similar apparatus to that carried on the SLS missions. This work, which was supervised by physicians Thagard and Ellen Baker, significantly enhanced NASA's database on the adaptation of the body to the space environment. On 4 July, after miscellaneous cargo had been transferred, Atlantis undocked. To avoid Thagard, Dezhurov and Strekalov having to endure the discomfort of sitting upright during the return to Earth, couches were erected on the mid-deck. On landing on 7 July, Thagard staggered off the orbiter, but his Russian colleagues allowed themselves to be carried off recumbent to ensure that they would not jeopardise the follow-on biomedical tests.

Despite having postponed STS-70, it was not, as had been announced, put back after STS-69; instead STS-69 was slipped so that STS-70, which by then was ready, could be launched in record time after Atlantis's return, and when Discovery lifted off on 13 July this was achieved. After releasing the new TDRS satellite, this flight ran to its originally planned length to facilitate a variety of mid-deck experiments. At this point, an alarming discovery was made. Erosion of the inner lining of the SRBs was not uncommon, but the post-flight inspection of STS-71's boosters revealed the most severe O-ring scorching to date. As the nozzle-joint of an STS-60 SRB had also been damaged, it was decided to augment the joints of the already-mated boosters, which meant that STS-69 missed its 5 August count. A fuel cell that would not start properly prompted a T–8 hours scrub on 31 August, but Endeavour was launched on 7 September without incident carrying the SPARTAN-201-3 and WSF free-flyers. Although the WSF was able to be released on this occasion, an attitude-control fault interfered with its experiment. The SRBs from STS-69 were inspected as a matter of urgency to clear the way for Columbia, whose boosters had also had their nozzle-joint O-ring insulation augmented. One of the three SSMEs used by Discovery had tested a new oxygen pump with higher performance, and as this had functioned well Columbia had been fitted with two of these upgraded motors. Once this new oxidiser pump became standard, the effort would switch to improving the hydrogen pump to increase the Shuttle's capacity to deliver cargo to the 51-degree-inclination orbit in which the ISS was to be assembled. In the event, STS-73's count on 28 September had to be abandoned when the hydrogen pump on one of these new engines sprung a leak. Starting on 5 October, three successive counts were scrubbed due, respectively, to a storm, a hydraulics fault and an engine fault. After the weather pre-empted the count on 15 October, Columbia had to yield priority on the Eastern Test Range to an Atlas 2 rocket, but was finally launched on 20 October. By relying on telescience, the operations of the USML-2 experiments served as a high-fidelity rehearsal for the way in which science would be conducted on the ISS. Shortly before the de-orbit burn, a small piece of orbital débris struck one of the recently closed payload bay doors. After Columbia's safe return on 5 November, the focus was switched to Atlantis.

When the year's manifest had been drawn up, it had assigned STS-74's launch to 26 October, but the summer hiatus had compressed the post-STS-71 schedule, with missions being flown back to back. The progressive delays in dispatching Columbia had eaten up the slack and pushed back Atlantis. However, having demonstrated its ability to launch one flight just a week after its predecessor landed, NASA assigned STS-74 to the first available window. The 11 November attempt was pre-empted by poor weather at the TAL site, but Atlantis lifted off at the optimum point in the all-too-brief Mir window the following day. STS-74's mission was notable for the fact that this was the first time that a Shuttle was to add a module to a station. In keeping with the refreshing new spirit of international cooperation, this 5-metre-long airlock had been constructed by the Russians. Mounting the Docking Module (DM) on Kristall would, in future, guarantee the orbiter clearance from Mir's solar panels, thereby eliminating the need to swing Kristall onto the complex's axis (as had been done for the STS-71 docking) because repeatedly relocating Kristall would interfere with Mir's ongoing work. Since Thomas Reiter was serving his tour of duty, NASA was unable to leave an astronaut on the station. Atlantis's landing on 20 November brought operations for the year to a conclusion. As Endeavour was still being turned around after its repeatedly delayed STS-69 mission, it was impossible to launch it as STS-72 in December, and this slipped into the new year.

Its luck having returned, Endeavour lifted off on 11 January for STS-72 in the centre of the window for the rendezvous with the Space Flyer Unit (SFU) – which Japan had launched on its new H-2 rocket the previous March on the understanding that NASA would retrieve it. During its retrieval by Koichi Wakata using the RMS,

Table 6.14 Shuttle manifest, c. late 1995

Flight	Date	Orbiter	Objective
STS-72	Jan 1996	Endeavour	SPARTAN-206/OAST; SFU retrieval
STS-75	Feb 1996	Columbia	USMP-3; TSS-2
STS-76	Mar 1996	Atlantis	SMM-3
STS-77	May 1996	Endeavour	Spacehab 4; SPARTAN-207/IAE
STS-78	Jun 1996	Columbia	LMS-1
STS-79	Aug 1996	Atlantis	SMM-4
STS-80	Nov 1996	Columbia	WSF-3; ORFEUS–SPAS-2
STS-81	Dec 1996	Atlantis	SMM-5

The Tethered Satellite System is unreeled during the STS-75 mission.

the two long solar panels failed to retract, and were jettisoned. The 3-tonne package of experiments was then captured and placed on a cradle in the bay. The following day, the SPARTAN free-flyer was deployed, and while it performed its programme for NASA's Office of Aeronautics and Space Technology (OAST), two spacewalks rehearsed procedures for the ISS. A month later, Columbia was launched on time for STS-75. Its primary task was to repeat the TSS experiment, and on this occasion, although the tether unreeled smoothly, it broke shortly prior to attaining its full 20-kilometres and the satellite was drawn away by differential gravity. Nevertheless, just before this failure the act of trailing the conducting tether in the Earth's magnetic field had generated a low-power current with a potential difference of 3,500 volts, satisfying the scientific objective of the test. The rest of this cabin-confined EDO mission was devoted to the USMP experiments.

THE MILK RUN

Atlantis's count for STS-76 was slipped a day to 22 March due to excessive winds, but on that date it lifted off on time. This flight initiated a new phase in NASA's operations. Over the next two years a succession of astronauts were to serve tours on Mir, with Atlantis routinely making handover and resupply flights. Atlantis was the only orbiter equipped to carry the docking unit. The Shuttle–Mir flights were the fixed points on the manifest, and launching Atlantis on schedule was to take first priority. In addition, dedicating an orbiter to Mir for such a lengthy period meant that all the payloads that it would otherwise have carried had either to be deleted or reassigned to other orbiters. It was now evident that the rate could not be increased beyond the eight flights that was achievable in a good year. Rather than continue to fly a Spacelab – as had been done on the first docking in order to use its medical instruments – NASA had negotiated a contract with Spacehab Incorporated in which the company would bolt together two of its modules to form a bulk-cargo hauler. As this was not yet ready, Atlantis carried a standard module in its bay for STS-76. Although, previously, the Spacehab had been carried forward to allow the rear to be assigned to free-flyers, the fact that the ODS had to be up-front meant that the squat module had to be located further back. By placing Spacehab nearer the orbiter's centre of mass, it was possible to carry a much heavier cargo than would otherwise have been feasible. While Atlantis was docked, Rich Clifford and Linda Godwin spacewalked to remove superfluous apparatus from the DM and install an experiment package. They wore the SAFER backpacks in case their tethers broke. When it was time for Atlantis to depart, Shannon Lucid remained on the Mir side of the hatch. In an emergency, she would return to Earth with Yuri Onufrienko and Yuri Usachev in the Soyuz lifeboat, otherwise she would be retrieved by Atlantis when it returned four months later.

Endeavour's count for STS-77 was on schedule. In addition to the final Spacehab with NASA-funded lockers, the bay had a SPARTAN free-flyer with the innovative inflatable antenna experiment. The inflation of the antenna was a considerably more dynamic process than had been expected – the 1.2-tonne SPARTAN carrier tossed and turned, and was still pitching end over end when the 15-metre-diameter dish and a 28-metre tripod on which to mount a transponder rigidised. The antenna was later jettisoned to enable the carrier to be retrieved. On landing, Endeavour was returned to Rockwell for a refit in which it was to receive an ODS in preparation for missions to assemble the ISS.

The year's manifest had been unfolding as planned, and Columbia's launch as STS-78 on 20 June was on time. For the first time, a video camera recorded activity on the flight deck throughout the ascent. This Life and Microgravity Sciences (LMS) Spacelab took advantage of the EDO wafer and was another exercise in round-the-clock science with broad international participation to set the scene for the ISS. The post-flight inspection of STS-78's SRBs identified a significant degree of hot-gas erosion. Although this had not damaged the O-rings, it represented an unacceptable risk. Tests determined that the new sealant that had recently been introduced was faulty, and it was decided to revert to the older material. The new water-based putty replaced the old methyl-based type that had been prohibited by the Environmental Protection Agency. Because STS-79's boosters had already been integrated, the only option was to strip them down and start again. In order not to delay Atlantis for any longer than necessary, the segments intended for STS-80 were to be mated using the old sealant and reassigned to STS-79. Even so, this pushed the mission back by six weeks to 12 September. On Mir, Shannon Lucid, who was enjoying her tour, took this news in her stride. On 4 September, Atlantis had to retreat to the VAB to avoid a hurricane. Although this delayed the launch to 16 September, the countdown was faultless. Having spent six months on Mir, which the Russians now regarded as a normal tour of duty, Lucid handed over to John Blaha. It was the first time that NASA had had occasion to perform this ritual, and it marked a significant milestone in the programme. The double-sized Spacehab in Atlantis's bay carried the heaviest load of cargo to date, and it took several days to perform the transfers to and from the station. In addition, after Atlantis had undocked, a variety of experiments were performed in the module to test apparatus for later use on the ISS. On landing, Lucid expressed her surprise at how readily she readapted to the inexorable pull of gravity.

Despite having used the original type of sealant, STS-79's boosters had suffered significant erosion. In one case, the damage took the form of a series of 6-cm-deep grooves in the carbon cloth liner 'downstream' of small depressions carved in the throat of the nozzle. Such grooving had been observed earlier, but not on this scale. It had not reached the casing, but if the 3,100 °C gas had come into contact with the metal, the resulting breach could have produced a catastrophic failure. The effort to catch up on the schedule by dispatching STS-80 on 31 October was made contingent on a review of the recent changes to the booster's manufacturing process and, once the fault had been rectified, the count was recycled for a 15 November launch. Weather and competing calls on the Eastern Test Range slipped it to 19 November. On that date, after the count was held for a few minutes to allow a potential hydrogen leak to be investigated (and found to be spurious) Columbia finally lifted off. The WSF and the ORFEUS–SPAS free-flyers were deployed and retrieved as planned, but a series of spacewalks to test tools for the ISS had to be cancelled after it was found that the outer hatch of the airlock could not be opened because a screw in the mechanism had worked loose and jammed the ratchet. The rest of the mission passed off smoothly until it was time to return, and after two days for delays due to weather Columbia was ordered back to Edwards. At 18 days, this flight had stretched the EDO facility to its absolute limit. Apart from anything else, STS-80 was notable because it was astronaut Story Musgrave's final Shuttle mission – his sixth.

With Atlantis committed to Shuttle–Mir, its schedule revolved around visits to Mir every four or five months, so the delay in launching STS-79 was propagated to STS-81, pushing it into early 1997, but it was dispatched on 12 January on time. Those who had argued that the Shuttle would not be able to meet the tight launch windows for Mir had so far been proved to be unduly pessimistic. Shuttle–Mir had settled into a routine, with its own rituals. Blaha's handover to Jerry Linenger was uneventful. There is nothing like success for fostering a 'can do' attitude, so when Discovery took off to service the Hubble Space Telescope it all seemed very routine. After its repair by STS-61, the HST lived up to expectations by making startling discoveries, and the Space Telescope Science Institute was heavily over-subscribed with requests for observing time. This service call was to replace failed gyroscopes, Fine Guidance Sensors and tape recorders and to replace the spectrometers with two new instruments. The countdown was advanced two days, to 11 February, to allow for any minor slippage that might arise, because the Eastern Test Range was booked by various rockets for the week after 14 February, but there were no delays. Once again, the spacewalks went well. The only surprise was that some of the thermal insulation on the side of the satellite that faces the Sun had cracked and peeled off, entailing an extra spacewalk to tape improvised protective covers across the most damaged areas until the next service mission – scheduled for late 1999 or early 2000. Mark Lee and Steven Smith, and Greg Harbaugh and Joe Tanner conducted the spacewalks. Steven Hawley, the man who had originally deployed the HST, worked the RMS. The new instruments were the Near-Infrared Camera and Multi-Object Spectrometer (NICMOS), which was for extremely-high-resolution imagery, but also had a spectroscopic capability, and the Space Telescope Imaging Spectrometer (STIS). As the landing at the Kennedy Space Center was in darkness, the recently

Table 6.15 Shuttle manifest, c. late 1996

Flight	Date	Orbiter	Objective
STS-81	Jan 1997	Atlantis	SMM-5
STS-82	Feb 1997	Discovery	HST-2 service
STS-83	Apr 1997	Columbia	MSL-1
STS-84	May 1997	Atlantis	SMM-6
STS-85	Jul 1997	Discovery	CRISTA–SPAS-2
STS-86	Sep 1997	Atlantis	SMM-7
STS-87	Oct 1997	Columbia	USMP-4; SPARTAN-201-4
STS-88	Dec 1997	Endeavour	ISS-1

installed halogen lamps on the centreline of the SLF were illuminated to help Ken Bowersox to line up for the final approach.

After lifting off a day late, on 4 April, STS-83 rapidly deteriorated as a fuel cell that had never really settled down finally had to be switched off. Columbia was therefore recalled only four days into its assigned 16-day mission. The Materials Sciences Laboratory (MSL) was to have been the last in a series of eight Spacelabs, and was to have undertaken the most extensive investigation to date of combustion in space. This basic research was considered to be so important that, for the first time, NASA decided to turn an orbiter around and relaunch it with the same payload and crew. In the event, there was a convenient gap in the manifest in the summer, caused by the recent postponement of the first assembly flight of the ISS. Columbia was refurbished (a process that was eased by the fact that the payload did not have to be reintegrated), redesignated STS-94 and slipped into the schedule after STS-84, the next Mir mission. Atlantis was launched on time on 15 May and retrieved Jerry Linenger, who was succeeded by Michael Foale. After an uneventful countdown on 1 July, Columbia was relaunched and this time accomplished its 16-day MSL flight.

Slowly but surely, NASA was re-adopting dynamic scheduling as its standard operating procedure. This was just as well, because in April it had to slip STS-88 Endeavour by eight months to July 1998 because the Russians were behind schedule in constructing the module that was to be the ISS's habitat. The manifest for the rest of the year, and into 1998, was therefore revised.

STS-85 Discovery lifted off on 7 August as planned, as an international mission. In addition to the joint-US/German CRISTA–SPAS free-flyer, which was deployed to monitor the solar–terrestrial relationship for a week, the payload bay included a Hitchhiker Bridge, GAS canisters, and an MPESS carrying a prototype of the robotic arm that would be used to service apparatus mounted outside the Japanese laboratory of the ISS. The only unplanned moment of the mission occurred when the PAM motor that had failed to boost Westar-6 into geostationary transfer orbit so many years ago made a close pass by the free-flyer, but there was no danger of collision.

The next mission to Mir became a cliff-hanger even before it left the ground. Had Mir become too dangerous to risk another four-month tour of duty? NASA had

Table 6.16 Shuttle manifest, c. mid-1997

Flight	Date	Orbiter	Objective
STS-85	Aug 1997	Discovery	CRISTA–SPAS-1
STS-86	Sep 1997	Atlantis	SMM-7
STS-87	Nov 1997	Columbia	USMP-4; SPARTAN-201-4
STS-89	Jan 1998	Discovery	SMM-8
STS-90	Apr 1998	Columbia	Neurolab
STS-91	May 1998	Discovery	SMM-9
STS-88	Jul 1998	Endeavour	ISS-1

downplayed the fire during Linenger's tour without too much difficulty. However, the recent collision with a discarded Progress cargo ferry that was being manoeuvred around the station, and the crisis that this had precipitated, had led some members of Congress to call for Shuttle–Mir to be curtailed on the basis that Mir was dangerous and in any case, from NASA's point of view the joint programme had already served its purpose. A few hours prior to launch, Dan Goldin confirmed that NASA would honour its commitments. In fact, a change had already been made. Foale was to have handed over to Wendy Lawrence, but as she was too short to use the Russian EVA suit it was decided to send David Wolf because he was to make an excursion on the next mission, and would be able to help out in an emergency. Lawrence would still fly, because she had detailed knowledge of the cargo transfer procedures, but would not stay. Once again, Atlantis blasted off at precisely the optimum moment in the narrow window for a rendezvous with what was now effectively an international, rather than a Russian, space station. The STS-86 crew included Vladimir Titov and Jean-Loup Chrétien, the Frenchman who had already visited two of the Salyut stations. Scott Parazynski made an EVA with Titov to retrieve the experiments that had been placed on the DM by STS-76, and to test apparatus to be used on the ISS. After landing, Atlantis was mated with the SCA and flown to Rockwell for a refit. The final two Shuttle–Mir flights – which had been added to the initial seven-flight programme to help the Russians to extend Mir's utility until assembly of the ISS was well underway – were to be flown by Discovery. However, with the start of the ISS delayed, the newly refurbished Endeavour was left without a mission, and it was decided that Endeavour should fly the first of these Mir missions.

NASA's run of luck evaporated on STS-87. The SPARTAN-201-4 free-flyer was released but failed to start its sequencer, and could not stabilise itself and just drifted in space. This put NASA to the test. It had rescued satellites before, but by training a later crew to undertake the task using specially designed equipment. In this case, however, there was no convenient way of tacking the recovery onto a future mission. In the event, Winston Scott and Takao Doi, who had planned to make a spacewalk to test procedures for the ISS, set themselves up on each side of the bay and Kevin Kregel manoeuvred Columbia to position the free-flyer directly between them, to enable them to grasp the satellite and ease it down into its cradle using the universal tool – the human hand. The casual banter they maintained throughout the operation made it all seem so straightforward.

In January 1998 STS-89 retrieved Wolf from Mir, whose tour of duty had been refreshingly quiet, and delivered Andy Thomas for the final increment of NASA's

Table 6.17 Shuttle manifest, c. late 1997

Flight	Date	Orbiter	Objective
STS-89	Jan 1998	Endeavour	SMM-8
STS-90	Apr 1998	Columbia	Neurolab
STS-91	May 1998	Discovery	SMM-9
STS-88	Jul 1998	Endeavour	ISS-1

two-year-long continuous presence on Mir. When Endeavour developed a thruster problem, attitude-control authority was temporarily given to the station while the issue was investigated – when this was determined to be a faulty sensor, a software 'patch' was uplinked to by-pass the sensor and the Shuttle resumed control.

Columbia lifted off on 17 April for STS-90 with the Neurolab in its bay for the first mission devoted to studying how the human nervous system adapted to space. Its crew, which included veterinarian Richard Linnehan and medics Dave Williams, Jay Buckey and James Pawelczyk, worked around the clock for two weeks to carry out 26 experiments. The cause of the symptoms suffered by 66 per cent of people returning from a significant period in space was finally determined. It had been thought that orthostatic intolerance was a malfunction of the sympathetic nervous system, which maintains blood pressure by controlling the size of the arterial blood vessels, but, in fact, the light-headedness, dizziness, palpitations and difficulty in standing were caused by the heart shrinking and stiffening. It was a fitting conclusion to the final mission in the 15-year Spacelab programme. The module used was the one that the European Space Agency had contributed to Shuttle operations, and was first flown on the German-funded D1 mission. In April 1999 it was returned to the manufacturer for public exhibition. On being retired, the other flight unit, which NASA bought at cost price, was given to the Smithsonian Air and Space Museum in Washington, DC.

Table 6.18 Spacelab missions

Mission	Flight	Year	EDO	Objective
Spacelab 1	STS-9	1983	x	General microgravity research
Spacelab 3	STS-51B	1985	x	Life and microgravity sciences
Spacelab 2	STS-51F	1985	x	Solar physics
Spacelab D1	STS-61A	1985	x	Life and microgravity sciences
ASTRO-1	STS-35	1990	x	Astronomy
SLS-1	STS-40	1991	x	Life sciences
IML-1	STS-42	1992	x	Microgravity
ATLAS-1	STS-45	1992	x	Atmospheric studies
USML-1	STS-50	1992	1	Microgravity
Spacelab J1	STS-47	1992	x	Life and microgravity sciences
ATLAS-2	STS-56	1993	x	Atmospheric studies
Spacelab D2	STS-55	1993	x	Microgravity
SLS-2	STS-58	1993	2	Life sciences
IML-2	STS-65	1994	4	Microgravity
ATLAS-3	STS-66	1994	x	Atmospheric studies
ASTRO-2	STS-67	1995	5	Astronomy
Shuttle–Mir	STS-71	1995	x	Life sciences (long-duration Mir crew)
USML-2	STS-73	1995	6	Microgravity
LMS-1	STS-78	1996	8	Life and microgravity sciences
MSL-1	STS-83	1997	9	Materials science
MSL-1R	STS-94	1997	10	Materials science
Neurolab	STS-90	1998	13	Neurological life sciences

When STS-91 was launched on 2 June, it introduced the new Super-Lightweight External Tank, whose aluminium–lithium shell was both lighter and stronger than the earlier 'lightweight' version. A pressure sensor in one of the SSMEs failed at T + 20 seconds, and it was fortunate that this did not trigger an automatic SSME shutdown and prompt an abort. Subsequent investigation revealed that the sensor had been disabled by a broken piece of test equipment that had inadvertently been left in a pipe during Discovery's recent refit. Valeri Ryumin, a veteran of several tours on Salyut 6 and the Russian director of the Shuttle–Mir programme, was flying to make a thorough inspection of Mir to assess its continued use to host fee-paying visitors, and was surprised how cluttered it had become. Just before the hatches were finally closed, Mir's Talgat Musabayev and Nikolai Budarin gave Discovery's commander, Charlie Precourt, a memento to symbolise the passing of the 'batton' of human spaceflight to the ISS. Having retrieved Andy Thomas, Discovery made a fly-around and NASA said farewell to the venerable station.

DIFFICULT YEARS

It had been hoped to proceed with STS-88 in July and commence the assembly of the ISS, but in May, owing to a difficulty in funding the construction of the crew's habitat module, NASA was obliged to postpone it to "no earlier than 3 December". As a result of this slippage, the independent STS-95 mission was advanced to the autumn.

As the SSME's were ignited to launch STS-95 on 29 October, the hatch of the drag-chute compartment at the base of the trailing edge of Discovery's tail fell off – evidently because it was not installed properly – and struck the centre engine. Had the 5-kilogram aluminium slab severed the cooling lines that preheated the hydrogen prior to its being injected into the engine, the result could have been an uncontained failure of the engine leading to an explosion. While the launch sequencer would have inhibited the SRBs from igniting, Discovery would have been seriously damaged – possibly catastrophically. Shuttle programme manager Tommy Holloway described it as "a close call".

This mission had a public profile because its crew included John Glenn who, on 22 February 1962, became the first American to orbit the Earth when his Mercury 'Freedom 7' rode an Atlas missile. Fellow veteran Scott Carpenter was invited to reprise his exhortation at lift off: "Godspeed, John Glenn." Once he was safely in orbit, Glenn repeated his enthusiastic report: "Zero-g, and I feel fine!" Although criticised by some as a political junket because he had just retired as a senator for Ohio, Glenn – now aged 77, and the oldest person to venture into space – was to provide a unique data point for the investigation of how the human body adapts to weightlessness in orbit and to gravity on return to Earth. Aft of the Spacehab single module was a Hitchhiker with a variety of experiments, and the SPARTAN-201-5 free-flyer, which was to study the Sun. In addition, because the Hubble Space Telescope's NICMOS instrument had prematurely run out of coolant, Discovery was to test the radiator that was to be installed to restore it to service, and the material intended to

Table 6.19 Shuttle manifest, c. May 1998

Flight	Date	Orbiter	Objective
STS-95	Oct 1998	Discovery	HOST, TAS, IEH, PANSAT
STS-88	Dec 1998	Endeavour	ISS-01 (SSAF-2A)
STS-93	Jan 1999	Columbia	Chandra/IUS

be used to repair the telescope's insulation. As the cover of the compartment at the base of the vertical stabiliser had become detached at lift off, when Discovery landed on the SLF it did not deploy the drag-chute; fortunately, the crosswinds were minor.

Although a glitch in the hydraulic system caused Endeavour to miss its launch window by a day, STS-88 lifted off without incident on 4 December to commence the assembly of the ISS by mating NASA's first module to a vehicle which had been launched on a Russian rocket two weeks previously. Meanwhile, the fact that the Chandra X-Ray Observatory used chips from a batch that had proved to be faulty resulted in STS-93 being postponed for six months. With the next Russian module for the ISS slipping behind schedule, NASA decided to fly Discovery as STS-96 in May 1999 with a Spacehab double module of stores for the nascent station.

By April, STS-93 had been scheduled for launch in July. However, the plan to launch the ISS's habitat in September was considered optimistic, November being deemed more likely, and by the end of May this had been officially rescheduled for 12 November. As the Russian vehicle would take a fortnight to reach the ISS, the STS-101 mission that was to commission it was set for December. The prospect of improving the pace on the assembly was looking bleak, so Discovery, which had just returned from STS-96 in June, was reassigned from its STS-102 ISS mission to fly as STS-103 in October and service the Hubble Space Telescope, which had just lost a third gyroscope. After losing three of its original set of six gyroscopes, the HST had been fitted with new ones by STS-61 to restore the total to six. One had failed in 1997, another in 1998 and now it had lost a third. It required three to undertake its programme of scientific observations. The next service call had been scheduled for June 2000, but because the HST's redundancy had been lost NASA decided to divide this service over two missions, designated '3A' and '3B', with the second assigned to Columbia and tentatively scheduled for 2001.

The attempt to launch STS-93 on 20 July was scrubbed at T–7 seconds, barely half a second before the sequencer would have initiated the SSME start-up sequence. A sensor implied that hydrogen was leaking into the aft engine compartment but, frustratingly, the sensor was found to be faulty. The two-day recycle was scrubbed by the threat of thunderstorms, but Columbia was successfully launched on 23 July, its departure lighting up the night. Five seconds after lift off, Eileen Collins, the first female Shuttle commander, reported warning lights which meant that Columbia had suffered a major electrical failure. A sudden voltage drop on an electrical bus had disabled the primary controllers on two of the three SSMEs. The controllers were to operate the engines in response to commands issued by

Table 6.20 Shuttle manifest, c. late 1998

Flight	Date	Orbiter	Objective
STS-93	(Jan) 1999	Columbia	Chandra/IUS
STS-96	May 1999	Discovery	ISS-2A.1
STS-101	Aug 1999	Atlantis	ISS-2A.2
STS-99	Sep 1999	Endeavour	SRTM

Note that STS-93 was under review.

Table 6.21 Shuttle manifest, c. June 1999

Flight	Date	Orbiter	Objective
STS-93	9 Jul 1999	Columbia	Chandra/IUS
STS-99	16 Sep 1999	Endeavour	SRTM
STS-103	14 Oct 1999	Discovery	HST-3A service
STS-101	2 Dec 1999	Atlantis	ISS-2A.2

Columbia's main computers. (Collins later reported that the circuit breakers associated with the faulty bus had popped.) Although the flight was able to continue using the backup controllers on another bus, there was now no redundancy in those engines. Just as it appeared as if the scare was over, it was discovered that SSME #3 (one of the two affected by the electrical fault) was leaking hydrogen and therefore was operating about 100 degrees hotter than the others. If it became too hot, it would have to be shut down, which would, depending upon when it occurred, prompt either a RTLS abort back to the Kennedy Space Center or a TAL abort to Spain or North Africa. Collins's luck held. The leak was only about 2 kilograms per second. Nevertheless, with the 'hot' engine drawing more oxygen than normal, the question became whether Columbia could achieve low orbit before the oxygen ran out. In the event, SSME #3 leaked 1,500 kilograms of hydrogen and drew 1,800 kilograms more oxygen than its companions. When the emptying of the ET's oxygen tank resulted in the engines shutting down, Columbia was not only in orbit, but the orbit was only a few kilometres short of the one intended. For pilot Jeffrey Ashby it had been an exciting first launch. A veteran of three previous missions, Collins had proved that she had the 'right stuff'.

The remainder of the mission went to plan. Occupying all but 1 metre of the payload bay's length, the IUS/Chandra stack was the longest and heaviest payload a Shuttle had ever carried. The Chandra X-Ray Observatory – the third of the Great Observatories – is named after Subrahmanyan Chandrasekhar, a leading figure in the study of stellar structure. It was deployed without incident 7 hours after launch. It was the 15th (and final) IUS launched on the Shuttle. After several days devoted to miscellaneous mid-deck microgravity experiments, Columbia returned to Earth for its next refit.

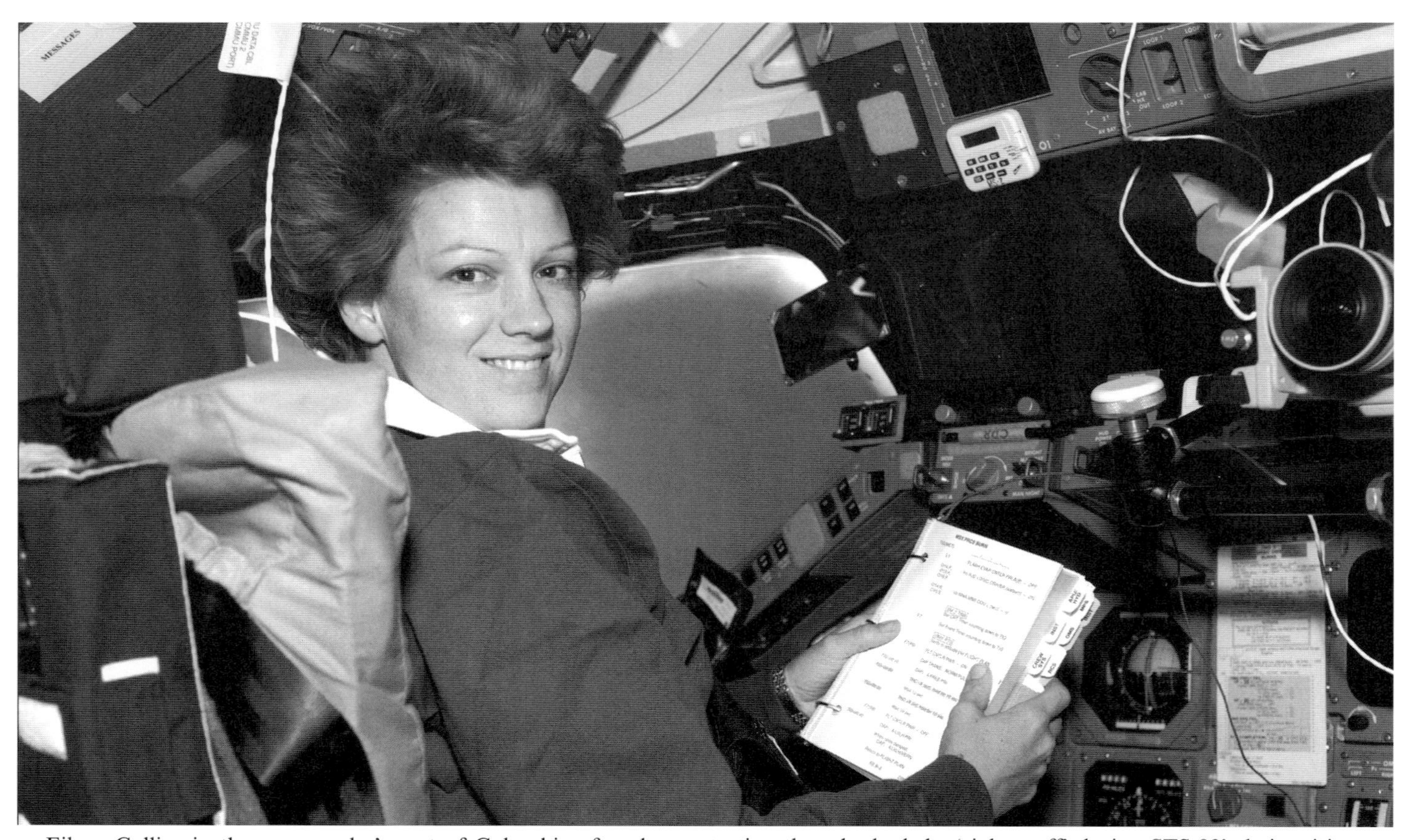

Eileen Collins in the commander's seat of Columbia after demonstrating that she had the 'right stuff' during STS-93's hair-raising launch.

Table 6.22 Shuttles with IUS payloads

#	Year	Flight	Orbiter	Payload
1	1983	STS-6	Challenger	TDRS
2	1985	STS-51C	Discovery	Magnum
3	1985	STS-51J	Atlantis	DSCS III
4	1986	STS-51L	Challenger	TDRS
5	1988	STS-26	Discovery	TDRS
6	1989	STS-29	Discovery	TDRS
7	1989	STS-30	Atlantis	Magellan
8	1989	STS-34	Atlantis	Galileo
9	1989	STS-33	Discovery	Magnum
10	1990	STS-41	Discovery	Ulysses
11	1991	STS-43	Atlantis	TDRS
12	1991	STS-44	Atlantis	DSP
13	1993	STS-54	Endeavour	TDRS
14	1995	STS-70	Discovery	TDRS
15	1999	STS-93	Columbia	Chandra

An investigation traced the hydrogen leak in SSME #3 to the thin pipes that are set within the engine's interior wall. Pumping hydrogen through these pipes prior to injecting it into the combustion chamber serves to warm the cryogenic fluid and to cool the engine bell and so protect it from the hot exhaust plume. Three of the 1,000 pipes that line the bell had been burst, and the fact that these were adjacent indicated that some object had struck the engine. Further investigation revealed the cause to be a repair to the injector that fed oxygen into the combustion chamber through 600 small holes. The procedure for dealing with a damaged injector line was to plug the appropriate hole by inserting a 2-centimetre-long pin. Two plugs had been installed on that particular engine, and one of them was missing. Evidently, as the engine was started, the pressure had ejected the pin, which had struck the wall of the bell. Tests indicated that the impact of the pin had weakened the pipes sufficiently for the pressure of the circulating hydrogen to burst them, and a study of the launch video confirmed that hydrogen was being dumped into the exhaust plume. If the impact had damaged a dozen or more of the pipes, the leak may well have been sufficient to prompt the engine to be shut down, resulting in a RTLS as soon as the spent SRBs had been jettisoned. When investigating the electrical problem that disabled two of the primary controllers, an examination of the cable trays running beneath the floor of the payload bay found damaged insulation on several sections of cable, and there was evidence of arcing between a bare wire and an adjacent screw. The other orbiters were also examined and found to have similar faults, with Endeavour being the worst affected with 20 degraded wires, some of which exposed bare wire. Some of the 350 kilometres of cable in an orbiter run through areas that are routinely accessed by technicians undertaking maintenance between flights, and it seemed that this activity had caused the wear and tear. In addition to repairing the damage, NASA revised the maintenance procedures.

In grounding the fleet to make repairs, NASA was obliged to ask the Russians to postpone the launch of their part of the ISS into the new year, which in turn meant slipping the STS-101 flight that was to commission it. By November, the repairs to Discovery were complete, Endeavour was almost ready, and work on Atlantis was about to commence. Meanwhile, on 9 September the manifest had been reshuffled: Discovery's Hubble service mission would fly in December and Endeavour's radar mapping mission would slip into the new year.

On 13 November the astronomical community was alarmed when the HST lost a fourth gyroscope and had to be put into hibernation. On 7 October the launch of STS-103 was tentatively set for 2 December, but owing to a succession of problems it was slipped to the 17th. On that date, however, a prolonged hold at T–9 seconds due to weather ended with a scrub, but Discovery launched on the 19th, ending the 149-day hiatus. The two-day delay had itself posed a problem. As NASA wanted Discovery back on the ground by year's end in order to avoid any 'Y2K bugs', its duration was cut and the number of EVAs was reduced to three. After the HST was grappled on the 21st and placed on the tilt-table at the rear of the bay, Steven Smith and John Grunsfeld, and Michael Foale and Claude Nicollier, replaced all six of its gyroscopes, a Fine Guidance Sensor, the main computer unit and an S-Band transmitter. They also fitted Voltage–Temperature Improvement Kits on the six batteries and a second Solid State Recorder to back up the one that had been installed in 1997 to replace the original reel-to-reel recorder. By extending their time outside – in the process making some of the longest EVAs to date – they managed to achieve all of their originally assigned objectives. NASA had allowed for two days of 'wave off' for weather, but Discovery managed to land at the first opportunity to draw to an end NASA's worst year for missions since the Shuttle resumed flying in 1988 after the loss of Challenger.

After STS-103, a routine examination of the fuel pump from one of Discovery's SSMEs found a flaw in the soft-metal plating employed to ensure that there was no gap between the tips of the blades and the pump's inner wall. The 5-centimetre-long debonding flaw was shown to be a perfect match for one in a seal that had been *rejected* during manufacture – in fact, that rejected seal section had somehow been assigned a new serial number. Although the flaw had not resulted in any problem, it was nevertheless an embarrassment for NASA's legendary procedures for process control. Further inspection of the records identified a further two rejected seals in an SSME that had been installed in Atlantis for STS-101, prompting the replacement of that engine.

Table 6.23 Shuttle manifest, c. October 1999

Flight	Date	Orbiter	Objective
STS-103	Dec 1999	Discovery	HST-3A service
STS-99	Jan 2000	Endeavour	SRTM
STS-101	Feb 2000	Atlantis	ISS-2A.2

During the Shuttle Radar Topographic Mapping mission that flew as STS-99 in February 2000, the 13.6-tonne SRL package mapped the ground-track in much the same way as the Magellan spacecraft had mapped the surface of Venus through its perpetual cloud cover. To provide the baseline for three-dimensional mapping by the interferometric technique on a single pass, a second receiver was mounted on the end of a 60-metre-long mast that was extended across the left wing. If the mast had failed to extend, two of the crew would have spacewalked to coax it out. In the event, however, it deployed perfectly. The crew ran two shifts to sustain round-the-clock operations for 11 days, during which they mapped 80 per cent of the Earth's land surface in unprecedented resolution.

At long last, NASA was able to turn its attention to the ISS – with Discovery, Atlantis and Endeavour flying 10 missions between May 2000 and December 2001. Since Columbia had not been equipped with an ODS during its refit, it was to fly the 'independent' missions, starting with STS-109 which was to make the '3B' service call of the HST. On 10 November 2001 one of the gyroscopes had 'glitched', and it had been decided to replace it before it failed completely. As the efficiency of solar arrays degrades in space, these were also to be replaced. The innovative NICMOS had consumed its coolant more rapidly than expected, and been shut down. It was to be fitted with a cryogenic cooling system that would last indefinitely. Also, in order to upgrade the telescope, the Advanced Camera for Surveys was to be installed in place of the Faint Object Camera, one of the original instruments which, although still functioning, had become obsolete. The count ran smoothly, and Columbia lifted off on 1 March 2002. On achieving orbit, the flow through one of the two freon coolant loops diminished, although not to such an extent as to require the curtailment of the flight. After the HST had been captured, John Grunsfeld and Richard Linnehan, and James Newman and Michael Massimino completed all the modifications during five spacewalks.

The Hubble Space Telescope has been called the "most important payload that the Shuttle ever carried". The HST should not be thought of as merely an expensive satellite that was delivered late and over budget, but rather as a unique, and hence priceless, asset. While it was not strictly necessary to have launched the HST on the Shuttle, it is also evident that (irrespective of the flaw in its primary mirror) the rate at which its gyroscopes failed would have made it unusable within five years – one-third of its planned life. Its continued operation was viable only

Table 6.24 Hubble missions

#	Year	Flight	Orbiter	Designation
1	1990	STS-31	Discovery	–
2	1993	STS-61	Endeavour	1
3	1997	STS-82	Discovery	2
4	1999	STS-103	Discovery	3A
5	2002	STS-109	Columbia	3B

Table 6.25 Shuttle manifest, c. March 2002

Flight	Date	Orbiter	Objective
STS-110	Apr 2002	Atlantis	ISS-8A
STS-111	May 2002	Endeavour	ISS-UF-1
STS-107	Jul 2002	Columbia	Spacehab
STS-112	Aug 2002	Atlantis	ISS-9A

because it had been designed to be serviced in orbit. The gyroscopes apart, advances in detector technology would have progressively reduced the HST's effectiveness, and by upgrading it with state-of-the-art instruments, the Shuttle has played a vital part in its overall life-cycle.

CRISIS

Columbia was to be turned around from STS-109 for STS-107, an independent science mission in July, but on 25 June the routine inspection of Atlantis following its STS-110 mission to the ISS identified cracks in a 'flow liner' designed to preclude turbulence as hydrogen is fed into an SSME. When all four orbiters were found to have such cracks, NASA had to ground the fleet. Although by August it was evident that flights would be able to resume in October, STS-107 was postponed to January 2003 in order to make way for two ISS missions.

On 1 February 2003, after an EDO mission of microgravity science experiments in the new Spacehab Research Module, Columbia was lost in re-entry – and its crew killed – because the thermal protection on the leading edge of its left wing had been severely damaged during the ascent to orbit. The grounding of the Shuttle fleet while Columbia's loss was investigated posed a considerable problem for the ISS.

Table 6.26 Shuttle manifest, c. September 2002

Flight	Date	Orbiter	Objective
STS-112	Oct 2002	Atlantis	ISS-9A
STS-113	Nov 2002	Endeavour	ISS-11A
STS-107	Jan 2003	Columbia	Spacehab

Table 6.27 Shuttle manifest, c. January 2003

Flight	Date	Orbiter	Objective
STS-107	Jan 2003	Columbia	Spacehab
STS-114	Mar 2003	Atlantis	ISS-ULF-1
STS-115	May 2003	Endeavour	ISS-12A

The Mir Space Station in its completed configuration.

Ken Bowersox, Nikolai Budarin and Don Pettit had been delivered to the ISS in November by STS-113 in the expectation of being retrieved by STS-114 in March. Contrary to press reports, as they had a Soyuz lifeboat they were not "stranded in orbit". As the Soyuz would not 'expire' until late May, there was time to plan. As it became clear that the remaining orbiters of the Shuttle fleet would be grounded until at least the end of the year, further assembly of the ISS would have to be indefinitely postponed. It was decided to continue to operate the ISS with a succession of two-person crews, this reduction being designed to optimise the use of consumables on the station, who would fly up and down in Soyuz 'taxis' and serve 6-month tours of duty – the maximum allowed by the Soyuz spacecraft's orbital service life. Until the Shuttle could be reintroduced, the task of sustaining the ISS fell to the Russians with their venerable Soyuz crew ferries and Progress cargo ships in a style reminiscent of the recently de-orbited Mir space station.

7

The art of spacewalking

VENTURING OUT

One of the rôles advocated for the Shuttle was that it would be able to rendezvous with a defective satellite so that 'spacewalking' astronauts could attempt to repair it. If an in-orbit repair proved to be impractical, then the satellite would be returned to Earth.

An important objective of the Gemini programme was to perfect rendezvous and docking. The Agena upper stage had been fitted with a special collar to enable a Gemini to mate with it, after which the Agena's powerful motor could be used for orbital manoeuvres. The first such mission was flown by Gemini 8. It docked easily, but when a thruster misfired, sending the spacecraft cartwheeling through space, it was forced to withdraw and make an emergency descent. Gemini 9 also attempted the experiment, but the rocket carrying its Agena target failed, and a secondary target, built for engineering tests, was hastily prepared and launched. Although this had no engine, it would serve as a docking target for Gemini 9. Unfortunately, the aerodynamic shroud protecting the collar failed to release, rendering impossible the docking. One option was to attempt to nudge off the shroud with the nose of their own spacecraft, but this was considered too dangerous. Gene Cernan was due to make an extravehicular activity (EVA) later in the mission to evaluate the Astronaut Manoeuvring Unit (AMU). Buzz Aldrin, a member of the backup crew, suggested that Cernan go out, don this jetpack, cross over to the Agena and release its shroud to clear the way for a docking, but this idea was rejected for fear that Cernan might become entangled in the shroud. The frustrated astronauts therefore took pictures of the fouled mechanism – whose appearance they likened to the jaws of an "angry alligator" – and withdrew. In fact, NASA's only experience of spacewalking was when Ed White ventured out from Gemini 4 for 20 minutes in June 1965. However, whereas White had enjoyed the sensation of weightlessness by simply floating on the end of a long umbilical, Cernan found that it was very difficult to maintain his position while he donned the backpack unit. As he floundered around, his exertions overwhelmed the suit's environmental system, fogging his visor, and he had to call a halt and make his way blindly back to the hatch. It was decided not to attempt to use the AMU again until basic spacewalking techniques had been perfected. Although simpler tasks were assigned to Michael Collins on Gemini 10 and Richard Gordon on Gemini 11, and both achieved their main objectives, their experiences confirmed that handholds were needed. Simply floating in space was one thing, but working in space was something else. Exploiting neutral-buoyancy in a water tank to simulate weightlessness, Aldrin rehearsed procedures for the use of handholds and footholds. As a result, on the final mission of the programme – Gemini 12 in November 1966 – he not only completed his assigned tasks, but was able to do so relatively easily. He used strategically placed handholds to reach a worksite, and used footholds to lock himself in position, leaving both hands free for the task. Also, as he did not have to struggle to maintain his position, he could pace himself so that he did not become exhausted. Ironically, he would have had no problem testing an AMU, if it had been carried. In light of this experience, NASA built the Weightless Environment Training Facility (WETF) at the Johnson Space Center in Houston.

After several spacewalks during Apollo missions, NASA faced a major challenge in 1973 when the astronauts rescued the Skylab space station, which had been badly damaged on its way to orbit. These impromptu spacewalks proved conclusively that astronauts could indeed work effectively outside a spacecraft. NASA was therefore confident that astronauts would be able to work outside the Shuttle.

EXPECTATION AND TRIUMPH

The objective of the first Shuttle flight was to demonstrate that the spacecraft could reach orbit and return to Earth, and the payload bay contained only instruments that recorded the status of the orbiter. On STS-2, however, Joe Engle was to don the new two-piece Extravehicular Mobility Unit (EMU), enter the airlock on the mid-deck, and conduct a depressurisation test. Unfortunately, one of the fuel cells flooded and had to be switched off, so the test was cancelled and the flight was curtailed. The depressurisation test was reassigned to STS-4. The cover of the classified CIRRIS sensor system in Columbia's bay had refused to open, and Ken Mattingly, who had made a spacewalk on his Apollo mission, and was to test the new suit in the airlock, suggested that he go out and release the cover manually, but permission was denied – the rules stated that astronauts should make spacewalks in pairs, to enable them to assist each other, and as there were only two people on board the controls could not be left unattended; a spacewalk during the test-flight phase of the programme would be sanctioned only in the event that the payload bay doors failed to close.

As soon as the National Space Transportation System was declared operational, two of STS-5's four-man crew were assigned the first excursion into the bay. This was postponed for a day because Bill Lenoir suffered a dose of 'space sickness'. The next day, however, he and Joe Allen spent four hours pre-breathing pure oxygen to purge nitrogen from their blood streams (such a long time was necessary because the orbiter had yet to be tested with a partial-pressure air mix). Problems arose even before they entered the airlock. In the process of testing the suit components, Allen found that a fan in his life-support backpack was running hot. Even although the excursion had to be cancelled, Lenoir donned his suit to test it but, unfortunately, its regulator would not maintain the necessary 4.3-psi pure oxygen pressure, so even the depressurisation test had to be cancelled; it was not a very promising start.[1]

The EVA goal to which NASA aspired was the capture, repair and release of SolarMax, a satellite crippled by a malfunction in its attitude-control system. This was a realistic option because SolarMax was the first satellite designed to be worked on in space. If this could be achieved, it would be a convincing demonstration of the Shuttle's unprecedented capability. On Challenger's first mission, Don Peterson and Story Musgrave became the first astronauts to venture outside, but only after the IUS/TDRS stack had been inserted into geostationary transfer orbit (GTO), *en route* to geostationary orbit (GSO). This deployment was accomplished on the first day. As initially manifested, Challenger was to have returned to Earth the next day, but in light of the previous frustrations, the STS-6 flight plan had been augmented to accommodate the evaluation of the spacesuit. On 7 April 1983, Peterson and Musgrave began their preparations with a full pre-breathing, as previously. As the suits raised no issues, they sealed themselves into the airlock. After opening the outer hatch, they used the handholds on the other side of the cabin wall and attached safety lines to a bar just above the circular opening. With this vital line in place, they each moved to a sidewall and hooked a tether with a slider to a cable running along each sill; they then released the hatch tethers. Each spacewalker could now traverse the full length of the bay without becoming entangled in the other's tether. If a tether broke, the orbiter would manoeuvre to retrieve the drifting astronaut. The primary objective on this first outing was to demonstrate a contingency procedure. A large annular cradle was mounted at the rear of the payload bay. It had held the IUS, and had been rotated up to about 60 degrees for payload deployment. It was meant to rotate down again afterwards, and it did, but in the event that it jammed, possibly even with the payload still in place, there was a procedure for rotating it manually. Similar to the procedure to close a stuck payload bay door, this was a contingency that NASA hoped would never need to be used, but it was wise to verify that it was feasible. Peterson and Musgrave made their way to the aft wall of the payload bay, set up a winch, ran a cable from it through a block on the tilting part of the cradle, drew it upright, and locked it into position. It was a good afternoon's work. Of the 4.3 hours they spent on their suits' environmental systems, 3.5 hours was spent in the bay.[2] There was much more to be done, however, before it would be practicable to tackle a job as complex as repairing a satellite.

On the next mission, the environmental unit was reset in orbit for the first time to adjust the standard 14.7-psi cabin atmosphere down to 10.2 psi while maintaining the mix of 21 per cent oxygen and 79 per cent nitrogen. Operating at this reduced pressure would slash the pre-breathing requirement in advance of a spacewalk. No external activity was assigned, but STS-7 was to carry out an important task in

[1] NASA measures pressure in terms of pounds per square inch (psi). The pressure at sea level is 14.7 psi.
[2] NASA measures EVA duration as the time spent on the suit's systems, rather than the time spent outside.

preparing for SolarMax. The retrievable Shuttle Pallet Applications Satellite (SPAS) had been deployed soon after reaching orbit. As Challenger returned to recover it, Bob Crippen flew the orbiter from the aft flight-deck station, looking 'up' through the overhead window at the satellite and used the recovery operation to evaluate the orbiter's manoeuvrability in proximity to another object – it proved to be extremely controllable, and the efflux from its Reaction Control System (RCS) thrusters did not disturb its much smaller companion.

The next step in preparing for SolarMax was assigned to STS-41B in February 1984. A manoeuvring unit had not been required for Apollo, but a redesigned form of the backpack that Gene Cernan had struggled to reach had been demonstrated inside Skylab's cavernous orbital workshop. According to the belief at that time, a spacewalker setting off to recover an errant satellite would require independent manoeuvrability, and the Shuttle variant was dubbed the Manned Manoeuvring Unit (MMU). By this time, NASA was providing spares for critical items. Only two spacesuits were required, but a third upper-torso (the part containing the systems) was carried to ensure that a spacewalk would not be cancelled if a suit was found to be faulty. Similarly, two MMUs were carried in case one proved unworkable. They were stowed in Flight Support Structures (FSS) mounted on the bay sidewalls, one on each side of the airlock.

On this occasion, two spacewalks were planned. As both were to be conducted by Bruce McCandless and Bob Stewart, a day of rest was scheduled in between. McCandless was first out. Because he had been the astronaut most involved in the MMU's development, he had been given the task of putting it through its paces. It was actually a pallet with a recess. When McCandless had reached the port FSS, he turned and eased the integrated pack of his suit into the recess. Two arm rests were then swung down. Each had a controller with push buttons and a joystick. The right arm specified rotation; the left specified translation. In all, the pilot had control of six degrees of freedom. Once he had verified all the MMU's systems, McCandless released his tether and swung the levers of the FSS's locking mechanism, and in so doing he became the first human to fly in space as an independent spacecraft. There must have been a temptation to zoom off and fly loops around the orbiter but, being its test pilot, McCandless worked methodically through a series of thruster firings to assess the manoeuvrability and stability of the MMU in each of its modes. With Stewart observing from the sidewall, McCandless moved up and down, left and right, forward and back, then pitched, yawed and rolled, all well within the payload bay. Once each of the degrees of freedom had been verified, he invoked a multi-axis rotation and activated the automatic stabiliser to verify that it could halt the motion and hold him stable. With this established, he began to move straight up out of the bay and did not stop until he was about 50 metres from the orbiter. After a few minutes, he returned. The MMU's nitrogen gas jets (24 in two independent sets of 12) delivered only 0.75 kilograms of thrust. The MMU had a mass of 150 kilograms, as did the suit, therefore the combined mass of an astronaut, the suit and the unit made all movement slow and deliberate. In fact, the trip had lasted almost 15 minutes. To follow up, McCandless went out again, this time to 100 metres. On his return, he stationed himself in front of the FSS, eased the MMU into its receptacle, and engaged the lock.

Bruce McCandless tests the Manned Manoeuvring Unit on STS-41B.

While McCandless had been playing Buck Rogers for the TV cameras – the downlink of which was broadcast live by the US networks – Stewart had been testing another piece of equipment that would be crucial in a satellite repair operation. The Manipulator Foot Restraint (MFR) was a platform that incorporated a pair of 'golden slippers' and was mounted on the end of the Remote Manipulator System (RMS) arm. Stewart retrieved the MFR from its sidewall stowage point and Ron McNair swung the end-effector of the RMS arm down to him so that it could be attached. With the restraint securely on the arm, Stewart climbed on board and allowed McNair to swing him around to determine how accurately the operator on the flight-deck's aft station could position an astronaut within the bay. Each time Stewart reached a notional work site, he would jiggle around to assess the extent to which his body's movement shook the arm. By the time Stewart finished the MFR trial, McCandless had returned with the MMU and they exchanged rôles. While Stewart flew the MMU, McCandless inserted a different foot-restraint, the Portable Foot Restraint (PFR), into the starboard sidewall. This placed him in front of a Stowage Assembly Box (SAB) in which were various tools and a mock-up of the electronics unit on the SolarMax satellite. His task was to establish whether the tools really worked, because apparatus designed for use in space sometimes failed. It was not simply a matter of making tools that would work in weightlessness; they also had to function in a vacuum and survive thermal stresses. Testing the SolarMax tools now would reduce the risk of a nasty surprise later. As Stewart had allowed the MMU to build up a higher speed than McCandless had, he soon returned. It was just as well, because, at a little over six hours, they were at the limit of their suits' nominal life-support capability. Having achieved all their objectives, they returned, highly satisfied, to the airlock.

Their second excursion was marred by a fault in the RMS. Although this ruled out a repetition of the 'swinging' test, that test had shown the arm to be an accurate and stable platform. As McCandless donned the second MMU, Stewart unstowed the Trunnion-Pin Attachment Device (TPAD), which was affixed to the front of the MMU. Projecting from the front of the TPAD was a stubby hollow cylinder that was a copy of the arm's end-effector. A SPAS deployable pallet was to have been hoisted out of the bay and slowly turned on the end of the arm to allow McCandless to attempt to dock with it by sliding the TPAD over a pin on the pallet to rehearse what an astronaut would hopefully do with the pin on the slowly turning SolarMax. Unfortunately, the electrical fault in the arm affected its wrist – the joint that was needed to rotate McCandless' target – so this part of the test had to be cancelled. It was nevertheless possible to verify the TPAD's locking mechanism with the SPAS mounted in the bay by positioning himself over the pin, and this he proceeded to do several times.

Stewart's main task this time was to work at another SAB – one that contained a replica of SolarMax's propellant system. He coupled hoses to valves, and pumped fluid from one tank to another. SolarMax was not in need of refuelling, but the basic spacecraft bus was of a multipurpose design that had been used on several satellites, and one – Landsat 4 – was in trouble as part of its communication system had failed. If the SolarMax repair proved successful, Landsat was likely to be the next to receive the astronauts' attention. However, because it was higher than a Shuttle could reach, it was planned to have the satellite lower itself, expending most of its propellant in doing so. Consequently, after it had been repaired it would require to be replenished. This Orbital Refuelling System (ORS) test was a step towards determining whether this would be practicable. The fluid Stewart pumped was not hydrazine rocket fuel; it was just a dye. If the valve and the pump worked, a later flight would test them under operational conditions. If it proved possible to pump fluid into satellites, then it would make sense to design orbital telescopes with cooling systems that could be replenished, thus extending their service life. There seemed to be a multitude of ways in which astronauts could make themselves useful in space.

When McCandless returned with the TPAD, the two astronauts exchanged rôles again. As Stewart was returning to stow the MMU, he noticed that the PFR had come loose and was drifting out of the bay. Even before Stewart could reverse his undoffing, McCandless was making his way hand-over-hand along the sill of the starboard sidewall. However, by the time McCandless reached the position of the PFR, it was too 'high' for him to reach while holding onto the orbiter. He then called steering commands to Vance Brand, on the flight-deck, who manoeuvred Challenger to enable McCandless's outstretched hand to retrieve the loose item in an excellent demonstration of teamwork. By the time the spacewalkers returned to the airlock, they had each accumulated 12 hours in EVAs. Events had gone very well. The only aspect of the SolarMax capture that they had not been able to simulate was its rotation, but this apart, the prospects were looking good. The activities had been caught on 35-mm film by cameras equipped with ultra-wide-angle lenses, in a format designed for projection on a planetarium dome to give a 360-degree wraparound view. An IMAX camera was also carried for the first time; its 70-mm film combined a large format with high resolution.

The astronauts' satisfaction was tempered by the fact that the other goal of the mission – the deployment of two commercial satellites – had turned sour by motor failures that had prevented them from reaching their assigned orbits. Nevertheless, they had cleared the way for the attempt to fix SolarMax two months later.

In November 1980, only a few months after it had been launched, SolarMax had lost its attitude-control system; a power spike in the electronics had blown the fuses. Its mission was to follow up the Orbiting Solar Observatory satellites by monitoring the Sun during the most active part of the 11-year sunspot cycle. Most of its seven instruments had to be aimed precisely at the Sun. The satellite was put into a spin so that the few instruments that did not require a steady view could carry out a restricted programme. The Sun had passed its peak and was nearing its minimum level of activity when STS-41C finally drew up alongside the stricken satellite. In addition to restoring its attitude-control capability, the astronauts' list of tasks included repairing an instrument that had been a little flakey from the very start. If SolarMax could be made fully functional, it stood every chance of surviving long enough to follow the Sun through its minimum and up to the next maximum, and thereby carry out its original mission.

The rendezvous with SolarMax required Challenger to climb to 560 kilometres – higher than the Shuttle had been before. On the way, it dropped off its primary payload, the Long-Duration Exposure Facility (LDEF). By the time the orbiter assumed station about 100 metres away, the controllers at the Goddard Space Flight Center had virtually cancelled SolarMax's spin and powered down most of its systems. Its fate rested with George 'Pinky' Nelson and James 'Ox' van Hoften.

Nelson donned the MMU, collected the TPAD, and set off towards SolarMax. As had his predecessors, he flew very slowly, and it took him almost 10 minutes to draw up alongside his target. SolarMax was actually two units bolted together. At the base was the spacecraft bus, and above it was the scientific package. A pair of broad solar panels projected from each side of the interface ring. SolarMax was not actually static; it was rotating on a stable axis at about one revolution per minute. As the pin on which the TPAD was to lock was on the bus, Nelson had to manoeuvre under the level of the solar panels to ensure that they did not strike the top of the MMU as they passed slowly over his head. Everything looked good, so when the trunnion pin came into view Nelson moved in to drive the collar of the TPAD over it. Apart from the spin, this was exactly what McCandless had done. On this occasion, however, when Nelson activated the mechanism, the TPAD refused to engage the pin. He withdrew, waited for SolarMax to present the pin again, then re-docked with the same result: the end-effector in the TPAD refused to lock onto the pin that it had been specifically designed to grip. In frustration, Nelson withdrew a few metres to consider what to do. But what, indeed, could he do?

Almost as an afterthought, Nelson reached out and grabbed hold of the tip of one of the solar panels as it drifted by. If he could cancel the residual spin, Challenger's RMS might be able to do what his TPAD had not. Unfortunately, before the Goddard controllers could intervene to warn Nelson off, his action had transformed the axial spin into a crazy pitching roll that threatened to incapacitate SolarMax. No longer able to keep its solar panels facing the Sun, it began to draw on its battery. In normal circumstances, this chemical battery was used only during the 30 minutes or so that SolarMax was in the Earth's shadow, and it would then be immediately recharged when the spacecraft passed back into sunlight. Even with most of its systems powered down, SolarMax would rapidly exhaust its battery. Nelson attempted to overcome the tumbling motion, but the conservation of angular momentum prevented him from doing so, and he only made matters worse. With his MMU running low on nitrogen propellant, Nelson was recalled to the orbiter. With shocking suddenness, it seemed that the satellite that he had set out to rescue had been written off! Challenger manoeuvred clear, and the astronauts did their best to sleep. It was a close call, but Goddard managed to orient SolarMax so that the solar panels provided enough power to operate its core systems. This provided the time needed to stabilise the spacecraft, and Challenger returned two days later. This time it was decided to attempt to snatch SolarMax with the RMS. This had not been the favoured means of capture, because it required the arm to absorb the energy of any residual spin, risking damage to the arm's delicate joints. In the event, it was easy for Terry Hart to station the end-effector and slide it over the pin on SolarMax's bus when it rotated into view, and the arm absorbed the energy without difficulty. It was immediately deposited onto the FSS at the rear of the payload bay so that it could draw power from the orbiter via an umbilical. This structure incorporated a large U-shaped cradle and a tilt-table. It had been built to enable the Shuttle to launch new satellites using the same type of Fairchild multipurpose bus. If it proved impossible to fix SolarMax, it was to be returned to Earth. Glad to have recovered the situation, the astronauts took the rest of the day off and reviewed the procedures for the repair operation the following day.

While Nelson retrieved the PFR, van Hoften attached the MFR to the arm and mounted it. With Nelson tagging along, Hart then transferred them to the rear of the bay where Nelson set his foot rest into a socket near the sill of the sidewall. As the tilt-table could rotate as well as tilt, it was positioned to give the repairmen access to SolarMax's attitude-control system. The bus comprised three boxes, each about 1 metre square, set 120 degrees apart. In fact, because the nature of the fault was not understood, it had been decided to replace the entire controller. In practice, this meant replacing one of these boxes. As the bus had been designed to be serviced, the astronauts had only to undo a set of bolts to release the box. The new controller was in a container on the FSS, and there was also a container to take the old controller. It was a simple matter for Hart to swing van Hoften down with his bulky package for the swap. The heavy lifting work would have been feasible without the RMS, but the arm made it easier, and faster, and time was the limiting factor on a spacewalk.

Installing the new controller accomplished the main objective, because it restored SolarMax's ability to aim its instruments at the Sun. If they had been running late, the astronauts might have stopped working, but as they were ahead of schedule they set out to fix the coronagraph, which had never really worked properly. This was a much trickier task, because they could not replace it, and they had to take it apart to repair it. First, however, they had to gain access to the instrument. The satellite was repositioned on the FSS to present the instrument section. The first task was to slice through the thermal insulation blanket with a pair of heavy-duty scissors; the loose flaps were then folded back and held in place by duct tape. Six screws had to be undone to release the underlying hatch. The screws could damage the instrument if they found their way into it, so they had to be retrieved – an awkward procedure for gloved hands. As it would not be possible to reuse the screws, van Hoften attached another strip of tape to serve as a hinge. The astronauts were making good progress, which was just as well because they were required to undo a total of 22 screws to disconnect the instrument's controller! The task of wiring up the new controller fell to Nelson. Fast-action spring clips were used to re-establish the 11 circuits. This done, the hatch was closed, the insulation was flattened, and both were taped in place. Although this tape would alter the thermal properties of the instrument unit, when the spacecraft was operating this surface would not face the Sun. Having completed both their primary and secondary tasks, van Hoften and Nelson retreated to the airlock. Their excursion had lasted a record 7.3 hours. SolarMax was released the next day, 12 April, the third anniversary of the first Shuttle launch.

Finally able to control its orientation, SolarMax aimed its instruments at the Sun and the coronagraph proved to be operating better than ever. This was a well-timed repair, as a fortnight later, even although the Sun was well past its peak, SolarMax recorded the largest flare since 1978. As the solar cycle slowly attained its new peak, it excited the upper layer of the Earth's atmosphere, which increased the rate of

George Nelson and James van Hoften repair the SolarMax satellite on STS-41C.

decay of the satellite's orbit. A propellant replenishment was tentatively assigned as a secondary objective on the 1990 mission to retrieve LDEF (which was also suffering from this increased drag) but SolarMax re-entered the atmosphere in December 1989, having maintained its watch on the Sun up to the end. The retrieved attitude-control system was returned to Fairchild, repaired and reused for the Upper Atmosphere Research Satellite (UARS). To the critics who said that it would have been cheaper to build another SolarMax and buy a Delta rocket, NASA pointed out that the repair had been undertaken as a secondary task after Challenger had deployed its primary payload. Others argued that as a dexterous robot could have performed the repair, such autonomous robotic spacecraft should be developed, but NASA said that a human presence made the vital difference between overcoming the unexpected and being overcome by it. It was a long-standing difference of opinion.

In any event, while the repair of SolarMax had been underway, other astronauts were rehearsing a procedure to recover the two satellites lost by STS-41B. As in any sales organisation, NASA could not afford to stand still, it had to up the *ante*. Thus, while Joe Allen and Dale Gardner trained in the WETF for the next ambitious EVA, two manifests were merged, and STS-41D became the first mission to carry three communications satellites. Although no spacewalk had been planned for this flight, a build-up of ice outside the waste water vent on the port side of the orbiter created an icicle half a metre long. A similar icicle had formed on STS-41B and been ignored, and then it had evidently snapped off in re-entry, collided with the OMS pod and damaged the thermal insulation. Therefore, it was decided to knock this icicle off. The water vent was in the sidewall, about level with and aft of the crew hatch. Henry 'Hank' Hartsfield was able to contort the RMS to reach the side of the orbiter. Taking care not to damage the thermal insulation on the leading edge of the orbiter's wing, which was just beyond the icicle, he nudged the icicle several times, but it refused to budge. Each crew includes two EVA-trained members whose rôle is to perform 'contingency' tasks, and Richard Mullane and Steven Hawley donned up in the airlock, but before they could enter the bay Hartsfield announced that he had finally managed to snap the icicle off at its root. The vent's nozzle was redesigned, and it gave no further trouble.

The next planned spacewalk was on STS-41G in October 1984, when Kathryn Sullivan and David Leestma took the ORS a stage further. This time it was mounted on an MPESS installed at the rear of the payload bay to hold a large format terrain-mapping camera. Their objective was to use seven special tools to insert a valve into the plumbing mock-up. As it could be done only once, they worked as a team. Once they had left the bay, 90 kilograms of hydrazine was pumped through the valve for a high-fidelity simulation of replenishing a satellite. This had not been performed while the astronauts were outside in case the highly toxic fluid leaked, and because there was no simple way to decontaminate suits in the airlock. Bob Crippen, commander of this flight, had been reluctant to use hydrazine, but the flow was completed without incident. This test led to the replenishment of Landsat 4 being approved. However, because the satellite's polar orbit was unreachable from Florida, the flight would have to depart from Vandenberg in California. The task was assigned to a mission which, at that time, was scheduled for 1987.

TWO UP AND TWO DOWN

The next spacewalking drama – the recovery of the two lost satellites – took place in November. Again, it was done on the fly as STS-51A's objective was to deploy a pair of communications satellites. Better yet, the insurer was paying NASA to bring back the old satellites so that they could be refurbished. The solid-rocket motor of a PAM was supposed to burn for 83 seconds, but a common fault had caused both of these motors to fail after just 10 seconds, sufficient only to raise the apogees to 1,000 kilometres. After the useless motors had been jettisoned, the satellites had been lowered using their own small thrusters. As both satellites were now in similar orbits, spaced a few hundred kilometres apart, this recovery mission involved much more manoeuvring than any previous flight. Discovery drew to a halt barely 10 metres away from Palapa. The satellite's 60-rpm spin had been virtually cancelled, and its slow residual roll ensured axial stability. As there was no time to waste, Joe Allen and Dale Gardner were soon in the bay preparing their tools. The fact that Palapa and Westar were both HS-376s was fortunate, as it enabled the same tools to be used in each case.

Because these satellites had been designed to operate in GSO, and no one had expected them to be seen again after deployment, they had no trunnion pin. It was thus not possible for the RMS to reach out and snatch them as it had SolarMax, so this time the participation of the spacewalkers was crucial. As the absence of a pin also meant that the TPAD could not be utilised, a new capture procedure had to be devised, and the necessary equipment produced. To capture the satellites without damaging the solar cells wrapped around their sides, it had been decided to develop a device to engage the ring at the base that had mated with the PAM stage. Although its designers referred to this as the capture device, its shape had inevitably resulted in the astronauts dubbing it the 'stinger'. It was really several devices in one. Like the TPAD, it had a clamp to attach to the arms of the MMU. A long rod (the stinger itself) projecting in front was to be inserted into the throat of the satellite's apogee kick-motor. A ring at the base of the rod was to mate with the adapter on the satellite. There was also a trunnion pin on the right-hand side to enable the RMS to take the satellite once it had been stabilised. This was ingenious improvisation, but would it work?

Allen manoeuvred into position directly behind Palapa, waited a few minutes, then eased forward. The stinger's success was evident even before Allen reported it, because his audience saw him suddenly adopt the satellite's spin. The MMU soon cancelled this, and once Allen had manoeuvred into a convenient orientation, Anna Fisher brought the RMS in over his shoulder to snare the stinger's trunnion pin. With Allen still attached, Fisher pitched the satellite to a vertical position and eased it into the bay, where Gardner was waiting with shears to cut off the omni antenna rod that projected out from the top of the satellite. The only way to stow the satellite in the bay was to turn it upside down. A bracket had been built to run over the top of the satellite, but it could not be attached. It had been tested on similar satellites at Hughes, but as the HS-376 was tailored to its customer's requirements the feed horn on the folded antenna on top of Palapa projected a little further than the tool's designers had been told, and prevented the clamps from engaging. In the original plan, Gardner would attach the bracket, Fisher would raise Allen back out of the bay and release him, Allen would then pitch the satellite over, and Fisher would snatch the grapple fixture on the bridge that Gardner had installed. Allen would then disengage the stinger and Fisher would rotate the satellite, in the process lowering it onto the waiting pallet. However, accepting that the common bracket clamp (as the bridging bracket for the top of the satellite was called) would not fit, the astronauts set out to improvise.

In the revised plan, Allen disengaged from the stinger, which remained on Palapa, with the arm holding it. While Allen stowed the MMU, Gardner moved the PFR to the sidewall. Allen then stood on the PFR, and took hold of the satellite by its still-folded main antenna. When Fisher withdrew the arm, Allen did his best to hold the satellite steady while Gardner disengaged the stinger. Gardner placed a 'shower cap' over the rocket nozzle to prevent flakes of solid propellant from falling out into the bay when the orbiter landed. Gardner then lifted a mount from the floor of the bay. This was an A-frame with a ring matching that on the base of the satellite. Mating the rings fastened the frame to the satellite. All this time (well over an hour) Allen held the satellite steady then, together, Allen and Gardner pitched Palapa down into the bay and Gardner fastened the A-frame to its floor mount. Despite the failure of the common bracket clamp, the two men had managed to lock the errant satellite into the bay. By the time they closed the airlock, they had been out for six hours.

Two days later, Discovery pulled up alongside Westar 6, and Allen and Gardner went out to retrieve it. This time Gardner used the MMU. Rather than waste time testing to determine whether the bracket would fit this satellite, it had been decided to adopt a revision of the impromptu procedure. This time the arm carried the MFR, and Allen rode on it. The satellite was captured in the same way, and Gardner flew it over to Allen, who took hold of it using the antenna, as previously. The stinger was immediately disengaged and stowed, together with the MMU. With Gardner on the bay floor, Fisher slowly brought the arm down. The fact that Allen had Westar by its antenna meant that the satellite arrived the right way up, so Gardner was able rather more easily to fit the A-frame and mount it on the bay floor. The final act was to sever the omni antenna so that it would not interfere with the closing of the bay doors. Before they came in, they perched on the arm and displayed a placard to the cameras announcing that they had two satellites *For Sale*. Later, they displayed another card, echoing that shown by the crew which had deployed the Shuttle's first commercial satellites. This one identified the Shuttle as operating for the *Ace Repo Company* whose corporate motto was *The Sky's No Limit*. Discovery's crew were subsequently awarded the Silver Medal by Lloyds of London, the insurers. This award was issued to persons whose "extraordinary exertions contributed to the preservation of property". Both satellites were subsequently sold and relaunched under new names.

Once again, astronauts had overcome equipment failure and used improvisation and sheer muscle to complete their objective. It is also important to note that the satellites had run into trouble only nine months earlier. Although NASA had almost immediately agreed to try to recover Palapa, Westar had been added to the plan only

On mission STS-51A, Dale Gardner prepares to capture Westar 6 by inserting his 'stinger' into the nozzle of its kick motor.

After having recovered Palapa B2 and Westar 6, Dale Gardner and Joe Allen hold up their '*For Sale*' placard.

two months before the flight; so in contrast to the SolarMax repair, into which years of preparation had gone, this double repossession mission had posed a different organisational challenge. In addition to achieving its immediate objectives, NASA had eased into a dynamic planning mode in which subsequent flights rectified the failings of earlier ones. Considering the complexity of mounting each flight, this dynamic planning was extremely impressive.

HOT-WIRING A SATELLITE

As the Leasat-3 satellite emerged from STS-51D's bay, its release was meant to trigger the switch that activated it. With the deployment seemingly accomplished, Discovery withdrew to leave the satellite to run through its start-up sequence. When the satellite did not send any telemetry to its ground station, it was realised that the infallible switch had somehow failed. Rather than leave the dormant satellite to be rescued by a later mission, it was decided to extend the flight, and rendezvous for a contingency EVA. STS-51D had a notable passenger, Senator Jake Garn, who was 'observing' in his capacity as chairman of the subcommittee that oversaw NASA's budget. Surely all that was required was to throw the switch! This was a great opportunity to show that the most effective backup system in space was a human presence.

The initial plan called for spacewalkers David Griggs and Jeff Hoffman to mount the RMS and gingerly throw the switch. Ideally, that would start the timer of the sequencer that controlled the engine firings during the GTO manoeuvres. But what if the satellite started up in a disturbed state and fired its motor immediately? It was reluctantly decided that it would be too risky to let the astronauts throw the switch manually, and a method would have to be devised that did not require an astronaut to be near the satellite when it came to life. The crew was told to get some sleep while the support staff on the ground considered the matter, but the astronauts stayed up half the night devising a scheme of their own and the next day a plastic document cover was cut to construct a flexible 'fly swatter' with a narrow slit near its tip. Griggs and Hoffman took this into the bay and taped it to the 40-cm wide cylindrical end-effector. Once they had retreated to the airlock, Rhea Seddon brushed the tip of this masterpiece of improvisation against the solar cells coating the surface of the slowly rotating drum, then waited for the switch to come around. After several near misses, the fly swatter's slit caught the switch and threw it, as intended. Discovery withdrew 100 metres to observe the satellite's activation, but when the rod of the omni antenna did not swing up on time it was acknowledged that the satellite was lifeless, and had to be abandoned. It was evident that the problem was not a simple switch fault; if it had been, the fly swatter would have overcome it. If a way could be devised to jump-start the derelict satellite it might still be rescued by a later mission.

A detailed analysis of the likely failure mode revealed just such a possibility, and several months later, after deploying its trio of brand new satellites, STS-51I drew up next to Leasat-3. James van Hoften was waiting on the RMS, and Mike Lounge eased him out to the satellite, which was still axially stable due to its slow roll. Van

Hoften had taken a specially built bar to span the gap between two of the sockets used to hoist the satellite on the ground. He waited until the attachment points rotated into view, then pushed the capture bar into position. When it next came around, he grabbed it and manually cancelled the spin, using his arms to absorb the energy. This was achieved despite a fault that had impaired the movement of the RMS. Meanwhile, Bill Fisher had set up station on a PFR on the starboard sill. The arm slowly moved van Hoften down towards the bay, and he dragged the 7-tonne satellite with him. Once it was in place, just over the bay, Fisher attached a second handling bar and took the satellite. Van Hoften replaced his bar with a trunnion pin assembly, dismounted the arm, and removed the MFR from the end-effector. The RMS then relieved Fisher of the satellite. It was teamwork all the way. With Leasat in place directly above the bay, both men were free for the really tricky work – the installation of the Spun-Bypass Unit (SBU). This piece of apparatus, specially built for this mission, contained the timers required to activate the satellite. All they had to do was install it so that it would bypass the faulty switch. It was to be attached to a point on the side of the drum, and its cables strung out to the circuitry. Although the system had been tested on a satellite in the Hughes factory, doing it in space was not expected to be easy. It was made more difficult because the HS-381 had most definitely not been designed to be worked on by astronauts.

The first task was to safe the satellite's pyrotechnics to ensure that there was no risk of their triggering some circuit while disabling the seemingly dead Post-Ejection Sequencer and installing the SBU. All the complex rewiring was completed without incident, and the SBU diagnostic display gave the welcome news that the satellite's battery was in good condition – there had been concern that the battery might have degraded. The next task was to install a Relay Power Unit (RPU) to directly engage the relay to deploy the satellite's omni antenna and, to everyone's delight, this swung up. This also started the telemetry downlink and enabled the satellite to accept radio commands. By this point, the astronauts had effectively restored Leasat to a state equivalent to that following a successful deployment. As the two men had been out for almost seven hours, there was no time for redeployment. It was decided to leave this task to a second excursion, which meant extending the flight. Because there were consumables for only one extra day in space, there could be no day of rest, so the two spacewalks took place back-to-back.

Leaving Leasat on the RMS overnight was not really desirable with the arm in a partially degraded state, but breaking the repair had left no option. There were some uncertainties in activating the Leasat after so long in a dormant state. If the solid propellant in its integrated perigee-kick motor had grown cold enough to enable cracks to form, it was likely to explode on ignition. On the second excursion, a probe was inserted into the nozzle of the engine to enable its temperature to be measured. The probe was equipped with a telemetry link so that it could monitor later attempts to warm the propellant. Once this was verified to be functioning, the arming pins were pulled from the SBU to start the various timers. Fisher resumed his station on the sidewall and held the satellite, then the arm disengaged and van Hoften removed the trunnion pin. After slipping the MFR back on the arm, van Hoften mounted it. He then retrieved a third handling bar. Rather smaller than the capture bar, this was the spin-up bar. Once van Hoften had hold of the satellite, Fisher removed his bar, Lounge drew the arm up out of the bay, and van Hoften dragged the satellite with him. However, it could not just be released. As it had first emerged from the bay, six months ago, it had picked up a 2-rpm roll. That had been cancelled when van Hoften had grabbed it. Now he had to restore it. The HS-381 was the first satellite built by Hughes specifically to exploit the great width of the Shuttle's payload bay. This squat 4-metre-diameter, 3-metre-tall drum was the largest satellite so far encountered by a spacewalking astronaut. Even a mighty heave barely made it spin. Two further boosts were required to build it up to the required rate. The task was complicated by the fact that the initial impulse also set the satellite drifting away, and the arm had to chase it to position van Hoften to grab the bar each time it came around. The sight of one of its astronauts manually spinning up this enormous satellite was tailor-made for NASA's Public Affairs Office. The bypass performed flawlessly. The solid motor was left to warm for six weeks before being commanded to fire, and it propelled the satellite up to GSO as intended.

This Leasat repair is an excellent case study of how NASA added activities to planned flights. STS-51D had been forced to abandon the satellite and depart, but it had later been determined that it should be possible to perform a rescue. STS-51B was a Spacelab flight. As this module took up the entire bay, there was no room to work on a satellite. In any case, the RMS was seldom carried with a Spacelab. The job could have been given to STS-51G, which was to deploy satellites, but it would have allowed little time for that crew to prepare, and STS-51F was another Spacelab. Thus, by a process of elimination, STS-51I was the first available mission to offer a clear bay in which to work and sufficient time to train. The dynamic planning cycle-time enabled mistakes to be rectified about six months later. In fact, it is amazing just how flexible the Shuttle was proving to be. Deploying three satellites and hot-wiring another was pretty good for a single flight by any standard, and the fact that it had been pulled off at short notice indicates that, as an operational system, the Shuttle had matured rapidly. Spacewalkers were being assigned ever more complicated tasks, and were succeeding – and with each success they became more adventurous. While they were always careful to employ a step-by-step methodology in developing the state of their art, they were definitely taking increasingly 'longer strides'. If a placard were to have been shown following the Leasat rescue, surely it would have read: *No Mission Too Difficult*.

ASTRONAUT HARD-HATS

In early 1984, President Reagan authorised NASA to build a space station. At the start of the 1970s, when the Shuttle had first been proposed, it had been as part of a project to build a space station; the Shuttle was to facilitate assembly, and the station was to provide the Shuttle with a community in space to serve. The projected cost had been so great, however, that NASA had been obliged to delete the station. The cost depended not only on what was to be built, but also on how it would be constructed. In particular, it depended on whether complex semi-automatic

On STS-51I, James van Hoften rides the Remote Manipulator System and manually 'spins up' the refurbished Leasat 3 satellite.

deployment systems would have to be developed, or whether most of the assembly work could to be undertaken by astronauts.

The fact that spacewalkers were not only performing so much work, but were doing so *by hand*, encouraged NASA to believe that it would be able to rely on spacewalkers to put together the structures that would form the space station. It was important to assess fabrication techniques as soon as possible. With the results injected into the design process, it would be possible to select between different station strategies, and so refine time and cost projections. An experiment to test two means of erecting trusses was assigned to STS-61B. This was no 'fly off' to choose between competing techniques; the objective was to ensure that the problem was understood. No tools were required, as both structures were to be built with rods and individually shaped nodes that snapped together by hand.

On the first of two excursions, Jerry Ross and Sherwood Spring began by going to the MPESS situated conveniently at the front of the bay. They elevated a platform alongside the sidewall sill. The first experiment was to erect a 15-metre-tall triangular cross-section trusswork on this platform. For this Assembly Concept for Construction of Erectable Space Structures (ACCESS) test, the MPESS was loaded with 100 tubular rods and a stock of tongue-in-groove nodes using sliding-sleeve locks. It would take only a few seconds to make each joint, although the astronauts had to unstow, transport, orient and attach each element in turn. At first, Ross took up station at the base and observed while Spring built the structure, doing all his own fetching. At the half way point they exchanged places, with Spring passing successive pieces up to Ross. In this way they worked as a team, with each performing his own task, and it took only half an hour to build the 10 frames required for the tower's full length. It was immediately disassembled to enable them to move on to the other apparatus, known as Experimental Assembly of Structures in EVA (EASE) that was to assess a different construction strategy. As this apparatus used rods twice as long as those of ACCESS, it was impractical to construct a full truss. In fact, only one 2-metre triangular segment was erected. Many such segments would require to be joined together to create a truss. It was built on top of the MPESS, in inverted form. Ross and Spring worked as a team by coordinating to deal with opposite ends of each element in turn. Because they each had to hold onto the structure with one hand, they had to locate a rod for connection and engage it using only their other hand. This was an acquired art. As soon as it was finished, the structure was disassembled. After the two structures had been built and stripped several times, the astronauts called it a day. Although at 5.5 hours this was not the longest spacewalk, it had been the most energetic to date. It had also been the most dexterous. Both men reported that their fingers ached.

Nevertheless, after a day's rest, they went back out to do it all again. This time, Ross stood on the MFR to build the tower, which required him to coordinate with Mary Cleave, the RMS operator. Once it was complete, Ross took a cable from within the MPESS and ran it up the length of the tower, clipping it into place as might later have to be done to embed an umbilical in the space station's truss. Once this had been stripped out again, Spring released the tower from its base and Ross took it with him when Cleave lifted him high above the bay. Once he was clear, Ross

On STS-61B Jerry Ross manipulates the ACCESS truss high above the bay.

swung the truss level to test the ability of an astronaut to manipulate a large object. After he had rotated the structure several times, Ross reoriented it vertically, and Cleave lowered him to allow Spring to reattach the truss to the base. As a final test before stripping it down, Spring replaced an element half way up the tower to simulate repairing damage to an erect structure. As a variation on the EASE test, the triangular structure was erected with one astronaut on the RMS and the other on a PFR, in order to compare their efficiency with that when they had been free. Not unexpectedly, having both hands free made the insertion of the rods into the nodes very simple, but it required close coordination with the arm operator to move around to fetch the elements, which was another factor that had to be considered when assessing working efficiency.

The objective had been to determine whether it would be practicable to rely on astronauts or whether it would be necessary to develop self-deploying structures. Stereoscopic cameras recorded the entire exercise so that their activities could be modelled in 3-D and subjected to time-and-motion analysis. By comparing work efficiency with that achieved in the WETF, the fidelity of training methods would be able to be improved. It was evident that although neutral buoyancy offered effective simulation of the *experience* of weightlessness, the pace at which astronauts could work had been underestimated because the drag of the water significantly inhibited *movement* – in some tasks they were twice as fast as in the water tank. This reality check also showed that they could maintain a set pace without becoming exhausted. Lest this appear insignificant, compare the degree of dexterity with that undertaken on the Gemini spacewalks, when simpler tasks had involved a major effort that had led to early exhaustion.

Concerning the issue of how the space station should be assembled, this simple test was sufficient to confirm that astronauts could erect structures, and there was no need to develop self-deploying structures. NASA's long-standing advocacy for the rôle of the human being in space seemed to have been vindicated. With a space station to be built, it seemed that there would soon be plenty of work for astronaut 'hard-hats'.

HIATUS

In a sober reassessment of operating procedures following the loss of Challenger in 1986, heightened concern for crew safety prompted the decision that the objectives for each mission would be set well in advance – there would be no improvisation. In particular, there would be no more last-minute building up of the flight plan to make spacewalks for which there was little or no time to train – in fact, EVA was to be minimised. Also, the MMU was to be grounded – astronauts would be restricted to the payload bay, and be tethered at all times. The agency's institutional confidence had been sapped, and its gung-ho spirit reversed. Now any mission was difficult if it involved unprecedented risk. Buck Rogers had evidently perished along with the Challenger Seven.

WORKING IN THE BAY

Although the Shuttle resumed flying in September 1988, astronauts did not return to the bay until April 1991. When Atlantis was preparing to deploy the Compton Gamma-Ray Observatory, the satellite's antenna stuck in its stowed position. Jay Apt and Jerry Ross already had a spacewalk scheduled for the following day, but they went out early to attend to the boom – a single shake by a space-gloved hand released it from its clamp. Since they were out, it was decided to make a start on the activities set for the following day, which were to evaluate tools designed for use on the space station. The payload remained poised on the arm high above the heads of the two men as they worked in the bay, so they could pursue only tasks that did not involve the use of the arm. Ross ran a tether across the bay and proceeded to assess how easy it was to pull himself along it. Although it may appear trivial, the focus of their spacewalking assignment was to assess rapidity of movement because, even after the space station was assembled, spacewalkers would have to maintain it, and on such a large structure it would be essential to move *rapidly*, to save valuable EVA time. While Ross scooted back and forth, Apt stood on the PFR of the Crew Loads Instrumented Pallet (CLIP) and acted out various actions to enable sensors to record the forces that he imparted through his feet. Then, as was standard practice, they exchanged places and repeated the experiments. Underlying such mundane tests was a determination to ensure that every problem was thoroughly understood, to ensure there were no surprises. After about four hours outside, they returned to the airlock and the RMS released its payload.

The next day, Apt and Ross returned for their planned excursion, continuing the theme of rapid translation with the Crew and Equipment Translation Aids (CETA) experiment. This was to assess several prototype rail-carts under study for enabling astronauts to move themselves and their apparatus on the space station's truss. The evaluation rail ran the full length of one sidewall. In each type of cart, the astronaut stood in a PFR; it was the mode of locomotion that was being tested. In the simplest case he lay prone and used a hand-over-hand action to pull himself along. In another case he stood upright on the mechanical cart and pumped a lever – in much the same manner as was done on a manual railroad buggy. In the most sophisticated case he operated a rotary dynamo to drive the electric cart. After evaluating how a solitary astronaut could travel, everything was repeated with one astronaut serving as the bulky equipment in need of rapid transfer. Once the cart tests were finished, Apt mounted the arm and was swung about at much higher speed than usual to assess the loads on the arm's structure. This kept them fully occupied for six hours. Compared with the drama of previous excursions, it was rather dull, but the aim was to acquire engineering data. The favoured translation technique proved to be a manually drawn cart. It seemed that in weightlessness there was no need for the mechanical advantage provided by more sophisticated devices. In an operational system, the cargo will be attached to a mount behind the foot restraint. Some of the tests were purely passive: Ross, for example, tested new gloves designed to facilitate greater dexterity. Having participated in the ACCESS and EASE assessments, he was well qualified to judge their effectiveness, but perfecting the gloves would prove to be a major project in its own right.

TRY, TRY AND TRY AGAIN

STS-49 in May 1992, which was Endeavour's first mission, marked NASA's return to satellite rescue. This time, however, it was not a Shuttle that had lost one of its satellites, and it was no last-minute affair. In fact, NASA had played no part in losing the satellite; the agency had become involved only when it was asked to attempt a rescue, and had spent *two years* assessing the feasibility of the task and working out how best to perform it. For NASA, there was a rich irony in the fact that it had been called upon to rescue this particular satellite.

As with the Leasat series, the HS-393 had been designed specifically to take advantage of the width of the Shuttle's payload bay. On the outside, therefore, it was a taller version of the HS-381 drum. Its upgraded HS-376-type transponders could relay 120,000 telephone calls simultaneously. Intelsat had ordered half a dozen and soon after it had booked them on the Shuttle, Challenger was lost and it was left to compete for the limited stock of rockets. The problem was that because it was the largest commercial communications satellite ever built, only a 'heavyweight' rocket with an outsized aerodynamic shroud could launch it, and in the scramble for such launchers, the government agencies had priority. As Intelsat had been forced to wait for either the upgraded Ariane or the new 'commercial' Titan, it covered itself by booking both. The Ariane became available first, so the first satellite (Intelsat 602; designations can be arcane) was launched in October 1989, and the second (Intelsat 603) followed six months later on a Titan. It was only the second flight of this version of the Titan; the first had carried two smaller satellites, and the deployment system was new – it was supposed to be able to accommodate variable numbers of payloads, but it was improperly wired and the payload did not separate from the upper stage of the rocket. To prevent the satellite from being dragged back into the atmosphere along with the spent stage, it was ordered to separate from its perigee-kick motor and fire its own thrusters to raise its orbit. Although stranded in low orbit, at least the satellite was in space, and if a Shuttle crew could fit a new kick-motor it would be able to continue up to GSO. NASA was thus invited to rescue a satellite that it had originally contracted to deploy.

The task had been assigned to Endeavour by a process of elimination. There had to be time to train; it had to be a flight on which the bay would be clear for retrieval work; and it could not be a flight with a massive satellite for deployment, because the new kick-motor would require a support cradle. STS-49 was the first mission to satisfy the criteria and, since its bay was already given over to spacewalking, it was the obvious choice. Several motors were considered. Because the satellite had used a substantial fraction of its propellant to stave off orbital decay – which would restrict its operating life if ever it reached GSO, cutting it from 15 years to 12 years – an IUS was an attractive option because the two-stage vehicle would be able to execute the circularisation burn in addition to GTO insertion. However, bolting an HS-393 onto an IUS would not be easy. The second option was to employ the Orbital Science Corporation's Transfer Orbit Stage (TOS), which had been developed specifically to place heavy Shuttle payloads in GTO. Although it had never been used in this way, it was cleared for flight on the Shuttle. The third option was to fit the same type of motor – an Orbus – as originally used. In fact, the TOS and the first stage of the IUS used variants of this same motor. It was decided to carry a new Orbus in a modified TOS cradle and mate it with the satellite.

The challenge was to find a way to capture Intelsat 603, which was considerably larger than anything astronauts had ever worked with before. Since the support ring at the base of the satellite was 3 metres in diameter, a stinger was impractical. The only option was to develop a capture bar which would have a trunnion pin on it to enable the RMS to take the captive satellite and manoeuvre it into position directly above the motor. If the motor had a suitable adapter, the satellite could be connected to the motor using the bar, the RMS could be withdrawn, the trunnion pin removed, and the payload released. The tricky part, of course, would be capturing the satellite. The astronaut would position the bar across the diameter of the base of the satellite and fire a mechanism that would engage both of its ends simultaneously. The complication was that because the HS-393 was a spinner, the astronaut would have to deal with a moving target. It was only half a revolution per minute, however, and WETF trials had indicated that it could be done.

Bruce Melnick swung Pierre Thuot on the RMS up to the slowly rotating satellite while Rick Hieb remained in the bay. After Thuot had carefully aligned the bar across the ring on the bottom of the enormous cylinder, he edged the bar forward until it was in contact. An automatic mechanism was supposed to trigger latches to secure both ends at the same time, but Thuot had not struck the target with enough force to activate the mechanism. He withdrew, shoved the bar in harder, and this time the latches fired. Unfortunately, when Thuot attempted to use the bar to cancel the residual spin, it slipped off, and as soon as the satellite was free it reacquired its spin. Furthermore, his actions had made the satellite precess, making it much more difficult to align the bar. Not only did further attempts to capture the satellite fail, but each time he touched it the precession became more pronounced. After three hours he admitted defeat and Endeavour withdrew, leaving Intelsat 603 with a slow 50-degree nutation. One fact was evident. The WETF simulations had not accurately represented the degree to which the satellite would be disturbed by a light contact. The clue was the fact that it had reacquired its spin as soon as the capture bar had disengaged. It was a certain sign that the liquid propellant was retaining sufficient momentum to spin the satellite as soon as it was free. This also explained the precession – the liquid was sloshing around. If he ever managed to capture the satellite, Thuot would have to hold it steady until the liquid had shed its kinetic energy, and take care each time he moved it. Despite training in a water tank, they had been caught out by the physical properties of fluids in weightlessness.

Overnight, the satellite's handlers eliminated both the precession and the residual spin, and it was completely stable when Endeavour returned the next day, Thuot went out again, confident that he would be able to capture it. This time he eased the bar against the ring and activated the latches manually, but they failed to engage. He persevered for five hours, but was obliged to abandon the attempt to capture the satellite. This time he left it in a flat spin. After Endeavour withdrew, the controllers once again set out to stabilise the satellite, but it was beginning to look as if *homo spacewalker* had met his match, and that the satellite would have to be abandoned.

Just one excursion had been allocated for its repair; the second attempt had used the first of two spacewalks intended for other activities. Should this other programme be sacrificed for a third attempt on the satellite?

While the crew slept, their colleagues in the support team back room and in the WETF tried to find an alternative way of capturing the rogue satellite. It was all too evident that the capture bar would have to be abandoned. To design, build and certify the bar for flight had cost in excess of $2 million, and it almost certainly met its specifications. Rehearsing an activity on the ground and doing it in space can be significantly different, and this was one such example, but how could an astronaut retrieve the 4-metre-diameter, 6-metre-tall drum without using the bar? And in any case, the bar would have to be in place to enable the RMS to manoeuvre the satellite onto the motor. Could Thuot capture the satellite by hand? Could he hold it in place while Hieb fitted the bar? If Thuot stood on the arm to hold the satellite, Hieb would not be able to reach it unless the orbiter manoeuvred really close in. If Thuot lost control of the satellite, it might strike the orbiter and damage the payload bay doors or the vertical stabiliser. Meanwhile, the crew of Endeavour had also come to the conclusion that the ideal tool for grappling a satellite was the human hand, and they added a twist of their own. The Shuttle carried a spare suit, and if Tom Akers, who was scheduled for a later excursion, was also to come out, then Thuot could grab the satellite and slowly swing it down into the bay, where Hieb and Akers would be waiting for it. They could then hold it steady while Thuot attached the bar, free of any concern about disturbing the satellite's motion. As soon as the bar was in place, they could revert to the original plan. When this idea was put to Houston, a WETF test was ordered to find out if the airlock could take three spacesuited astronauts – it was a tight fit, but feasible, so in the absence of a better idea the plan was approved.

Endeavour had propellant for just one more rendezvous. If Intelsat 603 eluded them this time, it would have to be written off even though it was perfectly healthy. If they failed, it would not be for lack of ambition. Dan Brandenstein, chief of the Astronaut Office, manoeuvred the orbiter to position the satellite a mere 2 metres above the bay, with its base facing down. Hieb had taken up station on a PFR on the starboard sill, and Akers was on an MPESS in the centre of the bay. The satellite was just above their heads. Melnick, who had already demonstrated his prowess on the arm, positioned Thuot around the far side. For 10 minutes they just observed the satellite's motion. It had a slight nutation with a 3-minute cycle. They waited until it was vertical with respect to the bay, and then six hands simultaneously grasped the base ring. They held the satellite still for several minutes to allow its fluids to settle. Whilst they had captured the satellite in a vertical attitude, it was not conveniently oriented for fitting the capture bar, which Hieb had placed on the sidewall below his station. Working in concert, and a little at a time, they gingerly rotated the massive drum through 120 degrees. When it was stabilised, Hieb let go, retrieved the bar, and placed it just beneath the ring. From his position on the sill, however, he could not reach the controller in the centre of the bar to engage the latches, so he held the bar in position with one hand and the ring with the other. Akers then acted as an anchor while Thuot manoeuvred beneath the satellite, indicating his instructions to Melnick by way of hand movements. With the latches of the bar engaged, Thuot tightened a

On STS-49 Pierre Thuot riding the Remote Manipulator System attempts to use a specially-designed 'capture bar' to retrieve the Intelsat 603 satellite.

In the first (and so far only) spacewalk by a trio of astronauts, Pierre Thuot (riding the arm), Richard Hieb (on the starboard sill) and Tom Akers (in the bay) manually capture and manipulate the enormous Intelsat 603 satellite.

number of bolts to complete the attachment, and then dismounted the arm. Melnick dispensed with the MFR and took hold of the satellite, and Akers, who had held the satellite in a vice-like grip for an hour and a half, was finally able to let go. It was now a simple matter to reconfigure the bar by deleting various appendages used during the capture phase, and installing a clamp at each end. Once the RMS had positioned the satellite on the motor, the clamps were engaged and the job was done. The shock came when Kathryn Thornton sent the command to eject the stack from its cradle and nothing happened, but the backup circuit worked and Intelsat 603 set off on the first stage of its journey to GSO.

BACK TO BASICS

The spacewalkers had triumphed again, but only after switching tactics. Perhaps the most significant conclusion was that the human hand ought to be the *tool of choice*, not a last resort. Clearly, if they had set out to do it the way it was eventually done, the STS-49 crew could have rescued Intelsat's satellite on the first day. In the event, the Assembly of Station by EVA Methods (ASEM) was sufficiently important for the mission to be extended for Akers and Thornton to perform a truncated version of it on a single spacewalk. The first task was to construct a truss segment with both astronauts floating freely. In fact, as in the case of EASE, only one section was built, but it represented half of a 14-metre cube that would form one element of a truss. As with EASE too, it was found that the best way to handle a rod was to set it free a few centimetres from a node, rapidly swap hands on the structure, prepare the node with the now-free supporting hand, then retrieve the rod and pop it in. This worked even when the far end of the rod was not anchored, but it took some time to become accustomed to letting the rod go, because the WETF training had not lent itself to this technique. It was therefore somewhat easier to assemble the joints in space than in training. A key objective of the test was to uncover such empirical data. When they were finished, the MPESS on which the rods had been stowed (and on which Akers had stood the day before) was hoisted by the RMS until it was just above the bay. The spacewalkers extended legs from the triangular structure and then – using only voice commands – directed Melnick to lower it into position to mate with the assembled truss segment. This tested the team's ability to coordinate to the degree needed for the arm-operator to work 'in the blind' on construction tasks. The entire assembly was immediately stripped down to leave time to test the Crew Propulsive Device (CPD). When Ed White floated out of Gemini 4, he had taken a simple hand-held gas jet with which to control his movements. Although it worked, he had barely familiarised himself with its use before it exhausted its supply of oxygen propellant. Michael Collins had tried it on Gemini 10 and found it difficult to use. An improved version had been evaluated inside Skylab. Now another such gas jet had been developed, and Akers briefly tested it in the front of the bay. The problem with each of the hand-held devices was that it was difficult to aim the impulse through the centre of mass, so it tended to impart an unwanted rotation in addition to a desired translation. The MMU could isolate each degree of freedom, making it simple to use, but it had been deemed impractical to assign MMUs to astronauts venturing outside the space station, and the hand-held gas jet was one of several devices that was being considered to allow an astronaut with a broken tether to return to the nearest part of the structure. Akers and Thornton crammed as much of the ASEM programme into their 7-hour walk as they could, but some tests – such as an inflatable-rod, intended for their second excursion – had to be deleted.

Endeavour's maiden voyage had set EVA records. In particular, it saw the first three-person spacewalk. This had been tough on Thuot and Hieb. The flight plan had called for a total of three excursions, with Thuot and Hieb making the first and third, and Akers and Thornton taking the second to allow the others a day's rest. In the event, Thuot and Hieb had made three walks, two of which were back-to-back, thereby providing further data on human endurance. NASA had learned a great deal about spacewalking from this mission. The ASEM rods had been easier to assemble than expected, which was good news, but the sensitivity of the big satellite had come as an unpleasant surprise. The WETF was excellent for preparing astronauts for the sensation of weightlessness, but the water-drag and the residual action of gravity in the tank was impairing training for manipulating objects. This prompted concern for the practicality of relying on spacewalkers to build the space station. To minimise this technical risk, it was decided that the contingency-EVA astronauts on forthcoming missions should venture into the bay to rehearse basic procedures. No activity was to be taken for granted. In the event, however, the big problem turned out to be *inactivity*.

Endeavour flew the first of these generic spacewalks as STS-54. After it had deployed its TDRS satellite, the bay was clear for the tests to begin. They started in the airlock, with Mario Runco and Greg Harbaugh taking turns to move one another so that their backpacks slotted into the storage brackets on the wall. This may sound trivial, but the astronaut doing the work could not see the bracket. Once in the bay, they slipped into and out of foot restraints, without bending down to see their feet. Even on the ground, many everyday tasks become difficult if direct vision is denied. It was prudent to establish now that such tasks could be achieved in space, so as to preclude the possibility of finding out later that they could not. Another deceptively trivial assignment was the attempt to move purposefully in the bay. Certain items located outside the space station's pressurised modules will have a finite operational life, and must occasionally be replaced. Chemical storage batteries, for example, are to be based at the far end of the truss, near the solar panels, and astronauts will have to take fresh batteries out and return with the old ones. To simulate manipulating an Orbital Replacement Unit (ORU), Runco and Harbaugh dragged one another along the sill, employing a single-handed translation technique. They found this rather tiring. The inefficiency of handling a bulky object while floating freely argued that it would be wise to fit foot restraints at all sites on the station where astronauts were intended to carry out such work.

When Endeavour flew next, as STS-57, David Low and Jeff Wisoff rehearsed specific procedures for the repair of the HST. To rectify the flawed optics, a package as bulky as a telephone booth would have to be very accurately inserted into the telescope's instrument bay, and it was crucial that this be done without causing any damage. Their first task, however, was to stow the antennae of the just-retrieved

EURECA satellite (its retraction mechanism had failed). As this proved to be more difficult than expected, they were unable to begin their assigned programme for two hours. Wisoff played the rôle of a bulky package, Low rode the RMS, and Nancy Sherlock operated it. Low picked up Wisoff, then instructed Sherlock to move him to a specific location. Such tests had been performed previously to evaluate the stresses on the arm when making large-scale movements, but this time it was the ability of the arm to make *fine* adjustments that was on trial. After retrieval, the HST was to be stationed on a FSS at the rear of the bay, so these tests were conducted with the 15-metre RMS fully extended. Finally, Wisoff assessed a new torque-wrench. Working on bolts had been awkward in the WETF, but they posed no difficulty in space.

A few months later, Jim Newman and Carl Walz continued the tool tests in Discovery's bay. Story Musgrave was to head the HST repair operation, but while rehearsing in a vacuum chamber he had suffered mild frostbite in his fingers. In an effort to determine whether an astronaut could overcome chilled fingers – and thus preclude having to curtail an excursion – Newman held the palm of his hand against one of the high-powered floodlamps that illuminated the bay while it was in shadow. Although the heat radiated by the lamp was able to penetrate the fingers of his glove, this action demonstrated that the glove was not as effective a thermal barrier as had been thought. This prompted concern that the suit was likely to lose heat if subjected to prolonged shadow – as it had in the vacuum chamber. A better glove was ordered, but because this would not be available in time for the HST repair, an over-glove was added as an interim measure in the hope that the additional layer would reduce heat loss without impairing the glove's already marginal dexterity.

The equipment tested this time included a power ratchet, a semi-rigid tether designed to stabilise the operator of a torque-inducing tool, and a new PFR that incorporated a platform that could be raised and tilted by activating pedals with the feet. The only problem was that the SAB steadfastly refused to shut. The fact that closing the SAB took 45 minutes was a timely reminder that even a familiar object could slip a schedule. To assess the risk to the HST from any contaminants emitted by a spacesuit, sensors monitored the environment in the bay while the astronauts were outside. Absolutely nothing was being left to chance. The rendezvous with the HST two days into STS-61 was to be followed by an unprecedented five consecutive days of spacewalks, with Story Musgrave and Jeff Hoffman alternating with Kathryn Thornton and Tom Akers. Additional excursions could be made, but Endeavour's maximum endurance was 11 days, and at that time all outstanding tasks would have to be elided. Endeavour had 7 tonnes of ancillary apparatus in its bay. In 400 hours of WETF-time, Musgrave and his colleagues had choreographed each action that they were to make in space. After mastering each component task, they had made a series of 10-hour simulations to check that the individual tasks followed on from one another seamlessly, and that there were no omissions. No crew had ever trained more single-mindedly for external activity. They *had* to succeed – there was far more resting on the outcome than the future of the telescope; their actions would also determine the prospects for the space station.

Claude Nicollier snatched the HST with the RMS, and set it down on the FSS at the rear of the bay. This could rotate and tilt the HST to enable the astronauts to work on any part of it and remain directly visible to the arm operator. Power was supplied by an umbilical, so the orbiter immediately oriented itself to put its captive in shadow, where it would remain for most of the repair activity.

Three of the six gyroscopes in the HST's attitude-control system had already failed, and another was showing intermittent signs of wear. If (or rather, when) this also failed, the telescope would have to be withdrawn from service. The primary task of Musgrave and Hoffman on the first outing was to restore this capacity. The only problem they encountered was at the end, when two of the four bolts proved reluctant to re-engage. By the time they managed to shut the cover and tidy up, they had been out for almost eight hours. The following day, the Goddard Space Flight Center controllers commanded the solar arrays to roll up so that they could be dismounted and replaced. The starboard array was so badly distorted that it jammed, but this did not really pose a problem. Akers and Thornton retrieved the other array and stowed it in a container in the bay for carriage back to Earth where it could be carefully examined. After Thornton had detached the stuck array, Nicollier raised her well clear of the HST on the end of the arm so that she could cast the twisted array adrift. The replacement arrays were affixed to the HST, but were not extended.

For their second outing, Musgrave and Hoffman extracted the enormous wedge-shaped WF/PC instrument from the side of the HST, and slid in the new one. They had expected that they might encounter difficulties aligning the new unit to fit on the guide rails, but it slipped straight in – the HST had been *designed* for in-space servicing. With the main task complete, they moved onto the magnetometers, two of which had failed. These units were on the skin of the tube, but, as it was impractical to detach them, the new ones had to be placed on top of the old ones, and bypasses fitted. Akers and Thornton now had to remove the HSP instrument and replace it with COSTAR, the unit that was to provide the corrective optics for the telescope's three remaining on-axis instruments. The HSP and COSTAR were each the size of a telephone booth. An astronaut perched on the arm had to hold the 300-kilogram box of COSTAR and position it within a few millimetres of a guide rail that was not directly visible – this required a great deal of teamwork, but when it was properly positioned, COSTAR slipped in easily. For the follow-up task of this outing, the telescope's computer system was upgraded by installing a 386 coprocessor. On the final excursion, Musgrave and Hoffman replaced the controller for the solar array motors in order to overcome a fault that had limited the rate at which the arrays could be rotated, and the new arrays were then unrolled. All that remained was to install a crossover in the power supply of the GHRS in order to restore a failed data channel. And that was it! The HST – now in somewhat *better* condition than when it was initially deployed – was released the next day.

To everyone's relief, all the tasks had been accomplished more or less in the allotted times. Words can hardly express the concentration that must have been required to work on the world's most expensive telescope for up to eight hours at a stretch, in the knowledge that the slightest mistake could write it off. Nor can words readily convey the intricacy of the operation. Every iota of experience gained from previous spacewalks, and from the WETF training, had been exploited in devising

Perched on the Remote Manipulator System, Story Musgrave and Jeffrey Hoffman wrap up the last of five spacewalks to repair the Hubble Space Telescope during the STS-61 mission.

the individual operations, estimating their durations and integrating them into a viable sequence for each spacewalk. In reality, the estimated 30 hours of EVA time had stretched to 36 hours, but this was to overcome problems such as a cover that would not shut, and not because the astronauts had fallen behind schedule with one of the main tasks. In fact, the outcome proved that the repair operation had been thoroughly understood. All the tools had worked as intended, and the frostbite issue also appeared to have been resolved.

It was now self-evident that the training regime had been sound, but Musgrave's team had virtually monopolised the WETF for almost a year. Training at this level of intensity could not be made available to the astronauts who would perform the spacewalks to assemble and maintain the space station; these tasks would require several hundred hours of EVA effort per year. The supreme effort to fix the HST had paid off, but assembling the space station would require such intense activity to be made *routine*, and any apparatus that would ameliorate this workload was to be welcomed. STS-62 evaluated a prototype of the Dexterous End-Effector (DEE) that was being developed to enable the RMS on the space station to undertake delicate operations. It was able to sense the dynamic loads that it imparted on an object as it picked it up, thus giving its operator a degree of feedback that the arm could not offer. No spacewalk was necessary, however, because the RMS was able to retrieve the DEE from its stowage rack, test it and replace it. When the DEE enters service, it should reduce the need for astronauts to go outside to undertake basic installation and maintenance tasks. The next spacewalk, on STS-64, returned to the topic of how an astronaut with a loose tether could effect a hasty recovery. The new device was a compromise between a hand-held gas jet and a fully fledged backpack. This was a compact box that could be fitted to the base of the standard life-support backpack. Given its purpose, it was dubbed the Simplified Aid For EVA Rescue (SAFER).

The excursion undertaken by Carl Meade and Mark Lee was specifically to test this new propulsive device, so they were able to subject it to a thorough evaluation. In doing so, Lee made NASA's first untethered spacewalk since 1984. The control unit was on an umbilical stowed on the side of the pack and he had only to reach down to retrieve it. The unit had been designed to serve four functions in sequence: (1) to enable a drifting astronaut to rapidly cancel any rotational moments, (2) to turn to face the station, (3) to translate straight to it, and (4) to halt. The 38-kilogram unit was essentially a tank of nitrogen, with 24 jets positioned to deliver impulses to isolate all six degrees of freedom. While the 1.4 kilograms of nitrogen in its tank was just sufficient for its rôle as an emergency device, it did not provide much scope for a comprehensive trial, so a tank had been installed in the bay to enable the astronauts to top-up. The evaluation began with Lee methodically testing the SAFER's ability to control each degree of freedom. Then Meade, riding the arm, manhandled Lee to impart rolling, pitching and yawing rotations in order to assess the ability of the attitude-hold function to overcome them. They then swapped places, and Meade performed translation trials, moving out along the arm and back again. Finally, he headed straight for the airlock to demonstrate that the SAFER provided sufficient directional control to reach a specific point without the need to waste gas by making mid-course corrections. A data system recorded its performance, and later analysis confirmed the astronauts' enthusiastic assessment. Significantly, most of the training for this EVA had been conducted with a virtual reality system. Computer simulation already provided the basis for training to use the RMS. Training for spacewalking was more demanding, but as the fidelity of virtual reality simulations improved, it was hoped that they would reduce the demand for the WETF.

Whereas astronauts to date had tackled one spacewalking assignment at a time, and had been able to devote months to learning what was necessary to undertake a small number of specific tasks, on the space station they would be called upon to apply themselves to a much wider range of tasks. It would clearly be impractical for every spacewalker to know everything about every system they would encounter, so an Electronic Cuff-Checklist (ECC) had been devised. Worn on the forearm, this 1-kilogram system had an 8- by 10-cm screen with a 2-megabyte memory. In effect, it would enable each astronaut to look up procedures and checklists in a 500-page manual, the contents of which could be loaded from a laptop immediately prior to venturing out. Although Meade and Lee reported it to be quite effective, there was a problem on STS-63 a few months later when Michael Foale and Bernard Harris found that the device failed when subjected to the intense cold of a prolonged period in shadow. Finding out that the ECC did not like the –100 °C chill was a bonus of their excursion. STS-63 was testing the revised gloves. When astronauts are in sunlight the suit has to work hard to keep them cool, and in shadow it has to keep them warm. It had turned out that Musgrave's team had been so active during the HST servicing that their own metabolism had kept them warm during the extended time in shadow. To exacerbate the effect of heat-loss through the gloves, Foale and Harris tethered themselves to the end of the RMS high above the bay and proceeded to do *absolutely nothing*. The idea of having astronauts remain idle for several hours was anathema to planners accustomed to ensuring that each and every five-minute block on a flight plan was put to good use; but in this case, doing nothing was the point. After two hours, both men reported that their fingers felt as if they were touching ice, so they went on to a follow-on experiment in which they were to assess their ability to manipulate the 1.2-tonne SPARTAN free-flyer and accurately position it. However, the metal was cold, too – so cold, in fact, that it was painful to grip, and they were recalled. Jim Voss and Michael Gernhardt continued this trial on STS-69. The astronaut's undergarment had been modified to enable its coolant loop to be deactivated without also disabling the loop that kept the suit's electronics cool. This meant that when an astronaut spent any time in shadow the coolant loop could be deactivated. Voss and Gernhardt reported that their body heat kept them warm, even while they remained idle. To address the original glove problem, tiny electrical heaters were installed at the finger tips. These tests established that the problem of working in shadow had been solved, and even the improved ECC survived the cold this time. The back-to-basics strategy had paid a rich dividend and the prospects for the space station had improved with every flight. Voss and Gernhardt went on to undertake generic space station tasks such as removing protective blankets from simulated ORUs, using tools to unplug their umbilicals, swapping them, and then reconnecting and shielding them.

Two spacewalks on STS-72 were conducted by Dan Barry, Winston Scott and Leroy Chiao (who went out twice) for "the most extensive hardware evaluation ever". The 20 items that they tested for the space station included utility boxes, fluid connectors and a rigid umbilical to be used to run bundled electrical and fluid lines along the truss. To make the tests as realistic as possible, to enable the ergonomics of the tasks to be studied, parts of the bay had been rigged to resemble the relevant work sites on the station. It is important to note that the excursions designed to reduce the technical risk facing the assembly of the space station were secondary objectives on already planned flights, and this vital experience was purchased at minimal cost.

The next time astronauts ventured out, it was from Atlantis, docked with the Mir space station. Rich Clifford and Linda Godwin worked on the interface module that the Shuttle had attached to the station on its previous visit. They removed items such as a video camera that was no longer required, and then installed the four exposure cassettes of the Mir Environmental Effects Package (MEEP). As the orbiter would not be able to undock to chase them down if a tether broke, both wore SAFERs. One year later, Jerry Linenger joined Vasili Tsibliev in spacewalking in the Russian Orlan spacesuit. He was swung on the end of a crane so that he could install another experiment on Mir. Michael Foale and Anatoli Solovyov went out a few months later to inspect the damage to the Spektr module caused by a collision with a Progress cargo ferry. And a month after that, Vladimir Titov joined Scott Parazynski in spacewalking from Atlantis to retrieve the packages that NASA had emplaced on Mir's exterior.

The plan for STS-80 to assess a 5-metre crane designed for moving ORUs was frustrated when Columbia's outer hatch could not be opened. When the mechanism was stripped after the mission, a tiny screw was found to have worked loose and fouled the ratchet. Tamara Jernigan and Tom Jones, clearly frustrated, had had to repressurise the airlock and re-enter the cabin. The real concern over the stuck hatch was that if the TDRS antenna ever jammed, preventing the payload bay doors from closing, or if the doors themselves failed to close, astronauts would have to go out and rectify the fault. But if a stuck hatch blocked their way into the bay, the orbiter would be stranded in space. This was another valuable lesson. NASA installed a set of tools so that, if necessary, the ratchet mechanism could be dismantled and the hatch released manually.

Table 7.1 Spacewalks in the Shuttle era *not* involving the ISS

#	Mission	Date	Hours	Astronauts	Objective
1	STS-6	7 Apr 1983	3.5	Peterson and Musgrave	Demonstration
2	STS-41B	7 Feb 1984	6	McCandless and Stewart	SolarMax rehearsal
3	STS-41B	9 Feb 1984	6	McCandless and Stewart	SolarMax rehearsal
4	STS-41C	8 Apr 1984	6	Nelson and van Hoften	SolarMax capture
5	STS-41C	11 Apr 1984	7.3	Nelson and van Hoften	SolarMax repair
6	STS-41G	11 Oct 1984	3.5	Sullivan and Leestma	ORS test
7	STS-51A	12 Nov 1984	6	Allen and Gardner	Palapa-B2 retrieval

Table 7.1 (cont)

#	Mission	Date	Hours	Astronauts	Objective
8	STS-51A	14 Nov 1984	6	Allen and Gardner	Westar-6 retrieval
9	STS-51D	17 Apr 1985	3.1	Hoffman and Griggs	To attach fly-swatter
10	STS-51I	1 Sep 1985	7.2	Fisher and van Hoften	Leasat-3 capture
11	STS-51I	2 Sep 1985	4	Fisher and van Hoften	Leasat-3 release
12	STS-61B	30 Nov 1985	5.5	Ross and Spring	ACCESS test
13	STS-61B	2 Nov 1985	6.6	Ross and Spring	EASE test
14	STS-37	7 Apr 1991	4.6	Ross and Apt	CETA test
15	STS-37	8 Apr 1991	6.2	Ross and Apt	CETA test
16	STS-49	10 May 1992	3.7	Thuot and Hieb	Intelsat capture
17	STS-49	11 May 1992	5.5	Thuot and Hieb	Intelsat capture
18	STS-49	13 May 1992	8.5	Thuot, Hieb and Akers	Intelsat 603
19	STS-49	14 May 1992	7.7	Akers and Thornton	ASEM test
20	STS-54	17 Jan 1993	4.5	Runco and Harbaugh	ISS preparation
21	STS-57	25 Jun 1993	5.8	Low and Wisoff	HST preparation
22	STS-51	16 Sep 1993	7.1	Newman and Walz	HST preparation
23	STS-61	5 Dec 1993	7.9	Musgrave and Hoffman	HST service
24	STS-61	6 Dec 1993	6.6	Akers and Thornton	HST service
25	STS-61	7 Dec 1993	6.8	Musgrave and Hoffman	HST service
26	STS-61	8 Dec 1993	6.9	Akers and Thornton	HST service
27	STS-61	9 Dec 1993	7.3	Musgrave and Hoffman	HST service
28	STS-64	16 Sep 1994	6.8	Lee and Meade	SAFER test
29	STS-63	9 Feb 1995	4.6	Foale and Harris	ISS preparation
30	STS-69	16 Sep 1995	6.8	Voss and Gernhardt	ISS preparation
31	STS-72	15 Jan 1996	6.2	Chiao and Barry	ISS preparation
32	STS-72	17 Jan 1996	6.9	Chiao and Scott	ISS preparation
33	STS-76	27 Mar 1996	6.1	Clifford and Godwin	Mir experiments
34	STS-82	13 Feb 1997	6.7	Smith and Lee	HST service
35	STS-82	14 Feb 1997	7.5	Harbaugh and Tanner	HST service
36	STS-82	15 Feb 1997	7.2	Smith and Lee	HST service
37	STS-82	16 Feb 1997	6.6	Harbaugh and Tanner	HST service
38	STS-82	17 Feb 1997	5.3	Smith and Lee	HST service
39	Mir	29 Apr 1997	5.0	Tsibliev and Linenger	Experiments
40	Mir	6 Sep 1997	6	Solovyov and Foale	Experiments
42	STS-87	24 Nov 1997	7.5	Scott and Doi	Rescue SPARTAN
43	STS-87	3 Dec 1997	5	Scott and Doi	ISS preparation
44	Mir	14 Jan 1998	3.8	Solovyov and Wolf	Kvant 2 inspection
45	STS-103	22 Dec 1999	8.3	Smith and Grunsfeld	HST service
46	STS-103	23 Dec 1999	8.2	Foale and Nicollier	HST service
47	STS-103	24 Dec 1999	8.1	Smith and Grunsfeld	HST service
48	STS-109	4 Mar 2002	7.0	Grunsfeld and Linnehan	HST service
49	STS-109	5 Mar 2002	7.3	Newman and Massimino	HST service
50	STS-109	6 Mar 2002	6.8	Grunsfeld and Linnehan	HST service
51	STS-109	7 Mar 2002	7.5	Newman and Massimino	HST service
52	STS-109	8 Mar 2002	7.4	Grunsfeld and Linnehan	HST service

THE TASK AHEAD

Cosmonauts working on the Mir space station had demonstrated that construction in orbit is feasible. Their achievements are evident in pictures of Mir taken at various times during its long life. Yet as the assembly of the International Space Station got underway, NASA was conscious that at least 600 hours of external activity – fully thrice all of its spacewalking experience to date – would be required to complete it, which would really put *homo spacewalker* to the test. Nevertheless, the astronauts and cosmonauts who would do it had the comfort of knowing that their high-fidelity training procedures and tools would break the assembly into a series of manageable tasks.

8

Microgravity

FACILITIES FOR COMMERCIAL RESEARCH

As soon as the Shuttle began commercial operations it became apparent that there was scope for small microgravity payloads. The basic package was the Small Self-Contained Payload, more commonly referred to as the Get-Away Special (GAS). These 0.5-metre-diameter canisters were carried in the payload bay, mounted either individually on the sidewall, or a dozen at a time on the GAS Bridge Assembly, which was an MPESS running across the bay. The cost of the GAS payloads was deliberately held down, but different categories of user were offered different rates – for example, an educational institution could book a canister for a few thousand dollars. When the Shuttle first flew, there was already a long waiting list, which was serviced on a 'first come, first served' basis. However, when Challenger was lost, only 53 GAS had been flown, and the 100th did not fly until 1994, on STS-60. The GAS even offered possibilities for a commercial operator to provide a service in which it would configure an experiment to fit into a canister and prepare and certify it fit-for-flight, because NASA had strict regulations for such matters, the documentation for which often weighed more than the experiment.

In order to provide its experiments with a higher quality of microgravity than was possible within the bay of a Shuttle that was manoeuvring in space, the German company Messerschmitt-Bolkow-Blohm eagerly seized the opportunity to service the European market by developing a bridge structure that could be lifted out of the bay by the RMS and released as a short-duration free-flyer. MBB named its platform the Shuttle Pallet Applications Satellite (SPAS). This bridge was attached to a triangular support truss of carbon-fibre tubing. It had a three-axis control system and nitrogen gas jets for attitude control, and 60 per cent of its 1.5-tonne mass was available for payload. When it first flew on STS-7 in June 1983 it became the first payload to be released and later retrieved by the Shuttle. NASA had charged a minimal fee for flying the SPAS in return for the opportunity to mount a camera on it to provide the first views of an orbiter in space. To mark the occasion, Sally Ride fixed the RMS in the shape of a '7'. The SPAS had seven GAS cans for the European Space Agency and Germany's Ministry of Research and Technology, and the varied payload included two materials-processing experiments, a solar-transducer test, a heat pipe, a pneumatic conveyor, a remote-sensing camera, and a spectrometer to note the gaseous environment as the orbiter manoeuvred close alongside. The SPAS flew again on STS-41B, but a problem with the RMS meant that it could not be deployed, and the experiments had to be conducted in the bay.

The SPAS was not flown again prior to the loss of Challenger, and in 1988 it was sold to the US Department of Defense. The German Space Research Agency built a more capable free-flyer. In effect, the European Retrievable Carrier (EURECA) was a double SPAS bridge without the triangular support truss. Unlike the SPAS, which had to be retrieved within a few hours because it drew its power from a battery, EURECA had solar panels for extended operations. It was to be released by one Shuttle and retrieved by another, months later. This new free-flyer was much more than a scaled-up SPAS, and was aimed at a different class of microgravity payload – those that would take a long time to run their course. After being released by STS-46 in July 1992, EURECA was retrieved by STS-57 almost a year later. The European Space Agency's astronaut Claude Nicollier used the RMS to lift it off its cradle, and held it high above the bay while it unfolded its long solar panels. After it had been released, and to reduce orbital drag, EURECA fired a hydrazine thruster to boost its orbit to 500 kilometres. Over the next fortnight, its 15 experiments were checked out and started up. Most of the experiments had been completed by the end of the year, but because its retrieval had been pushed down the Shuttle manifest, those able to benefit from additional time were kept running. When retrieval was imminent, the payload was shut down, and the platform lowered its orbit back to 300 kilometres for pick-up. David Low grabbed it using the RMS and held it while it retracted its appendages. Although the solar panels folded away properly, the two long whip antennae remained extended. It transpired that the retrieval faced an unexpected time constraint. An electrical connector had been fitted into the end-effector of the RMS to enable EURECA to draw power for its thermal-regulation system while it was on the arm, but this feed had been incorrectly wired and it became imperative that the spacecraft be placed onto its cradle from which it would be able to draw power. Yet the flight rules prohibited placing the spacecraft in the bay with its whips extended, as they could easily strike the payload mounted on the bridge immediately in front. But unless this was done, it was possible that the propellant in EURECA's tank would freeze, split the tank, and contaminate the payload bay. The regulation was waived and the free-flyer was gingerly lowered into the bay. The following day, two astronauts went out and manually stowed the antennae. The 1,000 kilograms of payload included a pair of furnaces, a protein crystallisation facility, a radiation monitor, an instrument to measure the total energy output of the Sun, another to measure the energy spectrum of the Sun, a telescope for high-energy astrophysical observations, and a variety of technology demonstrations of apparatus intended for use on future spacecraft. In most cases the apparatus functioned satisfactorily. One experiment to assess a new type of thruster was crippled by a short circuit, but this did not occur until after some useful data had been acquired.

The operational concept for EURECA envisaged five flights, each of about six months duration, over a 10-year period. NASA allocated a second flight opportunity on the manifest for 1995, but the European Space Agency let this lapse due to lack of funding, and another flight opportunity in 1997 was also refused. The EURECA project serves to highlight an unfortunate trend in Shuttle applications – despite the carrier having been designed to be reusable, it was then used only once. A more constructive outlook, however, is that if ever there is an urgent need for a small free-flying platform, EURECA will be available.

One much-touted use for the Shuttle was the promotion of microgravity research and, through the perfection of methodologies, the commercial use of space. To prime this process, NASA, in partnership with industries and universities, sponsored a number of Centers for Commercial Development of Space (CCDS) in disciplines likely to lead to applications, and it offered to fly their experiments at discounted rates.

The mid-deck of the orbiter's cabin had 42 locker spaces, but 80 per cent of this capacity was routinely reserved for crew equipment. Experiments that would not fit into a GAS can and required some element of astronaut oversight could be flown in the free lockers, but the list of such experiments for educational and commercial users increased much faster than it could be addressed. Spacehab Incorporated was established in 1983 to service precisely this queue of payloads with a 'mid-deck augmentation module'. With NASA's enthusiastic support, Spacehab raised the venture capital required to place a contract with McDonnell–Douglas to develop a 5-tonne, 4-metre-diameter, 3-metre-long flat-topped pressurised module that was to be carried at the front of the payload bay and be connected to the mid-deck by a short tunnel. Ironically, by drawing the orbiter's centre of mass forward, the module, if it were to be filled to capacity, would jeopardise the vehicle's aerodynamics during the final approach to a landing. Although the module was limited to 1,360 kilograms of cargo, by accommodating an additional 60 lockers it so increased the capacity for experiments that a mission specialist was added to the crew specifically to supervise its operations.

Although Spacehab had been designed to service the commercial market, it had been assumed that the first customers would be government agencies. It was believed that the provision of ready access to space would in itself promote experimentation and, later, commercial exploitation of microgravity. The loss of Challenger in 1986 did not directly affect the development of the module, but it did little to encourage industry to place bookings. The economics of the module were similar to the Shuttle in that profitability required a high flight-rate. When Spacehab was devised, NASA had hoped to fly the Shuttle on at least a fortnightly basis. The company's business plan had envisaged the module making its debut in 1991, and rapidly ramp up to five flights per year. By the time the Shuttle resumed operations in 1988, it was evident that the flight rate would never exceed a dozen flights per annum, and, with so many high-priority payloads on this limited manifest, it was also apparent that Spacehab would be lucky to get *one* flight per year. To enable the company to firm up its existing commercial options NASA guaranteed the first flight as planned, and offered a lifeline in the shape of a $184-million booking for 200 lockers distributed over the first six flights. NASA intended to make these lockers available to the CCDS in order to remove the potential bottleneck in testing hardware for these projects. Although the module was not finished until early 1992, it was flown in June 1993 on STS-57, the mission that retrieved the EURECA free-flyer. Two dozen of the experiments were supplied by NASA, covering materials and life sciences, and an engineering test of a waste-water recycling system intended for the space station. The commercial experiments that the company had managed to secure filled most of the rest of the lockers, but it was already clear that this favourable payload mix would be difficult to repeat.

Despite Spacehab having been developed specifically to provide customers with *ready access* to microgravity, it took up to four years to run a mid-deck payload through NASA's fit-for-flight certification procedure, which was a real disincentive to experimentation. A service-provider such as Spacehab Incorporated, familiar with NASA's payload integration procedures, and already approved as a NASA supplier, was able to reduce this time to two years, but even that could not really be described as ready access. Also, since *time is money*, this delay prohibitively inflated the cost of what ought otherwise to have been small cost-effective low-budget experiments. And no matter what it carried, the company faced the problem that its income was extremely erratic – it did not receive its final payment until the payload had flown, and it was a hostage to slippage in the Shuttle's schedule. As a commercial venture, therefore, it was a rather risky business. In fact, NASA was carrying the module for free because the fee that it charged the company for each flight was exactly one-sixth of the value of the contract it had placed for lockers for the experiments supplied by the CCDS on six flights. NASA's rationale had been to establish a flight schedule to enable the company to generate revenue by leasing the *remaining* lockers on a fee-paying basis. By the time of Spacehab's second outing, in February 1994, NASA had renegotiated the contract. Because the company was not attracting sufficient commercial interest to use all of the lockers not assigned to the CCDS, NASA agreed to reassign *all* of its remaining utilisation to the next two flights, after which the company would be on its own.

Spacehab was saved by the transformation in the political situation following the collapse of the Soviet Union. When NASA decided to ferry supplies to astronauts on board the Mir space station for projects that would lead up to the International Space Station (ISS), it sought a logistics carrier – there were only two real options: Spacelab and Spacehab. Whereas Spacelab's laboratory-style instrument racks were inconvenient, the locker system was ideal for cargo. However, the factor that really made the decision was that the Spacehab module could be turned around within the planned four-monthly flight cycle. In winning this contract, the company secured its short-term future. The contract called for a double-length version of the module, for increased capacity, so the pressure hull of the engineering test article was affixed to the rear of one of its flight modules. As there were two operational modules, this left one for the already agreed science missions.

Fittingly, Spacehab's third flight was in February 1995 on the rendezvous with Mir that paved the way for the first docking that summer. In order to rehearse close-proximity manoeuvres alongside Mir, the module had been fitted with a roof window and a camera to offer the Shuttle pilot a viewpoint equivalent to the one that would be provided by the special docking unit. This Spacehab was most notable for

Charlotte, a spider-like tele-operator system. It was strung on wires stretching across the module, and could locate itself in front of any given locker. It had an arm with a gripper, and could flip switches and load samples and data cassettes. A video system allowed its ground supervisor to monitor its activities. It had been developed by McDonnell–Douglas for use on a space station. By its fourth flight, Spacehab was no longer newsworthy, it was part of the Shuttle infrastructure. Nevertheless, it kept the astronauts busy with a broad range of experiments supplied by the CCDS. On this mission, it was the inflatable antenna on the SPARTAN free-flyer which got the headlines.

Spacehab's first flight as a cargo hauler was on STS-76, in March 1996. As the double module was not yet ready, this was the standard version. However, because the docking system had to be at the front of the bay, the module was placed well-aft to restore the centre of mass. One tunnel ran from the mid-deck hatch to the base of the ODS, and another ran from this to the module. As all the hatches were on the same axis, this was a straightforward configuration. Placing the module further back had the advantage of increasing its permitted payload to 2,000 kilograms. The cargo included some 880 kilograms of Russian logistics for Mir and almost 750 kilograms for NASA's science activities. The Spacehab also carried a freezer in which to return the biomedical samples that had been collected on Mir. A total of 500 kilograms of scientific results and miscellaneous apparatus was retrieved. This was a significant milestone, because Mir crews served six-month tours and with only two Soyuz flights per year, each of which could return no more than 120 kilograms of compact items, Mir had become cluttered with apparatus for which there was no longer a use. The Shuttle, therefore, provided Mir with a return capacity that had previously been sorely missed. While it was ironic that a module developed to facilitate microgravity research had found its niche as a bulk-cargo hauler, it was *business* and, because it was genuinely selling its services to NASA, in 1995 the company turned in a profit for the first time.

It is clear that if it had not been for the Shuttle–Mir programme, Spacehab would have been a commercial failure. The basic concept was valid, but the raw commercial market did not develop as expected. Although Spacehab cleared the backlog of mid-deck payloads, enabling the semi-industrial CCDS to prepare their apparatus for use on the ISS, once the station is operational there will be little call for a microgravity module in the payload bay – the concept of 'mid-deck augmentation' evidently had built-in obsolescence.

BIOLOGICAL SYSTEMS

The long-term objective of NASA's life sciences programme is nothing less than to understand the fundamental nature and origin of life. In the shorter term, however, it has two imperatives: first, to understand the way in which life is adapted to the Earth – in particular the relationship between life and gravity – and, second, to develop the medical and biological systems required to enable humans to live and work in space. However, we exist in the context of a larger biological system, not just for food but also for environmental and waste management, and if we are to establish stations in space we will need to take crucial elements of this environment with us. It will therefore be necessary to develop the technology to sustain such an environment in space.

Flora

It is not really feasible to grow plants on the Shuttle because its orbital endurance is so limited, but it can be used to test apparatus as a preliminary to setting up a semi-permanent facility. As work on board the Salyut space stations showed, cultivators optimised for microgravity can fail for a multiplicity of reasons, so this opportunity to test apparatus was very welcome.

The Plant Growth Unit flown on STS-3 used a type of pine, mung-bean and oat to study the formation of lignin, the agent that stiffens the stem of a plant to enable it to withstand the pull of gravity. The objective was to determine whether a wood-plant would develop with less of this agent in microgravity. This test had long-term implications for building self-sustaining habitats in space. The human gut cannot digest a lignin–cellulose cell so, even although such a plant is rich in carbohydrate and protein, it cannot serve as a food. If a lignin-free variety could be grown in space, it would increase the range of foods available. This experiment, fully called Gravity-Induced Lignification In Higher Plants, was reflown on STS-51F. When the Shuttle resumed flying after the loss of Challenger, the Plant Growth Unit was used for the Chromex experiment, which was to determine whether the roots of a plant in microgravity developed in the same way as those on Earth. Root-free shoots were grown to measure the rate, frequency and patterning of cell-division in the root tips and the genetic make-up tested to determine whether this was upset by exposure to microgravity and radiation in the space environment. The plants grew well; indeed, the rate of root-tip growth was much faster than on Earth, but the roots became disoriented and grew in all directions. Of far greater significance for this experiment was the discovery of a substantially reduced cellular division in the root tips, and of increased chromosomal abnormalities such as breakage and fusion. Data from radiation monitors flown alongside suggested that the radiation dose was insufficient to cause the 10–30 per cent aberration rate seen in dividing cell chromosomes. The only way to separate the effects of microgravity and radiation would be to fly a centrifuge to reintroduce the effect of gravity in the radiation environment, but this has yet to be done.

The real challenge was to plant a seed, grow it, and from the flowering plant harvest a new seed that would then be planted to restart the cycle, because it is only by running through successive generations that genetic mutation is made manifest. Quite apart from the study of space genetics, the successful growth of plants will be crucial to the establishment of a fully self-sufficient orbital habitat, not just for food but also for atmospheric processing, as plants consume carbon dioxide and liberate oxygen. Unfortunately, even with its extended-duration facility, a Shuttle cannot remain in space long enough for a plant to complete its life cycle. This requires a semi-permanent orbital habitat. After many disappointments the Russians managed to run a weed called arabidopsis through its life cycle. As this had yet to be done for a staple food like wheat, when NASA was invited to have a succession of astronauts serve tours on Mir, it assigned them this task. It had taken the Russians many years to develop a successful cultivator by a process of trial and error on board the Salyut stations that preceded Mir. While the first flowering had been achieved in 1980, it took another two years of frustrating experiments before seed pods were produced – an event that greatly pleased the botanists.

The Wisconsin Center for Space Automation and Robotics, one of the CCDS, built the Astroculture cultivator to grow plants in a fluid-infused matrix – a process called hydroponics. The need to control fluids made this a technically demanding method. The initial flights in 1992 and 1993 were engineering tests of the water and nutrient delivery, the lighting system and the humidity control system. It was a self-contained unit within a mid-deck locker, and comprised two growth chambers. An inert material served as the root matrix. One porous steel tube was embedded in the matrix to supply water, and another to recover it. Although the water flow in the supply tube was pumped, the nutrient solution emerged from the tube by capillary action. The problem was to propagate the nutrients through the matrix without it becoming saturated. In this initial form, the apparatus was heavily instrumented to measure flow rates to define its operating parameters. When perfected, it was flown on a number of Shuttle missions in preparation for being sent to the ISS.

Rather than risk using one of its own experimental systems for growing wheat, NASA decided to use the Svet cultivator already on Mir, in which radish and lettuce had been successfully grown. It had a high-intensity lamp and grew the seed in a substrate infused with nutrient, and NASA instrumented it to monitor the physical parameters of the experiment. John Blaha was to have initiated this Greenhouse Experiment, but Shannon Lucid planted the first wheat seeds while waiting for the delayed Shuttle to retrieve her. Blaha harvested the crop just before he left, Jerry Linenger, replanted some of the seeds and tended them throughout their cycle, and Michael Foale continued the experiment. The growth of a staple food was a real milestone in space biology.

Fauna

As physiological changes run their course more rapidly in animals than in humans, on a space flight of limited duration more can be learned by studying animals. The first animal experiment on the Shuttle, flown on STS-3, was a sealed mid-deck locker containing bees and moths – species selected because of their very different ratios of body mass to wing area – and they were filmed for later analysis of their reaction to microgravity. A colony of ants was flown on STS-7 for further insight, and a hive of 12-day-old bees was carried on STS-41C to determine whether they would be able to make a honeycomb. On Earth, bees make a matrix with a hexagonal cross-section, which is structurally stronger than a matrix having a triangular or square section. The container mimicked the familiar diurnal light cycle and maintained a hot environment. The bees rapidly adjusted, and made a recognisable comb. This test was reminiscent of the one on Skylab in which the spider named Arabella had initially been confused by the strange environment, but had rapidly adapted and woven commendable webs. On the early Shuttle flights, such studies were secondary tasks, based on suggestions by school students.

The first mission devoted to life sciences research was the flight of Spacelab 3 on STS-51B in 1985. The Ames Research Center in California had developed several facilities for animal studies on the Shuttle. The Animal Holding Facility (AHF) was a cabinet-sized enclosure for monkeys. The two monkeys on this mission had been designated 'A' and 'B' in the hope that they would not become anthropomorphised by the public. Nevertheless, their carriage was criticised by Animal Rights activists. Although monkeys had flown in space before – notably in the early days of testing the Mercury capsule – this was the first time that humans and monkeys had flown together. The experiments were supervised on alternate shifts by the two medical specialists. As Norman Thagard's speciality was human medicine, Bill Thornton – who had worked with Ames on the development of the AHF – took the shift when the monkeys were awake. As the monkeys were rather withdrawn, and declined to eat from their food dispenser, Thornton took pity on them, opened their enclosure, and feed them by hand. This contravened the experimental schedule, which called for the animals to be fed only from their food trays. Opening the enclosure had been forbidden in order to preclude the escape of loose material, but when a food tray was removed for refilling loose material floated out. This débris was not simply uneaten food, it included pieces of excrement and droplets of urine. To retain particulates in the laboratory, the environmental system had been configured to ensure that the air-flow went from the orbiter's cabin into the Spacelab, but this proved deficient. One memorable scene on the video link was the look of disgust on Shuttle commander Robert Overmyer's face when some monkey droppings drifted in front of his face as he sat at his seat on the flight deck. Such material represented a health hazard to the crew, as it could be swallowed, inhaled, or trapped in the eye or ear. Worst of all for this pioneering crew was the pungent aroma of the animal enclosure. However, the animals were not the only source of unpleasant emissions. With an all-male crew, it had been a simple exercise to modify the toilet to collect urine for a human biology experiment, but this persistently malfunctioned and sprayed fluid into the cabin.

Spacelab 3 also carried two dozen rats in an Animal Enclosure Module (AEM), a unit that provided its occupants with food, water, air and light. It had been tested on STS-8 and STS-41B. The rats had been surgically implanted with sensors to enable biotelemetry to record physiological data during the flight. Immediately after landing, and before the process of readaptation to gravity could set in, they were dissected to extract tissue to preserve its state of adaptation to microgravity. The study of rats can provide a range of biomedically pertinent data. Skeletal muscles can be classified in terms of their rates of reaction to stimulus by nerve impulses. Those that react most rapidly are related to body motion, and those that react slowly are involved in maintaining posture in the presence of gravity. The response timing is governed by a specific protein, and by dissecting a rat the state of adaptation can be related to the production of this protein. After seven days in space, a rat will have lost 40 per cent of the mass in the muscles of its hind legs. These muscles would normally have been active in opposing gravity, but in space there is nothing to oppose, resulting in a marked decrease in the diameters of muscle fibres and an almost total lack of muscle tone. There are also biochemical changes. After seven days the process that creates energy in muscle cells is also almost totally absent. In space, the slow-acting load-bearing and postural muscles atrophy. On the other hand, the fast-acting motion-control muscles can be significantly enhanced. The musculo-skeletal system rapidly adapts to the environmental change on entering space and on returning to Earth. This confirmed observations by astronauts who had flown on Skylab for three months, and by the cosmonauts who made even longer flights on Mir. This long-term process is difficult to study in detail in the case of the human crew because they are required to exercise specifically to overcome such deterioration (and because they cannot be dissected upon their return), so the study of rats enabled the biochemical processes to be investigated.

Stamina is also decreased. It is suspected that in space the reduced muscle activity not only inhibits the protein that controls response time, but also enzymes associated with metabolism. Metabolism is the process by which food is broken down and converted into energy. In the absence of energy released by metabolism, the muscles rapidly exhaust their limited supply and begin to consume the glycogen that is stored in non-load-bearing muscle tissue, primarily in the liver. Glycogen is a polysaccharide stored as a reserve carbohydrate. It is a polysaccharide in that it is a mixture of monosaccharides, the most common of which is glucose – otherwise known simply as 'sugar' – and it plays a significant rôle in metabolism. The first Physiological and Anatomical Rodent Experiment focused on the rôle of glycogen. Specifically, it sought to find out how microgravity affects insulin control of glucose transport in the soleus, a calf muscle involved in overcoming the pull of gravity, as well as in movement. It was found that non-load-bearing tissue stores additional amounts of glycogen as a result of the modified regulation of glucose metabolism. As this test was done on STS-48 – the mission that deployed the Upper Atmosphere Research Satellite – the AEM was fitted on the mid-deck and could accommodate only eight rats. Also, because it was a 5-day mission, very young rats that would be undergoing development, and should therefore be more susceptible to this change in the short time available, were flown. This experiment was carried repeatedly, in each case focusing on a different aspect of adaptation. These biochemical studies showed that the body can adjust to a change in its gravitational environment in a remarkably short timescale.

It had long been known that bones atrophy in weightlessness. Data from rats indicated that this loss is worst in the case of load-bearing bones, that it results from a slowing in the rate of bone formation, and that the inhibition of the process of calcification is linked to a reduction of a protein secreted by bone-forming cells. But how did the cells detect the absence of gravity? Two Physiological and Anatomical Rodent Experiments on STS-56 set out to study bone development in detail. It was clear that gravity is essential for 'normal' skeletal development, because the rôle of bone is to bear the load imposed by the rest of the body, and posture reflects this. One experiment investigated the configuration that the body adopts in the absence of gravity by measuring the bone mass, mineralisation rates and bone strength. Earth-based experiments in unloading the skeleton of rats by hanging them by their tails throughout the formative stage in their development had shown that although bone continued to be produced, the cells formed a structure with little strength. In space, however, this process is inhibited. The aim was to determine whether this resulted indirectly from the reduced muscular stimulation, or was directly attributable to the bone. The second experiment studied the osteoblasts – the cells that make bone. The production of pre-osteoblast cells is mechanically sensitive, and the objectives of the experiment were to figure how such cells sense gravity, how production is switched off when gravity is removed, and how it is restarted immediately upon returning to Earth. Osteoblast proliferation was tracked by a marker for DNA synthesis, and the non-load-bearing jaws were studied as well as load-bearing shin bones to determine whether the process was selective. A student experiment on STS-29 had flown rats in which non-load-bearing bones had been fractured and then reset to study the effect of exposure to microgravity on the healing process. It transpired that the tiny amount of bone grown in space had little structural strength, and it began to look as if the duration that humans will be able to live in space will be limited by the issue of bone demineralisation.

In addition to rats, the first Space Life Sciences Spacelab had an aquarium with 2,500 baby jellyfish (*aurelia ephyra*) – one of the simplest organisms to possess a nervous system. In the sea, they maintain orientation by using a statolith sensor analogous to the mammalian vestibular apparatus. They swam about in circles, with spasmodic bursts of activity, so perhaps they, too, had low stamina. A variety of other aquatic species had already been tested. STS-5 had carried sponges, an organless animal that is characterised by the presence of a canal system and a series of chambers through which water is first drawn in and then expelled. Frogs flew on the first German Spacelab. A Japanese visitor to Mir took small tree frogs to see if the suckers on their feet enabled them to orient themselves in space, which they did. Tadpoles were observed on the Japanese Spacelab mission in 1992. Those hatched from eggs fertilised prior to launch were disoriented and corkscrewed, but those from eggs fertilised in space were content in microgravity. How did they know? Another experiment on this mission studied the adaptation of two carp, one of which had had its otolithic organs removed. All of these experiments provided valuable information for application to humans because the sensory apparatus is common; namely, tiny hairs in the inner ear with a calcified node at the far end, whose reaction to stimulus provides a sense of direction.

The follow-up Space Life Sciences mission employed the Extended Duration Orbiter (EDO) to sustain a 14-day programme. On this occasion, a bank of AEMs housed 48 rats which had been dosed with tracers to enable the process of adaptation to microgravity to be followed in greater detail. This time six of the rats were to be dissected in space so that their tissue and bone could be sampled *in situ*. Some of the others were to be killed immediately after landing, to allow the effects of descent to be isolated. The rest were to be killed at intervals thereafter to snapshot the process of readaptation. Martin Fettman, a veterinarian, was included in the crew to undertake this delicate task. He used a guillotine in a glovebox to kill the rats by decapitation, and as the blood oozed from the carcass it floated free, messing up the enclosure. If ever it proves necessary to carry out an invasive medical procedure on an injured astronaut, that will be the day when space medicine comes of age.

The US Army's Institute of Research based at the Walter Reed Hospital had developed the Space Tissue Loss experiment to study the effects of spaceflight on the body. Rather than study animals, it had built a mid-deck apparatus to grow cultures of live cells. The great advantage of this approach was that it was possible to study the cells directly, rather than as part of an organism. In addition, because the cells were in a culture, drugs could be injected. Tested as a secondary payload on STS-45, this automated apparatus grew both bone and muscle cells and determined a variety of parameters including cell shape and membrane integrity. The effects of enzymes were recorded *in situ*, and the cells then fixed to preserve them for return to Earth. On STS-53, in addition to the ongoing studies of muscle disintegration and bone demineralisation, this apparatus was used to investigate the response of white blood cells (associated with the immune system) to antigens produced by infectious agents and tumours. It was found that muscle growth in microgravity is impaired by disruption of the ability of precursor cells to fuse to make fibres. Just as the bone studies could ultimately lead to a treatment for osteoporosis, it was possible that an understanding of the process of muscle wastage would help to devise a treatment for muscular dystrophy.

The US National Institutes of Health (NIH) introduced a comprehensive suite of rat tests on STS-66 as NIH-R ('R' for rat). In fact, this was an international effort, and one of the experiments was undertaken for Moscow's Institute of Biomedical Problems. As each of the two AEM on the mid-deck housed five pregnant rats, this experiment pushed the study of the body's development back to the foetal stage. Immediately following the 10-day flight, all the rats were taken to a laboratory. The test exploited the fact that a rat has two wombs: one foetus was surgically removed and the other was allowed to run its full term. The mothers were then dissected. The aims included studies of placental development as a result of this prenatal exposure to microgravity, the rôle of gravity in the formation of the optic nerve, the vestibular otolithic apparatus, the immunological system, the regulation of body fluids, the tendon attachment process, and the formation of the muscle spindles which serve as sensory receptors. It had been shown by STS-29 and the first Japanese Spacelab that chicken eggs fertilised prior to launch gave chicks that developed weak bones. In NIH-C ('C' for chicken) on STS-66, cultures of cells taken from chicken embryos were grown. As cartilage is a connective fibrous tissue that transforms into bone if it ossifies, this study of cartilage mineralisation focused on the process of calcification to hone in on the problem.

The Center for Cell Research was established at Pennsylvania State University as one of the CCDS to focus on commercial projects involving physiological testing. It developed the Physiological Systems Experiment which first flew on STS-41. This used rats in AEMs to determine whether biological effects induced by microgravity mimicked terrestrial medical conditions sufficiently closely to facilitate testing of new pharmaceutical products. Tumour cells had already been flown to evaluate anti-cancer drugs, but this new apparatus was to address a wider range of ailments, including bone- and muscle-wasting diseases, organ tissue regeneration and immunological diseases. It became one of those mid-deck payloads that flew as a low-profile secondary payload several times a year. Its first flight was sponsored by Genentech, the Californian company that specialised in using recombinant DNA-based products to replicate natural proteins. The second flight was for the Merck

Company to assess a potential treatment for osteoporosis induced by oestrogen depletion in menopausal women. This could be extended to patients immobilised for months in bed, during which time, just as in space, the load-bearing bones atrophy. The NASA interest, of course, was in any possible application to inhibit the bone-loss that would afflict astronauts on tours of duty on the ISS.

There have been several notable cases in which a system is not fully appreciated until it is possible to view it whole, from outside. It was no coincidence that the vulnerability of the Earth was not recognised until it was observed from orbit, and the fact that an astronaut on the Moon could mask out the Earth with a thumb held at arm's length served only to reinforce this realisation. The shallowness of the atmosphere that harbours life is strikingly shown in the pictures of short arcs of the Earth's limb taken from low orbit. The vast scale of the planetary weather system was not fully appreciated until cameras were placed in geosynchronous orbit. The early history of the Earth remained a mystery until spacecraft revealed the cratered surfaces of the other bodies, prompting the realisation that the Earth must have undergone similar bombardment. It should come as no surprise, therefore, that an understanding of life's dependence on gravity, as a result of its evolution on the surface of a planet, is being revealed only now that we can observe life in the absence of gravity. As microgravity is highlighting the effects that take place on Earth when the body fails to accommodate gravity, this type of research is not simply intended to enable astronauts to remain in space for longer periods and return to Earth without debilitation, it is designed to improve our understanding of ailments and develop new treatments, and must be judged in this wider context.

Humans

In 1958, when the newly created NASA decided to send a man into space, medical specialists suggested numerous ways in which the astronaut might become disabled by the absence of gravity. Although Yuri Gagarin's single orbit robbed NASA of the glory of being first, the cosmonaut's survival was welcome news, because it silenced the prophets of doom and gloom. Clearly, the human body could function perfectly adequately in space, and Alan Shepard's all-too-brief suborbital hop a few months later revealed nothing to contradict this view. Gherman Titov, who spent an entire day in space, reported symptoms of nausea, but this was dismissed by NASA as an aberration, especially when Gordon Cooper made a similar flight with no ill effects on the final mission of the Mercury programme in May 1963. One of the primary objectives of the Gemini programme was to demonstrate that the human body had sufficient endurance to mount a lunar mission, and when Frank Borman and Jim Lovell spent almost two weeks in space on Gemini 7 in December 1965, NASA was sure that the medical risks of spaceflight had been exaggerated. It came as a shock, therefore, when the first Apollo crew suffered similar symptoms to those reported by Titov, and Borman vomited on his way to the Moon in Apollo 8. The Mercury and Gemini spacecraft had been too small for their occupants to move around, but Apollo had sufficient volume for its crew to stow their couches. As they performed weightless antics, the unfamiliar motions and viewpoints disturbed their vestibular organs, with the result that they experienced motion sickness. It transpired that, unlike Gagarin, Titov had unstrapped from his couch and floated within the confines of his 2.2-metre-diameter spherical Vostok capsule. Evidently the human body was more sensitive to weightlessness than NASA had believed. In 1964 the Soviets had sent up a physician, Boris Yegerov, but as he had been squeezed into such a capsule with two colleagues, the opportunity to study motion sickness was severely limited. On Skylab, NASA repeatedly set endurance records, with the final crew setting the record at 84 days. Skylab showed that although the human body was indeed initially disturbed, it adapted to the new environment. The simple instruments available gave tantalising clues into how the heart, muscles, bones and blood responded, but the fragmented data was insufficient to draw conclusions about the interrelationships.

The major trends were evident, however: the nausea derived from the brain receiving discordant visual and inner-ear otolithic signals. Only about 30 per cent of astronauts suffered any ill effects, but in the worst cases nausea during the first few hours resulted in a brief bout of vomiting. Susceptibility could generally be reduced by restricting sudden head motion, by avoiding orientations that offered views likely to induce vertigo, and by refraining from weightless gymnastics. This first phase of adaptation is complicated by the pooling of blood in the upper torso, as gravity is no longer drawing it to the legs, resulting in a sensation of lightheadedness and a puffing up of the face, especially around the eyes. It takes a few days for the vestibular disorientation to abate, doing so because the brain filters out the otolithic signals, and about a week to overcome the pooling of blood in the upper torso, during which time the body increases urination to shed what it believes to be excessive body fluid. During this time, the heart gradually adjusts to the fact that it is no longer pumping blood against the tug of gravity. Astronauts have to ingest extra water to preclude dehydration. As the body adjusts, the capacity of the cardiovascular system decreases, and the heart migrates up into the chest cavity. After a month, the body has settled into a state that is compatible with its new environment. It seemed that strict exercise ameliorated the atrophying of the load-bearing muscles, but the most alarming trend was that the slight loss of bone mass that had been detected following short flights evidently did not level out, with the result that after three months in space the Skylab astronauts displayed an effect comparable with a bone-wasting disease. Unlike a terrestrial osteoporosis sufferer, however, the astronauts' bones slowly recovered following their return to gravity.

As the Shuttle neared the point at which it would begin to fly, NASA picked up on what had become known as 'stomach awareness', relabelled it Space Adaptive Syndrome in order to encompass other factors influencing the body's adaptation to weightlessness, and formulated a plan to conduct a comprehensive study of the body as a system.

By this point, the endurance record was 185 days, set on board Salyut 6 in 1980. With the launch of Salyut 7 two years later, the record was extended to 211 days. It was then announced that a cosmonaut doctor would accompany the crew that would attempt to break this record, but it was 1984 before Oleg Atkov played his part in pushing the record to 237 days. Atkov had a multifunction biomedical test kit with which he was to track amenable parameters in unprecedented detail. Salyut crews on long missions hosted a succession of visitors for periods of a week. Tests were devised to note – both subjectively and objectively – every aspect of the early phase of adaptation. This programme recorded vestibular, hormonal, chromosomal and immunological changes; the heart's rhythm, structure and migration inside the chest; the capacity of the cardiovascular system; the composition and distribution of body fluid; the capacity of the respiratory system; bone loss; and muscle degradation. Other topics included: posture; skin sensitivity; sources of physical irritation; changes to the senses of hearing, taste and visual acuity; and aspects of brain activity and cognitive function. Psychological questionnaires tracked the self-assessment of working efficiency, and relationships among the crew and with the controllers on the ground. In addition, the video downlink was studied by behavioural psychologists to assess the state of mind of each member of a crew.

Meanwhile, NASA was gearing up the Shuttle flight rate and demonstrating its ability to deploy commercial satellites. As the maximum time a vehicle could spend in orbit at that time was about 10 days, the Space Adaptive Syndrome research was limited to the initial phase of adaptation to weightlessness. On reaching orbit on STS-3, Jack Lousma repeated his Skylab experience by vomiting. This was only a minor inconvenience, but when Bill Lenoir was sick on STS-5 and the first attempt to spacewalk had to be abandoned, NASA assigned physicians Norman Thagard and Bill Thornton to STS-7 and STS-8 respectively, to conduct their own tests. They were both subsequently assigned to the Spacelab 3 mission on which the rats and monkeys were flown. Although a Shuttle mission did not allow time for the body to adapt fully to microgravity, the tonnes of apparatus that it could carry in a Spacelab made it a veritable biomedical laboratory. The Space Adaptive Syndrome programme therefore developed parallel strands. The early subjective study of human reaction to space was supplemented with a detailed study of how the body regulated its physiological processes. As the tests became more sophisticated, testing started prior to launch, and continued after landing in order to study readaptation to gravity. The objective was to monitor the temporal variation of physiological parameters to track, in detail, the body's response to the absence of gravity.

Blood cells are carried along in the bloodstream by the plasma, which is the rich solution of proteins, nutrients, electrolytes, hormones and assorted metabolic wastes. In fact, there are two types of blood cell: the red blood cells (the erythrocytes) that contain the hæmoglobin that selectively absorbs and releases oxygen, and the white blood cells (the leukocytes) that provide the basis of the immunological system. The body contains about six litres of blood, of which the plasma constitutes more than half, and the sampling of blood is a very effective means of studying a whole range of factors affecting body chemistry. In the cardiovascular system, veins carry oxygen-deficient blood to the heart's right ventricle, which pumps it to the lungs where carbon dioxide is exchanged for oxygen. Pulmonary veins take the now oxygen-rich blood to the heart's left ventricle, which pumps it out through the aorta – the primary artery – so that it can refresh the body's tissue.

The migration of fluid to the upper torso upon entry to weightlessness causes the heart to expand to handle the increased blood flow. The lungs are also sensitive to gravity, which causes ventilation, blood flow, gas exchange and pressure to vary in different parts of these organs. In space, lung capacity decreases, possibly in response to the pooling of blood in the upper torso. As pressure in the arteries rises, nerve cells clustered in the heart, the aorta and the carotid artery, which is in the neck, signal the brain to adjust the heart rate to restore the blood pressure, but because the blood is no longer evenly distributed the only way to reduce the pressure is to reduce the capacity. The body's water balance is controlled by the kidneys, which regulate blood volume and electrolytes content, and remove the waste products discharged into the bloodstream by other organs. This action by the kidneys is known as the renal system. Hormone-secreting glands work with the kidneys to regulate the body's processes, and selectively secrete hormones into the bloodstream to control the rate of chemical reactions in other organs. This 'controlling' action is known as the endocrine system. In weightlessness the renal and endocrine systems increase urination to reduce the total body fluid and electrolytes and, thereby, the total blood volume. This adapted state is characterised by: a reduced heart volume; an overall reduction in blood volume (but with a greater proportion of that which remains present in the upper torso rather than in the legs); a slight increase in heart rate when at rest; and a reduction in the heart's ability to respond to strenuous exercise, with a consequent loss of stamina. However, none of these changes impairs cardiac function.

By measuring cardiovascular and cardiopulmonary parameters, and by sampling blood and urine to follow the endocrinic (hormone), erythropoietic (red blood cell) and immunological (white blood cell) evolution, this process can be tracked in detail. Analysis confirmed the earlier observations of an overall redistribution of body fluid and a reduction in plasma, red cell count and white cell count. Although there are several types of white blood cell, they all maintain a low concentration until they detect a foreign substance in the blood, and then they rapidly proliferate to create sufficient antibodies to attack the invader. In space, however, not only is the cell count reduced, when a blood sample was suitably stimulated, the lymphocytes (one type of such cell) proved barely able to proliferate, producing barely 3 per cent as much product as the control experiment conducted on the Earth. It is uncertain whether this really means that the body is more susceptible to infection in space. The only direct evidence is that none of the resident crews of the Mir space station has shown an unusual proclivity to infection by a member of a visiting crew.

In addition to the cardiopulmonary investigation, attention was focused on the effect of weightlessness on the musculo-skeletal system. The load-bearing muscles atrophy because they are not used, and there is an overall decrease in muscle strength. The loss of muscle mass is particularly pronounced in the calves. Cosmonauts have dubbed this deterioration 'chicken leg' syndrome. However, many cosmonauts return with greatly enhanced forearm muscles. Skylab data had shown that the most significant muscle-loss took place in the first month or so, and while rigorous exercise greatly slowed the rate of atrophying, it did not completely halt it. Load-bearing bones also atrophy in space. This takes the form of a gradual demineralisation in which calcium and phosphorus leach into the bloodstream, and the most significant loss of bone mass occurs in the legs and spine. As with the blood, bone is in a continual state of regeneration, and this departure from the terrestrial norm is actually the result of a change in the balance between the respective rates of production and destruction. (Rodent studies had linked this to reduced levels of osteocalcin – the protein that is secreted by bone-forming cells – which suggested that bone production is inhibited.) In the longer term, this demineralisation might pose the risk of kidney stones. The bone loss is similar to osteoporosis, but much more pronounced. On Earth, a sufferer typically loses bone mass at 2 per cent per annum, but it is an order of magnitude greater in space. Unlike most aspects of adaptation to weightlessness, this decalcification does not appear to have a limit – the longer an astronaut remains in a weightless state, the greater is the bone loss and the longer it takes to recover on returning to Earth. However, the loss rate is not a straightforward correlation with mission duration. One cosmonaut lost 8 per cent after six months, which was the same as the final Skylab crew in only three months. Also, whereas as one cosmonaut lost 20 per cent in five months, another lost only 15 per cent in seven months. Once decalcification is understood, it may be possible to develop a drug to either inhibit the loss rate or enhance the regeneration rate, and hopefully such a drug will also prove beneficial to osteoporosis patients.

Each of the various functions undergoes a phase of acute adaptation, then slowly settles down to its adapted state, but each takes a different time to peak. With so many functions simultaneously reacting to the onset of weightlessness, the body's adaptation is particularly acute during the first day. By the end of the first fortnight, most of the acute evolution has run its course. It is another few weeks, however, before the functions that are most sensitive to weightlessness achieve their fully adapted state, after which the slower-acting functions gradually depart from their terrestrial norm and become the dominant issue of concern. Thus, even although its orbital endurance was severely restricted, the Shuttle was able to make a significant contribution to understanding the body's adaptation by tracing the acute phase in unprecedented detail.

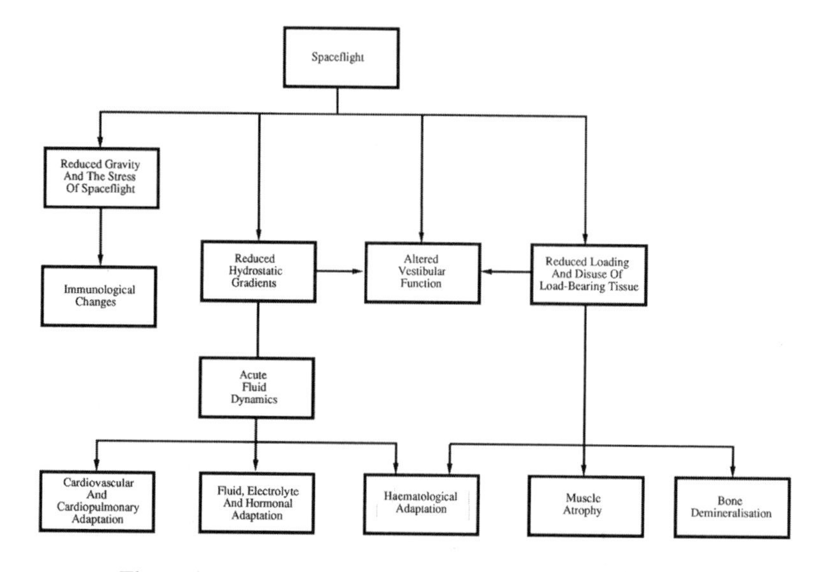

The various ways in which weightlessness affects the body.

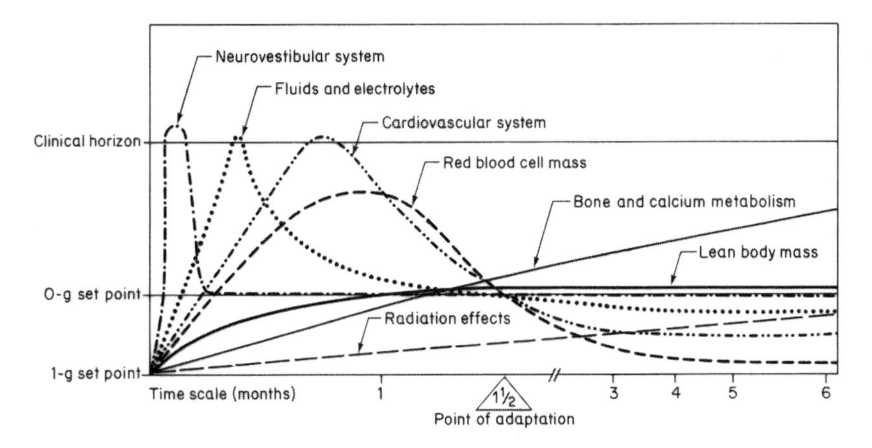

Timescales of how the mammalian body adapts to weightlessness.

The Shuttle–Mir programme provided an unexpected opportunity to study the fully adapted state. When Atlantis made its first visit to Mir, it carried a life sciences Spacelab to subject Norman Thagard and his two Russian colleagues to a battery of tests to record their state of adaptation after four months in space. Earlier work had shown that even the *g*-forces endured during re-entry of the atmosphere significantly upset the body's adapted state, so this *in-situ* snapshot was extremely valuable.

Many questions remain, however. For example, when 'excess' plasma is shed by the kidneys in the initial phase of adaptation, this increases the proportion of cellular material in the residue and the body gradually reduces the red and white cell count to compensate. How is this adjustment achieved? The red cell count is actually the result of an equilibrium of two processes, one creating erythrocytes and the other destroying them. Erythrocytes are continuously created by erythropoiesis – a process within bone marrow that is regulated by the hormone erythropoietin. Is the creation rate reduced, or is the destruction rate increased? If the creation rate is reduced, is this due to a signal sent via erythropoietin, or as a side-effect of changes affecting the bone? In space, biochemical reactions run at a reduced pace, so how is the regulatory system itself affected? There is evidence that hormone secretion is different, but the resultant effects on the kidneys, blood vessels, and heart have yet to be isolated. It is possible that a deeper understanding of the regulatory system could shed light on related terrestrial diseases, such as excessive blood pressure and heart failure.

In the final weeks of a six-month tour of duty on Mir, cosmonauts drank a saline solution laced with electrolytes to build up body fluids, spent an hour a day in their Lower-Body Negative Pressure chamber to draw blood into their legs in order to encourage the heart to increase the cardiovascular capacity, took stamina-enhancing drugs, and undertook rigorous exercise to rebuild their muscles and place an increased load on the heart. The first problem suffered on returning to Earth was that gravity drained blood into the legs. Even though the heart rate increased, the blood pressure *fell* due to the reduced cardiovascular capacity of the space-adapted state. The first result of this orthostatic intolerance was a tendency to black out. Dizziness resulting from diminished blood flow to the brain was compounded by the transition to an otolithically driven vestibular system, and by the generally weakened state of the load-bearing muscles. Although the heart and the sense of balance recovered within days, the muscles took rather longer to recover, and the cosmonauts walked with a distinctive gait. To help the heart to readjust, they wore inflatable leggings originally designed to enable pilots of fast jets to endure high-*g* turns. They also swam because neutral buoyancy relieved physical loading. Nevertheless, they were usually walking normally within a week. As cosmonaut-physician Valeri Poliakov pointed out to a press conference shortly after his return from spending 438 days (some 14 months) on board Mir, his very presence meant that the time required to readapt to gravity had been decoupled from the length of the flight.

As yet, no irreversible effects have been identified, and none of the observed effects is life-threatening, either during the course of a mission or on return to Earth.

Perhaps the most serious issue facing long-stay crews, or those venturing far from the Earth, will be the radiation environment. Apart from the risk of accumulating a lethal dose of radiation from solar flares, the most serious threat will be from cosmic rays – the heavily ionised atoms that travel at relativistic speeds. Astronauts travelling to the Moon reported 'seeing' flashes of light as these high-energy particles flew through their heads. If such a particle were to strike a cell's nucleus, it could damage the DNA within. The ultimate threat to space travellers could, therefore, be unpredictable genetic diseases, and the effects of weightlessness may turn out to have been but a minor inconvenience.

Table 8.1 International space endurance records

Astronaut/ Cosmonaut	Launch Spacecraft	Launch Date	Days In Space
Yuri Gagarin	Vostok 1	Apr 1961	0.07
Gherman Titov	Vostok 2	Aug 1961	1.05
Andrian Nikolayev	Vostok 3	Aug 1962	3.93
Valeri Bykovsky	Vostok 5	June 1963	4.97
Gordon Cooper	Gemini 5	Aug 1965	7.92
Pete Conrad			
Frank Borman	Gemini 7	Dec 1965	13.75
Jim Lovell			
Andrian Nikolayev	Soyuz 9	June 1970	17.71
Vitali Sevastyanov			
Georgi Dobrovolsky	Salyut 1	June 1971	23.76
Viktor Patsayev			
Vladislav Volkov			
Pete Conrad	Skylab	May 1973	28.04
Paul Weitz			
Joe Kerwin			
Al Bean	Skylab	July 1973	59.49
Jack Lousma			
Owen Garriott			

Table 8.1 (cont)

Astronaut/ Cosmonaut	Launch Spacecraft	Launch Date	Days In Space
Jerry Carr	Skylab	Nov 1973	84.04
Bill Pogue			
Ed Gibson			
Yuri Romanenko	Salyut 6	Dec 1977	96.42
Georgi Grechko			
Vladimir Kovalyonok	Salyut 6	June 1978	139.60
Alexander Ivanchenkov			
Vladimir Lyakhov	Salyut 6	Feb 1979	175.06
Valeri Ryumin			
Leonid Popov	Salyut 6	Apr 1980	184.84
Valeri Ryumin			
Anatoli Berezovoi	Salyut 7	May 1982	211.38
Valentin Lebedev			
Leonid Kizim	Salyut 7	Feb 1984	236.95
Vladimir Solovyov			
Oleg Atkov			
Yuri Romanenko	Mir	Feb 1987	326.48
Vladimir Titov	Mir	Dec 1987	365.95
Musa Manarov			
Valeri Poliakov	Mir	Jan 1994	437.75

MATERIALS PROCESSING

The first commercial payload to fly on the Shuttle was a mid-deck locker called the Monodisperse Latex Reactor. Its function was very simple: it was to produce large numbers of tiny latex spheres, all exactly the same size. On Earth, gravity interferes with the manufacturing process, and it was hoped that a more uniform product (a monodisperse product) would result from making them in microgravity. Even on its first flight on STS-3, the reactor created millions of perfect spheres 10 micrometres in diameter. Although all the output from each run was the same size, that size could be programmed anywhere between 1 and 100 micrometres. For spheres greater than 1 micrometre in diameter, gravity-induced distortions promote the formation of irregular lumps – a process called creaming – that ruins the terrestrial product. The ability to make perfect spheres 100 times larger than was otherwise possible represented a significant step forward. Since the attainable size was a factor of time, some of the product from one flight was used to seed a subsequent run in order to build up to the larger sizes without requiring overly-long flights. The apparatus comprised four 30-centimetre-tall reactor chambers and a large feeder tank, and was fully computerised. An astronaut was required only to switch the apparatus on; it turned itself off. Latex microspheres had a multitude of engineering and biotechnology applications, and the Bureau of Standards found a ready market for them at the rate of $400 per 15 million spheres. One mundane but important use was to calibrate scientific instruments such as electron microscopes. On the medical front, they could be used in cancer research to measure the pores in tumours so that spheres encapsulating anti-cancer drugs and primed for controlled release could be made to embed themselves on the surface of a malignant growth to deliver a concentrated dose of drugs straight to the cells most in need, yet lower the total dose received by the body. Because there is a significant difference in size between the pores in the lining of the stomach and the intestines, and in the tumours that grow there, such cancers were particularly susceptible to this kind of directed dosage. This project was undertaken jointly by the Marshall Space Flight Center and Dr John Vanderhoff, who developed the polymerisation process at the Dow Chemical Company in the 1950s. The space-borne reactor was successful, and its product was useful, but turning out little rubber spheres was hardly likely to excite the public.

Polymers

Early on, the 3M Company signed a 10-year contract to fly a series of experiments on the Shuttle. In addition to biologically active agents, the company was interested in polymers. A polymer is an organic material composed of very large 'long-chain' molecules which are formed by the union of smaller molecules, or monomers, in the process of polymerisation. The polymerisation of ethylene, for example, gives rise to the long-chain polyethylene molecule commonly known by its commercial name of polythene. The main mid-deck experiment on STS-51A was the Diffuse Mixing of Organic Solutions, developed in-house by 3M to crystallise organic material. Each of its six stainless-steel reactors enclosed three teflon-coated chemical chambers, and materials in the chambers were maintained in a separated state until electrically controlled gates were released to allow them to mix. These gates were opened slowly over a five-hour period so that their action would not to disturb the contents. On the test flight, one reactor contained urea in solution, and it diffused into an incompatible solvent that crystallised the organic material. In two other reactors, the end chambers contained tetraethylammonium oxonol and cyanine tosylate in solution, which were mixed in the central chamber. The other three reactors had proprietary materials. On its second flight, on STS-61B, two of the reactors grew crystals to study molecular growth, two grew crystals to study how the packing-density affected electro-optical properties, and two used dyes of different densities to investigate the physics of the mixing process. On Earth, the developing crystals settled on the base of the chamber and were distorted. In space, the apparatus yielded much larger and more regular crystals that were better suited to X-ray crystallography.

Another of 3M's projects was the Physical Vapour Transport of Organic Solids experiment. Although packaged in the same way as its earlier apparatus for Shuttle integration, it consisted of nine independent vacuum-insulated ampoules mounted within the container, each of which had an organic solid at one end, a silicon wafer on which a special metal film had been deposited at the other, and a buffer gas between. The organic material was vaporised, transported by the gas to the other end of the ampoule, and then condensed as a thin film on the temperature-controlled substrate. The thermal insulation was sufficient to allow internal temperatures as high as 400 °C without posing a problem for the mid-deck environmental system. The crystalline structure of each thin film was studied to determine its chemical, optical and electrical properties. This experiment flew on STS-51I, but the follow-up had to wait until the Shuttle resumed operations in the wake of the loss of Challenger, and was one of the suite of materials-processing experiments carried by STS-26. In the follow-up Polymer Morphology experiment on STS-34 some 20 samples, including a variety of polymer blends, polyethylene, and a form of nylon, were investigated.

3M began a new investigation on the first International Microgravity Laboratory carried by STS-42 in 1992. As this Gelation of Sols (GOS) study was part of the ongoing Applied Microgravity Research (AMR) programme, the experiment was called GOSAMR. A colloidal system is a mixture of two substances, one of which – the colloid, also called the dispersed phase – is uniformly distributed in a finely divided state through the dispersion medium. In a sol the dispersion medium may be gas, liquid or solid, but in a gel the medium is semi-solid. A gel is therefore a thick colloidal suspension involving particles smaller than 1 micrometre in diameter and gelation is the creation of a gel from a sol. The objective of the GOSAMR experiment was to investigate the influence of microgravity on the processing of gelled sols. Stoke's law of fluid dynamics predicted that there would be less settling out of the larger particulates; in other words, less sedimentation. Specifically, the company wanted to know whether composite ceramic precursors composed of large particulates and small colloidal sols could be produced in space with more structural uniformity, and if so, whether this would in turn result in a finer matrix with correspondingly superior physical properties. The aim was to determine the benefit to be derived from more regular structures, and to provide some degree of confidence that efforts to improve terrestrial processing would indeed lead to enhanced products. Advanced ceramic composites would improve the company's wide range of abrasive and metal-polishing products, so this exploitation of the Shuttle was directly related to the company's core business. GOSAMR contained five modules, each with two banks of eight double-barrelled syringes containing the various sol and gelling agent combinations. The electrically driven syringes were activated by switches on the control panel. Each module was programmed to shut down after a specific time, and the astronauts were required only to activate each module in turn. The apparatus was designed to create a precursor for an advanced ceramic material by chemical gelation – the process that forms a gel by disrupting a sol's stability. Although the precursor would have to be baked at a temperature of up to 1,650 °C to make a ceramic, this part of the process was not done on the Shuttle, it was done later, and the properties of the result examined. Some 80 samples were processed with a wide range of particulate sizes, sol sizes and gelation times. Colloidal silica sols with diamond particles, and colloidal alumina sols with zirconia particulates were processed, reflecting the products most likely to yield short-term commercial return (diamond ceramic is used in metal polishing, and zirconia-toughened alumina is a premium abrasive).

Other companies were also interested in microgravity polymers. The Battelle Laboratory in Columbus, Ohio, had set up the Center for Advanced Materials as a CCDS to investigate commercial opportunities for polymers, catalysts, electronic materials and superconductors. Its programme was delayed by the grounding of the Shuttle, but STS-31 marked the debut of its investigation into Polymer Membrane Processing. Polymer membranes are widely used in separation applications such as desalination of water, atmospheric purification, medicinal purification and medical dialysis. The standard method of making a polymer membrane is to deposit a thin film of polymer solution on a cast, and then let the solvent evaporate to leave the membrane in the required configuration. The membrane's porosity is determined by its structure, which is in turn influenced by the evaporation rate. Gravity-induced convection impairs the evaporation process. The aim was to investigate evaporation casting in microgravity, where such convection is absent. The apparatus comprised a pair of containers, one much larger than the other, connected by a valve. The smaller container was filled with the polymer solution, and the larger one was reduced to a near-vacuum so that when the valve was opened the sudden drop in pressure would induce the solvent to flash-evaporate. Two units were carried in a mid-deck locker. The experiment worked, but because the delicate membranes were impaired by the stresses of the landing and the subsequent handling, the apparatus was modified to better protect its contents. This time, the final step in the process was to inject water vapour into the container to quench the evaporation process. This eliminated further evolution of the membrane (as seemed to have happened on the first run) and served to insulate it from subsequent stress. The original polymers were reflown so that the results could be directly compared. While the improved apparatus worked, it was another year before the project was resumed. It then made five flights within a year, during which the timing of the quenching following the flash evaporation was varied. In one test the evaporation step was omitted, and water injection was used to precipitate the membrane directly from the concentrated solution. There was then a further pause before another series of experiments was conducted. The motivation for this research was to better understand membrane morphology, and so determine whether it would be cost-effective to try to improve the terrestrial manufacturing process.

Electrophoresis

The holy grail of the search for a microgravity application was the miraculous drug that would cure a deadly disease, and if it could be made only in space then so much the better, but in reality such hopes were little better than wishful thinking. A factor in the development of a *new* pharmaceutical product in the USA is that it must pass a comprehensive trial before the Federal Drug Administration can approve it. Even if a miracle drug were to be discovered as a result of microgravity research, and even if it could be manufactured on Earth, it might take a decade for the product to reach the market. Nevertheless, the pharmaceutical industry offered the best opportunity for developing a viable commercial process. What was required was an agent for which there was an *established market*, for which a small-volume product could be sold at a premium rate, and for which the standard manufacturing process was disrupted by gravity. Electrophoresis fitted this criterion; indeed, it seemed to be an application that had been waiting for microgravity processing.

Electrophoresis is a technique for separating material suspended in a fluid by exploiting the net molecular electric charge. If an electric field is applied across the fluid, it causes the molecules to move with an acceleration proportional to charge, separating out by molecular weight. If a mixture of constituents in suspension flows along a tray across which an electric field is applied, the molecules will gradually separate, and will emerge from the other end at different points, where the purified material can be drawn off. The difficulty with using electrophoresis on Earth is that gravity disturbs the flow and tiny convection currents cause mixing, which degrades the separation process. As convection does not take place in microgravity, it was hoped to improve both the volume and the purity of a material by processing it in space. The pharmaceutical industry had been trying to use electrophoresis to purify biologically active agents, so Ortho Pharmaceuticals, a subsidiary of Johnson and Johnson, teamed up with McDonnell–Douglas, one of the main developers of space hardware, to fabricate the Continuous-Flow Electrophoresis System (CFES). NASA was strongly supportive, and the Bioprocessing Laboratory at the Johnson Space Center assisted in certifying the experiment for flight. Unlike many of the mid-deck experiments, this was not a simple box with an on–off switch; it was what NASA termed a 'tended package', and required significant attention. Its primary element was the 2-metre-long rectangular-section electrophoresis flow tube, or canal, which was transparent to allow that the separation process to be photographed. The entire package of the canal, the pump, the refrigeration and the control electronics occupied a full cabinet and weighed 250 kilograms.

Starting with STS-3, NASA made most of the early Shuttle flights available to test this experimental unit as rapidly as possible so that a production unit could be built. Initially, it was used to purify raw protein, the amino acid compounds that perform numerous critical rôles in biochemical processes, but STS-8 marked a milestone by processing live cells. The cells were flown up in an incubator, purified, and replaced in the incubator for return. This was hand-carried onto the orbiter by a technician immediately prior to launch, and retrieved immediately on landing. Three types of cell were processed: human kidney, rat pituitary and canine pancreas.

In this first basic demonstration, kidney cells were separated from blood. An endocrine gland secretes a hormone directly into the bloodstream in order to exert a specific effect on some other part of the body. A hormone is a chemical messenger – usually a protein – that turns an organic chemical reaction on or off. An enzyme – also a protein – drives the chemical processes of life in a catalytic manner, so a hormone has to trigger the production of only a tiny amount of enzyme to prompt a significant effect. The kidney produces erythropoietin, the hormone that regulates erythropoiesis, the process that makes erythrocytes (i.e. red blood cells). On return to Earth, the purified cells were cultured to extract the enzyme that dissolves blood clots. It was hoped eventually to derive a treatment for anæmia, the ailment in which the kidney fails to produce erythropoietin at the required rate. As red blood cells live for only about 120 days, the body has to manufacture these cells at a rate matching the rate at which they die. In an anæmic patient the production rate falls behind. The pituitary is an endocrine gland that secretes a polypeptide hormone to promote growth. The beta cells in the pancreas secrete insulin, the hormone that regulates carbohydrate metabolism. Diabetes is the ailment whereby the body fails to make insulin. In principle, a single injection of a large quantity of beta cells would cure diabetes, but these cells were difficult to make. Transplanting cells had proved effective, but the procedure was not yet an authorised treatment.

In the tests, the CFES processed 700 times more sample in space than it had in ground trials, and produced a five-fold improvement in purity, which was very encouraging. Several dozen other agents (including hormones to stimulate bone growth that could be used to treat osteoporosis, epidermal growth agents that could be used to treat burns, and interferon protein used by the immunological system to combat viral infection) held out the prospect of a commercial industry in space. On all these tests, career astronauts had dutifully loaded the solutions and drawn off the product, but it was clear that if the system was to be scaled up it would be necessary to fly a specialist to look after it – which is how Charles Walker became the first payload specialist to fly on a Shuttle mission. The CFES had been conceived by McDonnell–Douglas in 1976 and pursued as a long-term research and development item so that in the event that microgravity materials-processing should prove to be commercially viable, this giant of the aerospace industry would lead the field. When Walker joined the company in 1977 he was appointed as chief test engineer for the CFES project. NASA charged the company a fee of only $40,000 to have one of its employees accompany its experiment on a mission, and Walker was the obvious candidate. In the 12-month gap between the engineering test on STS-8 and Walker's flight on STS-41D in August 1984, the apparatus was upgraded to enable it to sustain 100 hours of continuous operation to process 27 litres of raw material in an effort to make sufficient erythropoietin for Ortho to conduct all the clinical trials required to secure a licence from the Federal Drug Administration. Walker's knowledge of the system proved to be crucial, as he was able to spot problems at an early stage and intervene before the trial deteriorated. In particular, when a computerised degassing pump malfunctioned Walker modified the operating procedure in-flight. Unfortunately, his maintenance proved to be in vain. Although the canal had been sterilised, when he deactivated the system to attend to the pump the rise in fluid temperature allowed a contaminant bacterium to grow. Even although this rendered the product useless for a clinical trial, Walker was content that the new hardware had operated successfully from an engineering viewpoint. On his reflight in April 1985, he returned with a clean sample. However, in September Ortho backed out of the project in order to pursue a revolutionary means of making erythropoietin which exploited a breakthrough in genetic engineering – a small Californian start-up company called Amgen had found how to splice genes into fast-growing bacteria in a process that was not sensitive to gravity.

Undeterred, McDonnell–Douglas sent Walker up again in November 1985, and initiated talks with the 3M Company's Riker Laboratory to use the erythropoietin that it intended to make using an industrial-scale apparatus. Walker was now in the enviable position of having flown on the Shuttle more often than most of NASA's career astronauts. The new Electrophoresis Operations in Space apparatus was to be carried on an MPESS in the bay, and its 4-metre-long canal was to run for 100 hours to yield 24 times the output of the trials unit. Although it would eventually have to function autonomously, the apparatus was to be supervised on its proving flight in July 1986 by Robert Wood, Walker's colleague who had overseen its development, but the loss of Challenger in January 1986 obliged McDonnell–Douglas to place the project on hold. By the time flights resumed in September 1988, Amgen's protein, called epogen, had become the biotechnology industry's most successful product.

Although microgravity electrophoresis had been rendered obsolete even before it could achieve commercial viability, it represented a logical development of a known technology, and would have worked. The gene-splicing breakthrough could not have been predicted, and if it had not occurred McDonnell–Douglas would have been the first company to devise a manufacturing-in-space application.

Phase partitioning

The Phase-Partitioning Experiment was devised by an academic consortium led by the University of British Columbia in Vancouver, Canada, and managed by the Marshall Space Flight Center in Huntsville. Phase partitioning, which is used in the pharmaceutical industry, is a method of separating different types of biological cell. It is commonly used to separate bone marrow cells for cancer treatment, and also to purify transplant cells. Two immiscible liquids are mixed with the cells, and as they separate – or, in the vernacular 'demix' – the cells attach themselves to one solution or the other. On Earth the liquid with the lighter molecular weight floats to the top, but in microgravity the two phases take on a structure that can be likened to an egg, with the yoke floating inside the white. Because gravity-induced convection degrades the process on Earth, the goal of the experiment was to observe the process in the absence of gravity to understand the rôle of gravity in the separation process. The apparatus was a simple hand-held tray with a matrix of sample chambers, each of which was seeded with a two-phase system and shaken to mix the phases. As the cells demixed, the chamber was repeatedly photographed for a densitometer analysis of the images after the flight which would derived the rate and extent of the demixing, with each sample testing a specific physical parameter, such as viscosity. The trial on STS-51D in 1985 used two polymers in saline solutions. After a hiatus while the Shuttle was grounded by the loss of Challenger, the experiment flew on STS-26, and then again on the first International Microgravity Laboratory, but in this case with an electric field applied across the demixing samples.

Protein crystals

As a pharmaceutical is designed to induce or inhibit the chemical properties of a specific biological agent, the first step is to determine its properties. The best way to determine the properties of an organic substance is to crystallise it and use X-ray crystallography to expose its three-dimensional structure. However, it is difficult to crystallise organic material on Earth, and a flawed crystal is not only difficult to analyse, it is also not necessarily representative of the material's true structure. The crystallisation problem can be likened to a crowd filling a cinema: as soon as the door is opened the people rush in and, even though everyone may have a reserved seat, a few will inevitably sit in the wrong seats, causing confusion and disrupting the planned layout. Although humans might insist on sorting themselves out, this is not the case with protein molecules, which often latch onto the incorrect point, creating a flawed crystal. In microgravity the whole process runs more slowly, and as there is no disruptive convection a truly representative crystal can be more readily formed. In a protein crystal, it is not individual atoms that take their assigned slots, as is the case in an inorganic crystal such as salt; entire protein molecules join to form the matrix. The goal was to grow crystals the size of a salt grain. It was not necessary to produce them in bulk, as just a few would be sufficient to yield an insight into the structure of one of the lesser understood proteins. The pioneering work in 1953 by James Watson and Francis Crick that had revealed the molecular structure of DNA had employed a limited amount of high-quality X-ray diffraction data – although it required a great deal of insight to 'read' this data. Even with computer assistance, it took Max Perutz many years to fully understand the structure of haemoglobin, the erythropoietic agent responsible for taking up and later releasing oxygen. Progress in understanding proteins is dependent on the ability to crystallise them for analysis. Three methods of crystallising protein have been devised: (1) vapour diffusion, (2) liquid–liquid diffusion and (3) dialysis. Crystallographers were eager to find out how these processes operated in microgravity.

The first experiment to grow a protein crystal was made on Spacelab 1 by a team at the University of Freiburg in Germany, and the 60-hour session yielded a beta-galactosidase enzyme crystal that was *27 times* larger than any previously available for study. Further experiments were conducted on Spacelab 3 and the first German-sponsored mission. The most amazing result was a lysozyme crystal *1,000 times* the size of any seen before. Unfortunately, this programme was then interrupted by the loss of Challenger.

The Center for Macromolecular Crystallography was set up on the Birmingham campus of the University of Alabama as one of the CCDS. It developed the Protein Crystal-Growth experiment to investigate the potential for crystallising hormones, enzymes and other proteins. The principal investigator was Charles Bugg, the director of the Center, and the hardware was developed by McDonnell–Douglas. In addition to supervising the CFES on his STS-51D flight, Charles Walker tested this new experiment. In contrast to the massive electrophoresis unit, the crystal-growth apparatus was a small hand-held package that comprised a dialysis unit and a pair of vapour diffusion trays. Dialysis is a process of selective diffusion across a membrane. The dialysis experiment was a small block of lexan in which there was a central cavity containing membrane 'buttons' and glass ampoules with a precipitating agent. Shaking the apparatus would break the ampoules and activate the buttons. As the protein emerged from the membrane, it would be crystallised by the precipitant. Unfortunately, the vibration during the Shuttle's launch triggered the start of the experiment, spoiling it. The vapour diffusion apparatus survived. Each of the 35- by 8-centimetre rectangular trays incorporated 24 chambers with porous liners saturated by precipitant to crystallise a drop of protein solution injected by an overlying tray of syringes. To test whether they would resume growth, some of the chambers were seeded with a microscopic crystallised protein. The most significant product was a crystal of lysozyme 1.5 millimetres across. The vapour diffusion package was reflown several times prior to the programme being pre-empted by the loss of Challenger.

The simple hand-held apparatus had demonstrated that large crystals could be grown, but because the temperature had not been regulated and thermodynamic fluid effects had impaired the crystals, it was decided to develop an isothermal unit. This apparatus occupied a mid-deck locker and had a thermal regulator for a 20-chamber vapour diffusion tray. On STS-26 it grew 10 different proteins in 60 tests. Although not all of these samples produced usable crystals – either because they did not grow very large in the time available, or because they were damaged by the stresses of the return to Earth and subsequent handling – some were of exceptional size and quality. The vapour diffusion apparatus was flown on STS-26, STS-29, STS-32 and STS-31 to create crystals for pharmaceutical, biotechnology and agrichemical investigations. The prevailing mood was optimistic, because in the 1980s the proteins in interferon, insulin and human growth hormone were developed into successful products, and it was widely expected that cures would be found for cancer and the new scourge: the AIDS virus. And commercialisation was assisted by the fact that the Federal Drug Administration could approve drugs derived from pure proteins more readily than it could new artificial agents.

The Protein Crystallisation Facility introduced on STS-37 used temperature to activate and control the growth of the crystals, with the advantage that it eliminated thermally induced convection. The refrigeration and incubation apparatus had been altered to enable it to be programmed to follow a predefined temperature profile. It reduced the temperature from 40 °C to 22 °C (i.e. room temperature) over a 4-day period in order to enhance the regularity of the crystals. With four large chambers, this apparatus could process much larger amounts of material, in batches, and the results were encouraging. The vapour diffusion unit was tested on STS-48, but the plugs on the syringes failed to retract, preventing the raw material from penetrating the growth chamber; nevertheless, small crystals did form on the plugs. On the first International Microgravity Laboratory it functioned properly. It was then improved to enable a smaller sample to produce a greater yield, to make it more cost-effective, and flown on STS-49. When it was suspected that fluid effects were degrading the process, Larry DeLucas, Charles Bugg's deputy, flew as a payload specialist on the first US Microgravity Laboratory Spacelab to *study* the process of crystallisation, to determine the rates at which different proteins grew. To expose the crystallisation to inspection without jeopardising the conditions for growth, the apparatus was placed in a glovebox supplied by the European Space Agency. This pressurised workspace had facilities for both photo-documentation and video downlink. Since this was the first 14-day mission, there was time for DeLucas to vary the operational parameters to take account of his observations. This was more akin to laboratory science. Three of the standard vapour diffusion units were operated, one of them using the seeding technique. About 300 samples were processed in all. Of the 34 protein samples that were processed, 60 per cent had been tested before on shorter missions. Some of the resulting crystals were the largest ever seen. "Crystals from space are of such high quality", noted DeLucas afterwards, "that you can see details, like hydrogen atoms, that you wouldn't see with Earth crystals."

By STS-66 in 1994, the vapour diffusion apparatus had been transformed into the Crystal Observation System in a double-locker Thermal Enclosure System, and the crystallisation rate was regulated by adjusting the water evaporation to vary the precipitant concentration. The experiment was monitored by a video camera, both to enable an astronaut to inspect it without removing the apparatus from its enclosure and to record it for detailed post-flight analysis. The microgravity environment was monitored by the Space Acceleration Measurement System. A detailed study of the crystallisation would help to refine the laboratory technique. In addition to proteins, the commercial sponsors' interest had expanded the test subjects to include a variety of other macromolecules, such as viruses. The Protein Crystallisation Apparatus for Microgravity that made its debut a few months later was undergoing trials for scaling up. While it occupied only a single-locker Thermal Enclosure System, it comprised six cylinders, each of which contained nine trays, each with seven sample wells, and could therefore process 378 samples. When it accompanied the Crystal Observation System on STS-67, it utilised the tried and tested vapour diffusion technique. In its various guises, the Center for Macromolecular Crystallography's apparatus flew on more missions than any other commercial project, and genuine progress was being made. Of the first 100 proteins that these experiments tested, 25 per cent yielded crystals that were substantially larger and more regular than had ever been grown on Earth. Although this might at first appear to be a rather low success rate, it would be misleading to draw such a negative conclusion because the rate on Earth was much worse. The objective was to make crystals for analysis, rather than to turn them out in bulk for commercial sale, and a few crystals were sufficient for substantive analysis. Some of the proteins grew too slowly to form large crystals in the limited time available on a Shuttle mission, and while the introduction of the EDO helped, what was required was a growth period of several months. With little prospect of NASA building its space station in the immediate future, US companies approached the Soviets to send protein crystallisation experiments up to Mir. However, the US Government's ban on the export of technology to the Eastern bloc meant that even this proprietary apparatus could be 'exported' only if the package was sealed. Luckily, it was possible to encase a simple protein package in a box that had only an on/off switch. Payload Systems Incorporated served as agent for several pharmaceutical companies which opted to remain anonymous, but Boeing, which had never before used the Shuttle for such experiments, flew its package openly. In one case, two enzymes in 112 samples were processed by vapour and liquid–liquid diffusion techniques, and after two or three months yielded samples of the slower-growing proteins.

Generic processors

BioServe Space Technologies was established at the University of Colorado in 1987 to oversee the development of a suite of experiment facilities for the Shuttle. One of its affiliates, Instrumentation Technology Associates (ITA) of Exton, Pennsylvania, created the Materials Dispersion Apparatus (MDA) for biotechnology experiments employing the fluid mixing technique. As this could process a block of 100 materials samples at one time, it was meant for experiments requiring a large amount of data rather than limited proof-of-concept studies. Although conceived as a high-capacity protein crystal-growth unit, it was suitable for any process involving liquid–liquid diffusion, and the fact that it could be set up to mix as many as four fluids in each chamber made it a very flexible tool. It could be used for growth of alloy crystals, membrane casting and cell growth. BioServe was ITA's first customer, and the result was an integrated four-MDA mid-deck package called, obviously, BIMDA. It flew twice in 1991 and undertook a wide range of experiments growing both organic and inorganic crystals, germinating seeds, and fixing live cells. A few tests were aborted by mechanical failures on each occasion, but most were successful. The MDA was later flown as the Commercial MDA–ITA Experiment (CMIX) for the Consortium for Materials Development in Space, a CCDS on the Huntsville campus of the University of Alabama. It made its debut in 1992 on STS-52, and its 31 experiments covered the growth of protein crystals, thin-film membranes, and live cells. It flew again in 1993, then twice in 1995. In each case, ITA traded 50 per cent of the MDA capacity to the CCDS in return for the flight opportunity, and leased the remaining capacity to the biotechnology industry. It was modified to accommodate customers requiring chambers with 100 times the fluid volume to process living cell cultures. These Bioprocessing Modules obviously processed fewer samples than the standard MDA block. The even larger chamber of the Liquid Mixing Apparatus variant of the apparatus provided greater flexibility in the processing of samples, including options for laminar or turbulent flow when the fluids were injected.

It is worth summarising some of the projects undertaken with the MDA minilab in order to illustrate its tremendous scope. On a single flight in 1995, the Consortium for Materials Development in Space and its affiliates tackled the process of ageing, multi-drug effects on cells, neuro-muscular development, gravity-sensing and calcium metabolism, production of plant cell products and the usual protein crystal growth experiments; and ITA's commercial customers worked on a treatment for breast cancer and microencapsulation of drugs. In the case of ageing, it had been noted that cell growth slowed in the absence of gravity. To put this another way: many of the effects of ageing may ultimately derive from living in a gravity field. The experiment undertaken in this case monitored the growth of human lymphocytes, the primary type of cell flowing in the lymph system that distributes fluid around the body. The work on the effect of drugs on cells exploited this slowing down of cell metabolism. To be effective, a drug must cross a cell membrane, but after a while a resistance to drugs develops. In another case of membrane research, it had been noted that microgravity slows development of nerve and muscle cells. The nerve–muscle system is based on interactions between these two types of cells, by the transmission of chemical messengers across their membranes. When the process fails, neuro-muscular disorders result. Studying the slowed metabolism of cells from frogs provided welcome insight into these vital functions. Calcium regulates cellular activities leading to growth, which is why milk, rich in calcium, is an important aspect of a child's diet. It was believed that gravity played an active part in this process. To better understand the rôle of gravity, an experiment was conducted in its absence. While the protein urokinase was successfully crystallised on the first two CMIX missions, it was found to be a slow-growing type. When CMIX-3 flew on an

EDO mission, it was hoped that it would yield a urokinase crystal large enough to enable its three-dimensional structure to be determined to facilitate the development of a drug that would inhibit urokinase, and thereby inhibit breast cancer metastasis. Another commercial experiment tested a microencapsulated anti-tumour drug on mouse cells, preparatory to a programme of clinical trials. It is also worth noting that all of this research was done as a secondary objective on a mission that was devoted to astronomical observation.

BioServe built the Commercial Generic Bioprocessing Apparatus (CGBA) for growing micro-organism cultures and plants in fluid-processing chambers. It could accommodate 500 individual samples for a wide selection of experiments, controlled its own temperature, and incorporated an isolation glovebox for materials handling. It first flew in June 1992 on the first US Microgravity Laboratory. It became another package that flew regularly, and conducted a variety of experiments for commercial customers.

Fluids, melts and inorganic crystals

The multidisciplinary Spacelab 1 mission provided the first significant opportunity to investigate fluid dynamics on the Shuttle. One of the facilities in the materials-science 'double rack' was Italy's Fluid Physics Module. The focus of the study was the marangoni effect. The advantage of processing materials in microgravity was that there was no gravity-induced sedimentation and no convection, but other forms of convection were possible and one was tested in this case. The marangoni effect was a manifestation of surface tension that induced fluid mixing. It could be studied on Earth, but only in microscopic samples in which other effects were minimised – in microgravity, however, its *macroscopic* effects could be observed. It was studied in this case primarily to determine whether it was likely to disrupt later microgravity materials-processing experiments. A wide range of studies were pursued on the first German-sponsored Spacelab, and the Japanese Spacelab used a high-speed camera to record it for later analysis and computer modelling, and the second German Spacelab used a holographic camera to facilitate three-dimensional computer modelling. It turned out that the marangoni effect could be exploited in materials processing. Marangoni mixing within a fluid is driven by a temperature gradient over the surface layer, and it is possible to control the convection by controlling this gradient. Such fine control is not possible on Earth because, as noted, the effect is almost completely masked by gravity-induced effects. On the second International Microgravity Laboratory it was shown that different compounds of molten indium–gallium–antimony semiconductor can be mixed more uniformly using marangoni convection than is otherwise possible.

Another great advantage offered by microgravity is the ability to 'levitate' a sample in a crucible, enabling it to be processed without coming into contact with its reaction chamber. On Earth, it is possible to use a gas jet to levitate a sample, but the gas flow also acts on the sample, distorting it. Only in microgravity can a sample remain stationary without a force acting on it. Several containerless crucibles were built to investigate the manipulation of materials in microgravity. Once the sample had been moved to a specific point by some active control system, it would stay in a state of free drift. Such a crucible would not only permit a material to be heated to a temperature above that which would melt a conventional crucible, but also facilitate a highly corrosive reaction without risk of it being contaminated by the crucible, and without fear of it breaching the walls of the container.

The Drop Physics Module tested on Spacelab 3 in 1985 utilised ultrasonics to manipulate its contents. It was accompanied by Taylor Wang, who had designed the apparatus at the Jet Propulsion Laboratory in California. In the event, his presence proved fortunate, because no sooner was it activated than it broke down. As soon as Wang opened the cabinet, he realised that one of the power systems had failed. Given his familiarity with the apparatus, it would be a simple matter to strip it down and rewire it to run off the reserve power system. Surely this was an excellent opportunity to demonstrate the merit in flying investigators as payload specialists? However, it took the support team two days to reach the conclusion that Wang's proposed repair would not jeopardise the other Spacelab apparatus. While he waited patiently, Wang assisted with other experiments as best he could. Finally given the go-ahead, he by-passed the failed power system and for the remainder of the flight worked overtime to make up the lost time. When the apparatus flew on the first US Microgravity Laboratory it was operated by Eugene Trinh, another member of the development team, to study issues of fluid physics which could not be addressed in an experimental manner on Earth.

The 3-Axis Acoustic Levitator, which was also built by JPL, flew on STS-61C as part of the second Materials Science Laboratory. When the Acoustic Levitation Furnace built by Japan was tested on STS-47 it suffered fluctuations that allowed its contents to come into contact with the wall of the container. Germany's Tempus furnace, tested on STS-55, employed a magnetic field to manipulate its contents and studied solidification of metallic melts free of container effects, with a consequently ultra-high purity. It used an electric current flowing through coils of copper tubes to form a spot of minimum magnetic field strength in which a sample could be held. By varying the field, the sample could be moved, thus enabling it to be precisely located for processing in a thermal gradient. This field configuration had the advantage that the sample was maintained in the weakest part of the field, thereby ensuring that the field minimally influenced the solidification process. The Electromagnetic Levitator supplied by the Massachusetts Institute of Technology and flown on STS-61C also used an electromagnetic field, and the objective was to investigate the effects of flow during solidification of a material that was melted by induction heating as a result of being held in a cusp-shaped field. The advantage of an electromagnetic container over an acoustic container was that the former allowed processing in a vacuum, whereas the latter clearly required the presence of a gaseous medium to effect its control.

On the first International Microgravity Laboratory, in 1992, the European Space Agency introduced its Critical Point Facility for the optical study of the behaviour of fluids at, and near, their critical points. Any pure fluid possesses a liquid–vapour critical point that is uniquely defined by a thermodynamic temperature, pressure and density state. In a state in which temperature, pressure or density exceeds a critical value, liquid and vapour are indistinguishable. At the critical point, the fluid is highly compressible and small volumes fluctuate back and forth between liquid and gaseous phases. On Earth, observations are hampered by the fact that it is difficult to avoid compressing the critical region, because as soon as the vapour begins to liquefy and form droplets, gravity pulls the drops down and the critical region literally collapses. The closer the region is to the critical temperature, the more the fluid is compressible and the thinner is the critical zone, with the result that, at some temperature close to critical, the zone shrinks too small for any known experimental probe to measure its thermodynamic properties. Because the critical zone is broader in microgravity, it is possible to make measurements closer to the critical point. Intriguingly, physically different systems act very similarly near their critical points. The density and the heat-and-mass-transport processes of sulphur hexafluoride (a gas that can be made in extremely pure form) were measured close to its critical point in the Critical Point Facility by interferometric and light-scattering techniques. The Critical Fluid Light Scattering Experiment performed on STS-62, was particularly successful in this respect. Robert Gammon, the principal investigator, operated the experiment by telescience and was astounded by the data, which represented "a dream" that he had "spent his career waiting to see". Using a high-precision thermostat to control the temperature of a sample of exceedingly pure xenon to within a millionth of a degree, the onset of the state of equilibrium between the phases could be studied in detail using lasers to provide very accurate measurements of its density and pressure as it approached its critical point.

Another basic physics study concerned how helium transformed from a fluid to a superfluid phase at 2.2 K. As this temperature is known as the lambda point, the experiment performed on STS-52 was called the Lambda Point Experiment. The transition is easier to observe in microgravity because the process is very sensitive to pressure, and gravity causes the pressure at the bottom of a sample to be slightly higher than at the top, thereby imposing a temperature gradient across the sample and causing the process to creep. Before launch, the sample was chilled below the lambda point and kept in a cryostat. In orbit, the temperature was controlled to within a billionth of a degree as it was slowly raised back through the transition to enable the fluid's heat capacity to be measured as it changed state. Helium is the only element to display this superfluidity property.

Cosmonauts on the Mir space station (and, indeed, on its Salyut precursors) had shown that semiconductor crystallised in space was of very high quality. Knowing this, NASA set out to study the process to find ways to improve the technique, and sponsored the Crystal Vapour Transport Experiment that flew on STS-52 in 1992. NASA viewed this experiment as a precursor to the type of industrial research that would be undertaken on its space station – that is, assisting terrestrial manufacturing by studying fundamental processes in the absence of gravity. This furnace heated an ampoule to 850 °C in order to evaporate cadmium telluride and dissociate it into its constituents in the gaseous phase, and a temperature gradient caused the gases to be transported along a tube for recrystallisation. As the objective was to investigate the crystallisation process rather than to exploit it, the apparatus had a window to allow an astronaut to observe the effects of varying the operating conditions. Bill Shepherd and Mike Baker alternated in supervising it. They had been shown the process on

Earth to enable them to recognise any difference in behaviour in microgravity. Their visual observations were to be reinforced with video and still pictures. When one of the two furnaces shut down early in the flight, and the window on the other fogged over sufficiently to obscure the video camera's view, they resorted to sketching and displaying their sketches to the camera to enable their colleagues on the ground to follow the progress of the crystal's growth.

The Automated Directional Solidification Furnace that was tested for Grumman Aerospace on STS-51G and flown on several later missions, was to investigate the possibility of producing lightweight magnetic composite materials in space. Alloy samples were first melted and then solidified for examination post-flight. It operated by progressively working its way along the length of a sample – hence directional processing. The apparatus had been adapted from one previously flown on sounding rockets, and whereas it had previously been restricted to five minutes of operation, it could now process materials more slowly over a period of hours. Mephisto, carried on STS-52, was a French furnace to study directional solidification of metals. This study involved *in-situ* observation of the process and focused on the so-called solid–liquid interface – the point in the solidification process at which the solid and liquid phases were in contact. Directional solidification was also exploited on the first US Microgravity Laboratory mission to evaluate the extent to which, on Earth, gravitationally induced fluid-flow during crystallisation disrupts the microstructure of mercury–zinc telluride, an alloy that is highly prized for its sensitivity to infrared radiation. The intention here was not to develop a precursor to an in-space manufacturing application, but to determine the extent to which the terrestrial manufacturing process is flawed, and hence set a benchmark for its improvement. The experiment used the Crystal Growth Furnace, which was the first furnace the Shuttle had flown that was able to operate at a temperature of 1,350 °C. This reconfigurable apparatus could be used for a variety of processes. In one case the vapour transport technique produced mercury–cadmium telluride, which is another infrared-sensitive semiconductor. The apparatus made no provision for observing the crystallisation process, however; the results were examined post-flight.

As part of a Canadian suite of experiments flown on STS-52, the Experiment in Liquid Diffusion supplied by Queen's University in Ontario measured the diffusion coefficients of a variety of metals in their liquid phase. The samples were heated in a furnace until molten, allowed to diffuse, then rapidly solidified so that their state of diffusion could be determined after the flight. The experiment was performed in space because, on Earth, gravity-induced convection causes mixing that masks the degree of diffusion. The objective was to provide high-quality data with which to calibrate a model for the diffusion process, so that coefficients could be predicted for a wide range of metals, knowledge of which would assist industrial manufacturing processes. The apparatus was later sent to Mir for a more comprehensive study.

The Center for Commercial Crystal Growth in Space – a CCDS established at Clarkston University at Potsdam, New York – teamed up with Battelle's Advanced Materials Center to develop the Zeolite Crystal Facility. After a test on STS-50, the apparatus flew on STS-73 together with its principal investigator, Albert Sacco of the Worcester Polytechnic Institute. Zeolite, which is a mixture of silica and alumina, has the very useful property that it has an open crystalline structure that makes it selectively porous, and because it can function as a molecular sieve it is employed in catalysts, filters, absorbents and ion-exchange systems. For example, it is used by the petrochemical industry in the 'cracking' process to increase the gasoline yield by filtering out large hydrocarbon molecules. Unfortunately, zeolite crystals made on Earth are small and irregular, complicating its use in absorption, separation and ion-exchange processes. It was hoped that production in microgravity would yield larger and more regular crystals, and this proved to be the case. The first stage in exploiting this fact was to examine these new crystals to determine their properties, and assess whether it would be possible to improve the terrestrial manufacturing process, but the wider objective was to assess the possibility of commercial production in space, and subsequent flights therefore set out to determine the best procedure for doing so. The commercial viability of making zeolite in space derived from the facts that there is a ready market for the product, that natural crystal is rare, and that synthesis on Earth is difficult. Furthermore, since chemical additives would not be required when making zeolite in space, the first company to establish such production will be able to charge a significant premium for 'clean' membranes. Zeolite may well become one of the materials produced in bulk on a free-flyer serviced by the ISS.

Combustion

With STS-41 in 1990, NASA began to study the process of combustion in space. Although the outbreak of fire is a perennial danger facing a spacecraft, the absence of convection in microgravity reduces this threat because a flame will tend to burn itself out without spreading. The Solid Surface Combustion Experiment devised by Robert Altenkirch of Mississippi State University was the first of a comprehensive series of combustion experiments. On Earth, flame is strongly influenced by gravity; heated gas loses density, and it rises above a flame by buoyant convection. On the one hand, the cold gas that replaces it serves to cool the flame, but this circulating air flow also provides fresh oxygen which not only sustains the flame but also keeps it hot. Combustion, therefore, represents a balance between these heating and cooling factors and the outcome is determined by the speed of the airflow and the scale of the fire – a match, for example, will be snuffed out by a gust, but a camp fire will be strengthened. The specific objective of the experiment was to study the spreading of flame over solid fuels in the absence of buoyant convection. As air motion would be eliminated, combustion on a localised scale would soon burn itself out from lack of oxygen. Combustion could only be sustained if the flame spread away from the hot efflux into fresh air. With airflow absent, the balance between heating and cooling effects was calibrated by adjusting the oxygen content of the air in the container in which the experiment was done. In the orbiter's cabin, the oxygen fraction was 21 per cent, as at the surface of the Earth, but for the combustion tests it was raised and varied between 35 and 50 per cent. The process was recorded by video cameras, and the temperatures of the air and of the fuel (an ashless filter paper) were measured for detailed analysis. There was a brilliant glow when the paper first lit, but this diminished as the spherical flame slowly travelled along the length of the paper. In later tests, small samples of plexiglass were burned, with the initial bright orange flash giving way to the steady blue flame. These materials were selected because a comprehensive database for their combustion already existed. The experiment flew on eight Shuttle missions, and was later sent to Mir for further studies on a wider range of materials.

The influence of buoyant convection can be calibrated using controlled gusts, or forced convection. Because the forced flow can only *augment* the ambient flow, on Earth it is not possible to extend this calibration to speeds lower than that of the buoyant flow rate. However, it would be feasible to measure the effects of such minimal forced flow in the absence of gravity-induced convection. This provided a remarkable opportunity to increase our understanding of this fundamental process. Achieving mastery of fire stands as one of the most significant achievements of our early ancestors. Combustion engineering is the key to our modern way of living. The fact that it was possible to discover anything *new* about the process of combustion served to demonstrate that our view of the world is conditioned by the many and varied effects of gravity. The Forced Flow Flame Test – a refinement of the Solid Surface Combustion Experiment – was to assess how forced flow affected solid-fuel combustion. Once the apparatus had been tested, it was sent to Mir for extensive trials. The objective of the Candle Flame Experiment supplied by the Lewis Research Center was to determine whether a candle flame could be sustained in the absence of buoyant convection. The flame would require to rely on diffusive mixing of the essentially stationary surrounding air. Flame propagation was tested by using candles stationed at different spacings, and lighting one. Surprisingly, even when the candles were in close proximity the flame did not jump across the gap. Whereas the Solid Surface Combustion Experiment was performed in a sealed apparatus in order to vary the oxygen supply, the Candle Flame Experiment was done in the ambient atmosphere, using the glovebox only for physical isolation. It, too, was later sent to Mir. The Smoldering Combustion Experiment from by the University of California assessed the smoldering characteristics of polyurethane foam, both with and without a forced flow, and with different geometries, to identify the conditions that gave rise to the smoldering-to-flame and smoldering-to-extinction transitions. Having studied direct flame propagation, NASA turned its attention to radiative effects, in the form of the Radiative Ignition and Transition to Spread Investigation. The International Symposium on Combustion is the principal forum for research in the field. In 1996 some 10 per cent of the papers reported microgravity work.

The Materials Sciences Laboratory, flown in 1997, concentrated on combustion. The crew included Donald Thomas, an astronaut whose speciality was materials sciences, Roger Crouch, the chief of NASA's Microgravity Science and Applications Division, and Gregory Linteris, a combustion engineer at the US National Institute of Standards and Technology. The programme involved burning a wide variety of materials. Individual droplets of heptane and ethanol were ignited in the glovebox. The Droplet Combustion Apparatus was used for a University of California study of the ignition properties of a number of hydrocarbon-based fuels. The Combustion Module supplied by the Lewis Research Center incorporated a gas chromatograph, and this was used in a Californian experiment to study the process by which a flame, instead of moving over a linear front, could 'break away' as a sphere of flame. This phenomenon was very difficult to study on Earth. As the mechanisms that affect flame stability and extinction in mixed gases are poorly understood, despite the fact that they are fundamental to applications involving internal combustion engines, this research held out the prospect of a significant commercial return. A University of Michigan experiment used a laser-based instrument to study soot concentrations, as radiation from soot influences the durability of combustion. The soot from burning propane and ethylene in the Combustion Module was saved for post-flight analysis. Lest it be thought that little can be learned from soot, recall that it was only recently that carbon-60 Fullerenes were discovered in soot.

On Earth, overloaded electrical wiring is a common source of fire. To varying degrees, different types of wire resist passing an electrical current, and transform this energy to heat. The construction industry uses codes of practice based on extensive trials relating current to temperature in different wire configurations. Lewis focused on the development of systems for spacecraft, and the objective of its Wire Insulation Flammability Experiment was to assess the outgassing, flammability and flame-spreading characteristics of wires in order to identify any effects characteristic of microgravity. This data would form the basis of a 'building code' for outfitting the laboratories on the ISS – these facilities are to be generic so that the experimental suite can be revised, and to reduce the risk of fire breaking out on the station it was vital to know whether the electrical properties of power supplies were different in microgravity.

LIMITATIONS OF MICROGRAVITY

What is microgravity? Although, in general parlance, it is used synonymously with the terms zero-gravity and weightlessness, strictly speaking the 'micro' prefix refers to a gravitational force of 10^{-6} of Earth standard.

The quality of the ideal microgravity environment offered by flying in space is degraded by the vehicle's manoeuvres, by the crew's activities and by vibration from other apparatus. Ironically, firing the RCS thrusters to hold the orbiter in a stable condition can disturb very sensitive processes such as crystallisation. This can be overcome either by leaving the vehicle in free-drift or by placing it in an inherently stable attitude. As free-drift is inherently unstable, it is not often used. However, if a non-symmetrical vehicle is oriented with its major axis vertical, then it can exploit differential gravity – which tugs very slightly more on the end nearer the Earth – to maintain its attitude without firing thrusters. For the most sensitive experiments, therefore the orbiter is placed in this gravity-gradient attitude with its tail aimed towards the ground. As the best location for a sensitive experiment is at the orbiter's centre of mass, a Spacelab far aft in the bay is an inherently stable platform.

To assist researchers in interpreting the results of their experiments, NASA developed devices to *measure* the state of the microgravity environment. The first, the Space Acceleration Measurement System introduced on the first Life Sciences Spacelab, was a very sensitive three-axis accelerometer that could monitor vibrations affecting a nearby experiment. The Quasi-Steady Acceleration Measurement System built by Germany was tested on the second International Microgravity Laboratory. Having characterised the vibrations most disruptive to experiments, the next task was to attempt to remedy the situation. The stationary bicycle and treadmill float on springs to damp out the vibration as the crew exercises. In addition, isolation mounts were developed to damp out vibrations, to enable extremely sensitive experiments to be undertaken in a 'noisy' microgravity environment. As NASA built up its research facility on Mir, it installed the Microgravity Isolation Mount, a Canadian-supplied device that used a magnetic field to isolate an experiment from vibrations in the range 0.01–100 Hz. To test its effectiveness, the Space Acceleration Measurement System simultaneously measured the ambient vibrations while the vibrations that the mount transmitted were detected by their disturbance of the surface of a liquid in its test chamber in the glovebox. The Active Rack Isolation System designed by Boeing for use on the ISS, employed a mechanical suspension system to damp out micro-accelerations. On the second International Microgravity Laboratory, Japan tested the Vibration Isolation Box Experiment System that it had developed to use in its ISS laboratory.

Those opposed to a human presence in space argue that having people on board a space station leads to 'pollution' of the microgravity environment. Clearly, this is an oversimplification, as developing an ideal 'weightless' platform is a much more complex task than simply banishing astronauts from the facility. Although most of the experiments flown on the Shuttle rely on some degree of astronaut input, at the rate at which telescience techniques are being developed, it will soon be possible for apparatus carried on a free-flyer to be operated remotely, either from the ground or from the ISS. The best microgravity environment that a Shuttle orbiter can provide is about 10^{-5} *g*. A free-flyer can improve on this by an order of magnitude, but if it is to run a number of experiments simultaneously then the vibrations from one piece of apparatus may well disturb another. In addition, a host of subtle effects pose long-term problems. For example, calibration tests performed on Mir established that the gravity gradient has a measurable effect on a sensitive experiment. Although this is a low amplitude force, it acts continuously and its effect accumulates. These tests also showed that the atmospheric drag that causes a satellite's orbit to decay produces a detectable effect. These forces represent a quite different form of pollution of the microgravity environment, and will be very difficult to eliminate.

THE WAKE SHIELD FACILITY

In contrast to experiments to grow inorganic crystals by exploiting the microgravity of the space environment, the Space Vacuum Epitaxy Center which was established at the University of Houston in 1985 as a CCDS set out to use the *vacuum* of space to build crystals by epitaxy – the process by which a crystal is grown on a substrate in such a way that its structure is parallel to the substrate's lattice. In this case, the aim was to grow a crystal *atom-by-atom*, using a spray of raw material. It was hoped that in the 'hard' vacuum of space the crystallisation process would be more controlled, and yield a crystal that was both larger and more regular than was feasible on Earth.

Air pressure is traditionally measured in terms of millimetres of mercury in a capillary tube – a unit that is also known as the torr. On this scale, the pressure at sea level is 760 torr. Although a Shuttle travels in orbit, it is actually in the rarefied ionosphere. The pressure an altitude of 350 kilometres is a mere 10^{-7} torr, but this is prohibitive for epitaxy. However, if the orbiter were to orient itself belly-forward, calculations suggested that the pressure in its immediate 'wake' would be 10^{-14} torr. As the best that a terrestrial factory can achieve is 10^{-11} torr, vacuum epitaxy in a spacecraft's wake offered an opportunity to grow perfect crystals. Unfortunately, the orbiter pollutes its immediate environment with outgassing and thruster efflux, and the epitaxy apparatus could not be built as a pallet to operate in the payload bay. It had been hoped that if the RMS held the apparatus out of the bay the environment would be sufficiently clean, but tests demonstrated that this was not the case. The apparatus would therefore have to be made into a free-flyer that would be deployed to operate away from the orbiter and then retrieved. This made the apparatus rather more complex than originally intended, because it would have to run autonomously, but the prospect of achieving a 'hard' vacuum was enticing. The result was the Wake Shield Facility (WSF).

Although it was hoped to be able to manufacture a variety of thin-film products, it was decided to make semiconductors first, because exceedingly fine chips could be sold at a premium. Gallium arsenide was selected for the trial because it is difficult to manufacture large impurity-free gallium arsenide wafers. Other candidates were zinc selenide, indium phosphate and indium–gallium–antimonide. These exotic materials offered an order of magnitude improvement in performance over traditional silicon, and held out the prospect of not only integrating digital and analogue logic in a single material, but also microwave and photonic interfaces. The hardware was developed by Space Industries Incorporated in Houston, which was one of the Space Vacuum Epitaxy Center's affiliates. The 'shield' took the form of a 3.7-metre stainless-steel disk carrying the propulsion and attitude-control systems on its leading face and the epitaxy apparatus on its rear. Previously, Shuttles had withdrawn from free-flyers, but the WSF was to fire a nitrogen gas jet to initiate the separation to preclude the efflux from the Shuttle's thrusters contaminating the surface of its apparatus. When it was 75 kilometres away from the orbiter, ovens would evaporate cells of gallium and arsenic and spray this onto the substrate of the wafer. Throughout this processing, the orbiter would employ gravity-gradient stability in order not to fire its thrusters, and conduct microgravity research.

One of the early investigators on the project was Ron Sega, and his inclusion on the STS-60 crew proved fortunate. The WSF was to spend two days growing six wafers of up to 6 micrometres thickness. Instrumentation provided by the Air Force to measure conditions in the wake during the trial would also determine the degree to which the efflux from Discovery's thrusters impinged on the epitaxy package as the Shuttle returned. The experiment started well. Using the RMS, Jan Davis lifted the 1.8-tonne WSF from its cradle and held it above the bay while it was checked. It was heavily computerised in order to run autonomously, but could also be monitored by telemetry and commanded. Unfortunately, a firm radio-link could not be maintained. For two days, Sega worked in the hope that the test would be able to proceed, then a horizon sensor failed and deployment became impossible. In the circumstances, the best that could be hoped for was to hold the WSF high above the bay to verify the functionality of the epitaxy apparatus. The wafers made in this less-than-ideal test proved to be of comparable quality to the best made employing the same technique in terrestrial vacuum chambers. This was an encouraging start. As the next flight was scheduled for STS-69 in early 1995, the company set about modifying the WSF with some urgency. Extensive testing revealed that the sensor fault was due to electrical interference from a nearby power cable, so the wiring was rearranged. The radio-link issue was interference between the telemetry transmitter on the WSF and the relay system mounted on its payload bay cradle. The WSF was soon ready for its reflight, but the STS-69 mission slipped six months due to the hiatus in the summer of 1995 caused by uncertainty over when to send STS-71 for its historic link-up with Mir, problems with the SRBs, and a problem with woodpeckers attempting to nest on STS-70. Despite the rewiring of the communications system, it required two hours of trouble-shooting before Jim Newman was able to release the WSF. A few minutes after the separation manoeuvre, the telemetry link fell silent, and Endeavour had to risk polluting the surface of the apparatus by firing its thrusters to rotate to give the relay unit a better line of sight. For the first 16 hours of processing everything went as planned, then the cooling system malfunctioned and the attitude controller started to overheat. When the disk pitched over about 12 degrees and degraded the wake, the wafer-building underway at that time was abandoned. After the 'magnetorque' attitude control system re-oriented the disk, Houston allowed the apparatus to cool for 12 hours before resuming operations. To give the WSF time to complete its programme, the retrieval was put back 24 hours. It was not long before the platform pitched over again, however, but this time it was left to perform a full 360-degree flip. When the orbiter drew up alongside, the WSF was flying edge-on. However, it was stable, and was readily retrieved. Although the WSF had turned out to be difficult to operate, it had successfully produced four semiconductor wafers, and even though the wake had never bettered 10^{-13} torr, this was still 100 times better than could be attained in a terrestrial vacuum chamber, and the product was the purest gallium arsenide ever manufactured. The communications system was completely redesigned and, in an effort to overcome the heat build-up that had prompted the attitude-control failures, the thermal control system was upgraded. These modifications increased the mass to 2,150 kilograms. The test in November 1996 began with an alarming near-collision. As the end-effector of the RMS disengaged the pin on the front face of the WSF, it imparted a slow roll. The separation thruster could not be fired until this had been cancelled by the attitude-control system. This, however, took time, during which differential gravity caused the relative geometry of the two vehicles to change, with the result that when the thruster finally fired, the big disk skimmed low over the roof of Columbia's cabin as it departed. This incident prompted a review of the dangers of having self-separating payloads. This time everything went to plan, and wafers of aluminium–gallium arsenide and indium–gallium were produced in addition to gallium arsenide.

NASA had allowed the WSF three flight opportunities to demonstrate its capabilities. In effect, it had taken this entire allocation to achieve the objective set for the first flight. It had always been recognised that further flights would be contingent upon securing commercial backing. The troubled trials had hindered the timescale for commercial development, but it was also clear that even if the WSF was successful and the semiconductor industry did buy into the project, there would be limited opportunity to fly the package because the Shuttle's manifest would be dominated by flights devoted to the assembly of the ISS. This was the problem with a space-based manufacturing application: the Shuttle simply could not support any application that placed a heavy load on the manifest. In a commercial form, the WSF would have solar panels for extended operations and a carousel of raw materials for several hundred wafers. However, the WSF is not really a Shuttle payload. In its operational form it will be a platform to be flown alongside a permanent facility, from which it can be deployed, and to which it can return for servicing, and only then will it be able to turn out product at its optimum rate. As the semiconductor industry is a market that can set the price for a product of exceptional quality, it has been estimated that if that premium were to be 10 per cent, then a yield of only 200 kilograms of wafers per annum would be sufficient to render the venture profitable.

ALL DUE CREDIT

Critics have dismissed microgravity research as 'ivory tower' science having little commercial value, but much of the work has been undertaken by academics in concert with industry to ensure that the results stand the best chance of being applied. It is true that the potential for commercialisation has not lived up to the initial hype, but hype is exactly what the word implies. Much of the blame for this early overoptimistic expectation must be borne by NASA, for the way in which it promoted first the Shuttle and then, in the early 1980s, the rôle of a space station as a laboratory in space. Even if NASA did not originate some of the wilder concepts, it did little to correct such false promises. Nevertheless, companies in a wide variety of industries signed up as affiliates of the CCDS. To appreciate why, it is essential to understand that the result does not have to be space-based manufacturing to make the investment worth while. It is true that the goal of the McDonnell–Douglas's CFES project was a space factory, but the objective of protein crystal research is to increase our understanding of biological systems to enable drugs to be designed and manufactured on Earth, and the polymerisation and thin-film membrane work set out to seek insights to improve terrestrial manufacturing. If there was no commercial benefit to be derived from growing proteins in microgravity, then the biotechnology companies would long ago have stopped sending up crystallisation packages on the Shuttle, and they would not have seized the opportunity to use first the Mir space station and more lately the ISS – indeed, the first commercial microgravity payload to be sent to the ISS was to grow proteins.

Unfortunately for NASA, there is no public awareness of whether a given drug owes its design to knowledge derived from analysis of a crystal that was grown in space. In a sense, however, this ignorance is a direct result of the Shuttle's *success* in serving industry. Whereas the explorers of earlier centuries sailed on voyages of discovery and came back loaded up with looted baubles and carcinogenic materials, the Shuttle returns knowledge.

The Wake Shield Facility.

The Earth in space as viewed by Apollo 17.

9

Studying Earth

THE ATMOSPHERE

There are occasions when an understanding of a complex system can only be attained when the system can be observed in its entirety. In 1957–1958, teams of observers were dispersed to participate in the International Geophysical Year study, but even this enormous endeavour was unable to reveal very much information concerning the Earth as an ecosystem. Although one of the benefits of satellites is that they can scrutinise their home planet, this is not simply a matter of looking down from a great height, it is the fact that surveying the entire globe on an ongoing basis enables the various processes at work to be investigated in fine detail.

The atmosphere is structured in layers – troposphere, stratosphere, mesosphere and thermosphere – each with distinctive physical and dynamic properties, and it is convenient to classify these layers in terms of their temperature profiles. The lowest layer is the troposphere. Depending on latitude, it extends to an altitude of between 10 and 15 kilometres. It is strongly convective, and encompasses what has become known as the 'weather system'. Because the temperature of the lower atmosphere decreases with increasing altitude and decreasing pressure, it was naturally expected that this trend would continue all the way up to the boundary with space but, in fact, at an altitude of approximately 15 kilometres this trend reverses. This level – known as the tropopause – marks the base of the stratosphere. At an altitude of 50 kilometres the temperature is similar to that at sea level. This level – the stratopause – marks base of the mesosphere. The temperature then decreases to reach a minimum at an altitude of 80 kilometres. This level – the mesopause – marks the base of the thermosphere, in which the temperature soars. The existence of the stratosphere was noted in the nineteenth century, during which instruments were flown on balloons. As there is little vertical mixing, this is a distinct layer. It is the site of strong radiative, dynamic and chemical processes and the jet-stream winds are very strong. In 1930 Sidney Chapman finally explained the mystery of stratospheric warming: it is the result of ozone absorbing the ultraviolet rays from the Sun. This heating is strongest near the stratopause – the altitude at which the ozone is most concentrated. If it were not for the ozone, this thermal structure would not occur, and the atmosphere would progressively cool up to the mesopause. (As the gas density decreases, radiative cooling becomes dominant.) In the thermosphere, the temperature rises again because the rarefied atmosphere is dominated by atomic oxygen, which is excited by sunlight. Oxygen therefore plays a crucial rôle in determining the temperature profile of the upper atmosphere. In the lower atmosphere, oxygen exists as undissociated diatomic molecules (O_2). The intense sunlight at high altitude dissociates the oxygen, but recombination of atomic oxygen with undissociated O_2 leads to the formation of tri-atomic molecules of ozone (O_3). However, as it is a dynamic process, the ozone is in an equilibrium of formation and dissociation. At even higher altitudes, the recombination rate decreases, leaving only the atomic oxygen. A surprising discovery made on an early Shuttle flight was that a spacecraft glows as a result of its interaction with atomic oxygen – a phenomenon that has become known as 'Shuttle glow'.

LIGHTNING

When two centres of electrical charge become connected through a discharge channel the result is a bolt of lightning. As the high-current discharge generates very high temperatures, the surrounding air expands in a shock wave, producing thunder. Various mechanisms have been proposed for the electrical charge generation within a thunderstorm, but the process is stimulated when rapid convection transports water droplets, ice particles or specks of dust in a severe updraft. While the most readily observable lightning discharges are from a cloud to the ground (literally 'grounding' the charge), they also occur between neighbouring clouds, and since these are visible from both above and below they can be studied from space. An early experiment for the Shuttle was the Night–day Optical Survey of Lightning project. This employed a hand-held 16-mm film camera with a zoom lens, a diffraction grating on the lens for spectral characteristics, and an optical sensor for timing the flashes. The operator's commentary was recorded on audio tape. Although the crew of STS-2 were alerted on four occasions by meteorologists, they managed to observe only one storm and to record only one discharge, which was visible on four successive frames. Although this was a poor return for the time invested, later flights were more successful: indeed, on STS-4, in one particularly severe storm complex over South America, the astronauts noted that sympathetic lightning events progressively transferred the locus of activity across a distance of 600 kilometres. Such large-scale coordination is not readily observed from the ground because the field of view of any particular observer is so restricted; in fact, this was later defined as 'mesoscale' lightning. The objective of this study was to find a correlation between the degree of lightning and the convective state of a storm, to enable a satellite to identify storms likely to give rise to tornadoes and thereby enable specific warnings to be issued to the areas in its path.

The Mesoscale Lightning Experiment, introduced on STS-26, used a payload bay video camera equipped with a low-light imager, but unlike the earlier project this did not require crew participation as the camera was remotely controlled by Houston. It was a synoptic programme in the sense that it recorded the weather on the groundtrack while the orbiter flew through the Earth's shadow, but because it was so straightforward this could be done on every flight, and became a long-term study. Given knowledge of the orbiter's position, it was possible to determine the size of the lightning storm and measure its intensity. When the orbiter passed near a storm that was within observational range of ground-based lightning-detection stations those astronauts who had completed their normal duties were invited to make direct observations.

The magnetosphere is a consequence of the strong magnetic field created in the Earth's core. It contains an electrically conducting plasma of charged particles. Its lower region, which lies within the mesosphere, is the ionosphere, and its upper boundary, which is buffeted by the solar wind, is the magnetopause. It was found that thunderstorms can not only discharge to the ground, but can also launch intense bolts upwards into the ionosphere. Several distinct forms of such lightning (dubbed 'sprites') have been noted.

SOLAR–TERRESTRIAL RELATIONSHIP

The thermosphere has no specific boundary, it simply extends into space with ever-decreasing density. At 250 kilometres, which is the lower edge of the Shuttle's operating zone, the pressure is one 10-billionth of that at sea level, but this is still 10 million times the pressure out in interplanetary space. The gas density at the base of the thermosphere imposes sufficient drag to pull down a satellite in a matter of hours, but Explorer 51, which was launched in 1973, was manoeuvrable. It spent most of its time above 160 kilometres, but its motor was fired periodically to lower its perigee to make a few passes down near the mesopause and then to boost back to safety. As it is impractical to sample the mesosphere by this means, and because the stratosphere is totally unreachable, remote-sensing techniques have to be used.

Early results from Explorer 55 in 1975 prompted concern that ozone was being depleted by transient events such as plumes of ash from volcanoes, and interest immediately focused on aerosols – a fine spray of tiny droplets that are suspended in the rarefied stratosphere. In 1979 an Applications Explorer satellite introduced the Stratospheric Aerosol and Gas Experiment, and for a decade this photometer observed the Sun at orbital sunrise and sunset. By monitoring the spectral absorption features due to the passage of sunlight through the atmosphere, it measured ozone and aerosol concentrations, and because the light's trajectory was horizontal, it was able to measure this concentration as a function of both altitude and location, facilitating three-dimensional mapping of the distribution of aerosols in the atmosphere. Furthermore, over time it was possible to monitor the dynamics of the ozone layer. The Applications Explorer flew in an orbit of 55 degree inclination. Another satellite, Nimbus 7, was launched in 1978 to extend such observations into the polar regions. In addition to an improved form of the Stratospheric Aerosol and Gas Experiment called the Total Ozone Mapping Spectrometer, this larger satellite had the Earth Radiation Budget and Solar and

Backscattered-UltraViolet instruments and the Stratospheric Aerosol Monitor. A third satellite, the Solar And Mesospheric Explorer, was launched in 1981 to study the processes controlling the equilibrium in the ozone cycle.

The British Antarctic Survey based at Halley Bay had been monitoring overhead ozone concentrations since the International Geophysical Year. In the mid-1970s, it noted the appearance of a progressively more significant ozone depletion during the southern summer. Each year, the ozone concentration fell rapidly during the early spring, remained low throughout the summer, then slowly recovered. Even though the concentration in 1982 fell by 20 per cent, the data was not published because the Total Ozone Mapping Spectrometer data did not show anything unusual. In 1984, the ozone concentration fell by 30 per cent. When another site confirmed its reality, the results were written up, and a paper appeared in *Nature* in May 1985. On re-examining its Total Ozone Mapping Spectrometer output, NASA discovered that a 'filter' in the analysis programme had been set to reject 'spurious' data and had therefore failed to report this unexpectedly pronounced variation in Antarctic ozone. Fortunately, the raw data dating back to October 1978 had been archived and, when reprocessed, not only confirmed the Halley Bay discovery but also revealed the true scale of the phenomenon. The STS-41G mission of October 1984 was therefore well timed. Sally Ride hoisted the 2.275-tonne Earth Radiation Budget Satellite out of the payload bay using the RMS. When one of the two 7-metre-long solar panels refused to deploy, she rotated the satellite to face the Sun to heat the stuck mechanism, and then shook it until the panel unfolded and locked into position. It was a three-axis stabilised platform that kept its instruments facing the ground-track, and had a steerable TDRS antenna to relay data. Over the next 10 hours it used its thrusters to climb to 600 kilometres. At an inclination of 57 degrees, this satellite covered the same latitude range as the still-operating Applications Explorer. In addition to an improved Stratospheric Aerosol and Gas Experiment, it carried Nimbus 7's Earth Radiation Budget instrument. Similar Earth Radiation Budget instruments flew on NOAA 9 in December 1984 and NOAA 10 in September 1986. Because these latter satellites were to fly in polar orbit, they had to be launched on Atlas rockets instead of the Shuttle. They all measured the vertical distribution of stratospheric aerosols, ozone, water vapour, nitrous oxides, and other pollutants with an altitude resolution of 1 kilometre. The NOAA satellites were operated by the National Oceanic and Atmospheric Administration. All of this data was processed by the National Center for Climate Control. For purposes of analysis, the Earth was divided into a matrix of 1,000-kilometre-square 'cells'. The orbits of the mid-inclination satellites precessed round the equator in such a way as to yield a monthly average for each cell in their latitude range. As the Earth rotated on its axis beneath the polar satellites they were able to monitor the entire atmosphere on a daily basis, and since their orbits passed over each cell at a slightly later time each day they yielded detailed diurnal studies every month. Between them, these satellites mapped local, zonal, regional and global variations, and over time they enabled seasonal variations to be monitored. Such long-term worldwide data was basic to achieving a real understanding of the solar–terrestrial relationship.

The Earth's atmosphere is essentially transparent to light at visible wavelengths, but it absorbs longer wavelengths and is opaque over most of the infrared spectrum.

Some of the visible range of sunlight is reflected back into space by a veil of dust, ice crystals, stratospheric clouds and aerosols in the upper atmosphere. Some of the amount that penetrates the troposphere is reflected, but the remainder is diffused – which is why the sky appears blue – and much of the energy that reaches the lower atmosphere is absorbed at wavelengths characteristic of the chemical composition of its gases. Light is absorbed by electrons jumping straight to high states of excitation within individual atoms, but it is not re-radiated as visible light because the electrons drop back to their ground state by way of a series of steps and therefore produce a veritable cascade of low-energy photons. In the infrared region of the spectrum, the absorption of energy by gaseous molecules causes the constituent atoms to vibrate and rotate, and because these molecules are stable they lock this energy in the lower atmosphere. Water vapour and carbon dioxide are particularly effective at soaking up thermal energy – in fact, the Earth's surface would be 35 degrees cooler if it were not for the 0.3 per cent of carbon dioxide in the atmosphere. Carbon dioxide is therefore referred to as a 'greenhouse gas'. Only a small fraction of this thermal energy is able to leak away to space by radiative cooling. The key measure of the Sun's influence on the Earth is the 'radiation budget' – the net energy reaching the surface – which is dependent on the variation in solar energy reaching the Earth, on veiling of the lower atmosphere, and on greenhouse gases.

The first step in calculating the Earth's radiation budget is therefore to determine the energy spectrum of insolation, and any cyclic and longer-term divergent trends. Our orbit around the Sun is not circular, but slightly eccentric. On average, the Earth is some 150 million kilometres from the Sun, but it is a few million kilometres closer during the northern winter than during the southern winter. This introduces an annular variation in insolation. It is possible that the magnitude of the eccentricity in the orbit varies over a long timescale, in which case there will be a long-term cyclic variability in insolation. The output from the Sun is not constant, as sunspots cause the energy emitted to vary over an 11-year cycle. In fact, over very long timescales the Sun may actually be an irregular 'variable star'. Once the insolation is known, the energy that is reflected back into space is subtracted, together with that which leaks away by radiative cooling. The residual energy serves to heat the lower atmosphere, drive the weather system and sustain the biosphere. In the tropics, the Earth absorbs much more energy from the Sun than it radiates into space. At higher latitudes less energy is received, but the leakage continues. At the poles during an extended period of darkness the flow is entirely one way. The variance in temperature in different locations drives the weather system in which atmospheric and oceanic circulations try in vain to redistribute this energy to establish a state of thermal equilibrium. The oceans serve as a store for this energy. If it were not for the oceans the Earth would have a very different climate. The Earth Radiation Budget instrument comprised two parts. One aimed four cavity-radiometers at the nadir. Two of these had sufficiently wide fields of view to cover the entire disk of the Earth, while the other two had narrower fields for higher resolution. One radiometer in each pair measured reflected insolation and the other measured leaking thermal energy. A fifth radiometer stared at the Sun to measure the broadband insolation. The second part of the instrument slewed a set of narrow-field sensors back and forth to each side of the ground-track in order to scan a wide swath at high spatial resolution. One of its three thermistor-bolometers sensed reflected insolation, one sensed leaking thermal radiation, and the other sensed across the entire range to measure the Earth's total output. Overall, the instrument could measure solar insolation, visible and ultraviolet reflected radiation, and infrared leakage. When correlated with the concentrations of stratospheric agents determined by the Stratospheric Aerosol and Gas Experiment, the fluctuations in the radiation budget provided the basis for understanding the rôle of the stratosphere in the solar–terrestrial relationship. Such satellites provided sufficient data to calibrate computer models of the weather and climate. It had been difficult to compute the precise effects of cloud and carbon dioxide as factors affecting warming and cooling. As insolation is reflected primarily by clouds, the extent and location of cloud cover is a crucial factor. Furthermore, the components of the system are interconnected: significant warming affects oceanic and atmospheric circulation, which in turn determines the extent, type and distribution of cloud cover, which serves as a cooling agent.

Between them the Total Ozone Mapping Spectrometer on Nimbus 7 and the Stratospheric Aerosol and Gas Experiment on the Earth Radiation Budget Satellite showed that in 1987 the ozone above Antarctica depleted to 50 per cent, which was sufficient to define this as a 'hole' in the stratosphere. This hole was less serious in 1988, but grew progressively worse. The ozone is not a static layer; it is in a state of dynamic equilibrium between the creative and destructive processes. The depletion is the result of fluctuations in the rates of these processes in which the destructive process is stimulated by the presence of chemically active aerosols. It was ironic that although the industrialised nations were concentrated in the northern hemisphere, their pollution was having an influence at the south pole, and when a slight depletion of northern ozone was first noted in the winter of 1988–1989, it was clearly a global problem. In March 1982, STS-3 had tested an instrument to measure the ultraviolet part of insolation. This was the Solar Ultraviolet Spectral Irradiance Monitor (SUSIM), which had been developed by the Naval Research Laboratory. Beginning with the flight of Spacelab 2 in July 1985, SUSIM was flown as often as possible to determine whether the energy in this part of the spectrum varied over the solar cycle. This was important because ultraviolet radiation plays a vital rôle in the ozone cycle. Artificial chlorine compounds form a major constituent of the upper-atmospheric aerosols. The ionising radiation dissociates the compounds and the liberated chlorine was thought to be the active agent in ozone depletion. STS-34 in October 1989 carried a GAS canister containing the Shuttle–Solar and Backscattered UltraViolet (SSBUV). This was an improved version of the instrument carried by Nimbus 7, NOAA 9 and NOAA 11 (the most recent in this series, added in 1988). This radiometer measured the amount of ultraviolet reflected by the stratosphere. Its frequent flights on the Shuttle were not so much to collect data – which was already being done by the satellites – but to calibrate the instruments on those satellites, and thereby guard against the false identification of long-term trends.[1] Combining data

[1] Space is a harsh environment, and exposure to solar ultraviolet and atomic oxygen escaping from the Earth's atmosphere can degrade sensors.

from the Earth Radiation Budget Satellite with that from the Nimbus, NOAA and SolarMax satellites revealed that the solar output varied by a few tenths of a percent on a two-week timescale – that is, during half of a solar rotation. Measuring the variability in its output over a full 22-year magnetic cycle was therefore a key objective. It has been calculated that if the Sun varied over a period of hundreds of thousands of years with a systematic change of 0.5 per cent per century, it could account for all of the Earth's past climatic changes. Isolating any such a long-term trends will be difficult, but nothing could have greater significance for the future of humankind. However, no scanning system operates indefinitely. NOAA 9's Earth Radiation Budget scanner failed in January 1987, NOAA 10's in May 1989, and the one on the Earth Radiation Budget Satellite itself followed in February 1990. In each case, however, the nadir-viewing instruments continued to operate. SolarMax fell back into the atmosphere in 1989, and the Solar and Mesospheric Explorer fell silent when its battery failed in 1989. No satellite had done more to monitor stratospheric ozone depletion than Nimbus 7, which lost both its Earth Radiation Budget and Total Ozone Mapping Spectrometer instruments in early 1993. Fortunately, a few weeks later NASA launched the first satellite of its new Earth Observing System.

UPPER ATMOSPHERE RESEARCH SATELLITE

At 7 tonnes, the Upper Atmosphere Research Satellite (UARS) was the heaviest package specifically for atmospheric research. For STS-48, Discovery climbed to its maximum operating altitude to place the satellite in the required 600-kilometre orbit, which, at 57 degrees, was as highly inclined as the Shuttle could economically attain. The suite of 10 instruments was to conduct the first systematic study of the upper atmosphere's physical and chemical processes. After a thorough three-day check of the instruments, Mark Brown hoisted the UARS out of the bay using the RMS and deployed the single large solar panel and the boom with the TDRS antenna. Based on the experience with the Compton Gamma-Ray Observatory, where the boom had jammed, Sam Gemar and Jim Buchli were standing by in the airlock, but there were no problems and the satellite was released as planned.

The basic spacecraft systems were carried on the same type of Fairchild bus that was first used by SolarMax. Indeed, the attitude-controller had been retrieved from SolarMax and refurbished for reuse. However, the massive instrument module on the triangular bus bore no resemblance to the compact SolarMax. The UARS employed a combination of boxes, cylinders, antennae and booms – projecting at all angles – to measure vertical temperature profiles and map the concentrations of ozone, methane, water vapour and aerosols. In order to compile a comprehensive database on the mechanisms controlling the structure and variability of the upper atmosphere, the instruments were also to chart the upper atmospheric winds that shape the global distribution of chemicals filtering up from the lower atmosphere. Four of these instruments were specifically devoted to the ozone issue, four were devoted to measuring solar energy, and two measured the winds. A large part of the value of this package was its ability to perform coordinated observations. The goal of the UARS programme was to monitor two northern winters. It was confidently expected that the satellite would operate for at least five years and, hopefully, until later satellites of the Earth Observing System could take over.

The Halogen Occultation Experiment supplied by the Langley Research Center observed the Sun's passage across the Earth's limb to measure the distribution of hydrofluoric acid, hydrochloric acid and assorted nitrous agents, as well as methane, water vapour and carbon dioxide, so as to determine their regional and vertical distribution. The Microwave Limb Sounder was a radiometer for detecting thermal emission from vibrating molecules. Its antenna was pointed at the limb in order to integrate energy from the thickest possible air sample, but unlike the occultation measurements at sunrise and sunset, the fact that it directly sensed molecules by their emissions rather than indirectly by their absorption of sunlight enabled it to sample continuously. Built by the Jet Propulsion Laboratory in California, it used a new semiconductor transducer to turn the millimetre-range incident energy into radio for amplification. In practice, it was able to sense chlorine monoxide, hydrogen peroxide and water vapour at altitudes as low as 16 kilometres, with a 3-kilometre vertical resolution. It gave the first global survey of chlorine monoxide – the most important stratospheric aerosol. The Cryogenic Limb Array Etalon Spectrometer was built by the National Center for Atmospheric Research. This infrared emission spectrometer also measured the distribution of nitrogen and chlorine compounds, ozone, methane and water vapour. The Improved Stratospheric And Mesospheric Sounder built by the University of Oxford in England measured carbon dioxide, carbon monoxide, nitrous oxide and nitric acid, as well as methane, ozone and water vapour. The overlap in detector capabilities facilitated cross-checking to ensure that there was no instrumental bias. Methane and nitrous oxide were measured because, as with carbon dioxide, they act as greenhouse agents. In addition to determining chemical composition, these infrared detectors were able to determine temperature profiles.

Water vapour is an excellent indicator of atmospheric motion. The Total Ozone Mapping Spectrometer instrument on Nimbus 7 had revealed rivers of water vapour hundreds of kilometres long flowing in the lower atmosphere – some with as much water as contained in the Amazon – and it was suspected that other atmospheric constituents were formed into similar flows. The UARS was to make a follow-up study by interferometric analysis of Doppler variations in spectral data. The High-Resolution Doppler Imager supplied by the University of Michigan monitored the 10- to 45-kilometre altitude range and the Wind Imaging Interferometer from York University in Canada monitored above 80 kilometres. The result was the first global survey of upper atmospheric wind patterns. The UARS found that continent-sized windstorms rage in the mesosphere. Two patterns were discernible: migrating diurnal tides sweep around the world once a day, remaining in sunlight, while non-migrating tides are fixed over the major landmasses, and wax and wane with the diurnal cycle. Although these winds are very rapid, they harness very little energy owing to the low density of the air at that altitude. The upper atmosphere, nevertheless, was found to be much more dynamic than expected.

As solar energy is the driving force behind the Earth's weather system, it was essential to monitor this in parallel with the atmospheric observations. The

University of Colorado built the Solar–Stellar Irradiance Comparison Experiment to test a technique for checking the calibration of instruments such as SUSIM – which measured the ultraviolet component of insolation in the study of ozone. Hot blue stars emit strongly in the ultraviolet, and a selection of such stars were measured to create a reference against which variations in insolation could be checked. One of the instruments developed by the Jet Propulsion Laboratory and tested on Spacelab 1 in 1983 was the Active Cavity Radiometer Irradiance Monitor (ACRIM). The UARS carried an improved version of this instrument to measure the Sun's total output. In addition, the X-ray spectrometer of the Particle Environment Monitor supplied by the Southwest Research Institute in Texas observed the direct effects of solar wind plasma impinging on the upper atmosphere in the polar zones.

The stabilised UARS flew with its single 10-metre solar panel out to one side to give its chemistry and wind sensors a clear view of the nadir and the limb on the far side. The solar instruments were on a scan platform on the front of the bus so as to track the Sun. Although the satellite was not in polar orbit, its sensors could survey up to 80 degrees latitude, and it was able to follow the Antarctic ozone hole. With each passing year, the dynamics of ozone depletion became better understood. In winter, when insolation is at a minimum, the air over the continent forms an isolated vortex that locks in the aerosols. They are then frozen into droplets in stratospheric clouds, and thus serve as platforms for slow chemical reactions. On the Sun's return in the spring, photochemical processes tip the ozone equilibrium in the direction of depletion. Chlorine monoxide has a particularly avid affinity for ozone, and its abundance has been increased by human activity. This potentially runaway reaction tails off with the dissipation of the clouds in the summer, at which time the altered atmospheric circulation enables ozone to flow in from the subpolar latitudes, and it is this that makes the hole a cyclic feature. The situation in the northern hemisphere is less favourable to the formation of a distinct hole, but depletion can take place if local conditions are conducive. The UARS proved conclusively that the artificial compounds in the stratosphere were causing the ozone hole, and that, as had been suspected, chlorofluorocarbon (CFC) agents were the real culprits. Each year, the Antarctic hole not only deepened but also developed earlier, lasted longer, and increased in area. In 1992, it was almost as large as the continent itself and extended over the tip of South America, possibly because it was enhanced by particulates injected into the stratosphere by the eruption of Mount Pinatubo in the Philippines in June 1991. The ultraviolet component of insolation in the wavelength range nearest the visible spectrum – the so-called UV-A – is totally absorbed by the atmosphere, and is therefore harmless to the biosphere at the surface. In parts of the stratosphere undergoing severe ozone depletion, ultraviolet in the range 2900–3200 Ångstrom (UV-B) passes through and is a threat to life on the surface because even a brief exposure can promote skin cancer. The shorter-wavelength ionising radiation (UV-C) would be even more dangerous if it were not absorbed by molecular oxygen in the lower atmosphere. The extreme-ultraviolet region (at wavelengths shorter than 1800 Ångstrom) is absorbed by thermospheric atomic oxygen. Consequently, while ozone forms only a tenuous layer (a few parts per million) in the stratosphere, it is of vital importance to life on the surface. The good news was that the concentration of certain stratospheric aerosols was falling as a result of the limits imposed by the Montreal Protocol on ozone-depleting chemicals. Methyl chloroform, in particular, had risen 4 per cent per annum since records began in 1978, but from 1990 it fell progressively by 2 per cent per annum. Although this 'reversal' was encouraging, it was only one of many agents. Even if the emission of all destructive agents is terminated, it may take a century for the reservoir in the lower atmosphere to deplete.

In July 1992 the UARS was nearly 'knocked out' by a spring which locked its solar panel, but repeated cycling of the motor eventually shook the obstruction free and operations resumed. However, the Improved Stratospheric And Mesospheric Sounder failed a few months later, the Microwave Limb Sounder failed in April 1993 and the Cryogenic Limb Array Etalon Spectrometer exhausted its coolant a few weeks after that. The spacecraft's battery had deteriorated, but it was sufficient for the reduced payload to continue to make a significant long-term contribution, and although its 'design life' was a mere 3 years, by 2002 it had yielded a comprehensive record of the solar–terrestrial relationship during an 11-year solar cycle, and in doing so had provided a basis for the follow-on study by the more advanced instruments of the Solar Radiation and Climate Experiment satellite of the Earth Observing System launched in January 2003.

ATMOSPHERIC LABORATORY FOR APPLICATIONS AND SCIENCE

The Atmospheric Laboratory for Applications and Science (ATLAS) was developed to provide detailed snapshots to supplement the longer-term studies by the UARS. It was initially known as the Spacelab Earth Observation Mission. Its first flight, set for September 1986 on flight 61K, was cancelled following the loss of Challenger. In its original form it would have employed a short pressurised module and a pallet, but when it was reincarnated as ATLAS the module was omitted. On the first mission a second pallet was added to accommodate many of the instruments that were to have flown in 1986. To allow a free-flyer to be carried, subsequent missions had only one pallet, and just a core package on instruments.

The main atmospheric monitor had three instruments:

- The Atmospheric Trace Molecule Spectrometer (evaluated on Spacelab 3 in 1985) measured infrared absorption features at orbital sunrise and sunset to chart the concentration of atmospheric trace constituents.
- The Millimetre Atmospheric Sounder was flown for the Max Planck Institut in Germany. Like the Microwave Limb Sounder on the UARS, it detected microwave emission from vibrating molecules, and could therefore gather limb data continuously. It measured temperature profiles and concentrations of ozone and chlorine monoxide.
- The SSBUV was to check the long-term calibration of similar instruments on satellites.

All of the instruments devoted to measuring the Sun's contribution to the Earth's radiation budget had been used previously. An ACRIM was flown to recalibrate the instruments carried by the UARS.[2] The Solar Constant (SOLCON) and Solar Spectrum (SOLSPEC) instruments, which had been tested on Spacelab 1 in 1983, respectively measured the total output of the Sun and its energy spectrum across the infrared to ultraviolet range.[3] And SUSIM accurately measured the ultraviolet component to follow its variability over this cycle, because this radiation conditions many of the photochemical processes in the atmosphere. All of these instruments had been finely calibrated to enable them to test the underlying atmospheric models. The detailed snapshots complemented the long-term studies by the UARS and various NOAA satellites.

ATLAS-1 flew three other instruments to study the atmosphere:

- Atmospheric Lyman-Alpha Emission, supplied by the French space agency, measured absorption in the extreme-ultraviolet to determine the relative abundances of hydrogen to deuterium, the heavy form of hydrogen, because this ratio would shed light on how water is processed by the atmosphere.
- The Belgian Grille Spectrometer measured infrared absorption when the Sun was on the limb to chart stratospheric water vapour, methane and nitrogen compounds.[4]
- The Imaging Spectrometric Observatory measured ultraviolet emission lines resulting from photochemical reactions, to provide information on the energy transfer processes in the upper atmosphere.

In addition, two instruments investigated cause-and-effect relationships linking the ionosphere to the upper atmosphere.

Although the eruption of Mount Pinatubo occurred several months prior to the launch of the UARS, the satellite had been able to monitor the initial phase of the spread of the volcano's stratospheric veil. By the time that ATLAS-1 flew in March 1992, the concentrated stream that had girdled the globe at the volcano's latitude had been spread by the stratospheric winds into a thin haze that almost enveloped both hemispheres. ATLAS-1 provided a comprehensive chemical survey of the effects of this unusually intense volcanic plume, but one of the most significant results of the mission was the detection by the Atmospheric Trace Molecule Spectrometer of significantly higher concentrations of chlorine and fluorine – the photochemical decay products of CFCs. The results could be directly compared with those of seven years previously: chlorine had increased by 25 per cent, and fluorine by 70 per cent. It was no coincidence, therefore, that six months later, at its maximum, the Antarctic ozone hole covered the entire continent. During the following northern winter, total ozone fell by 10 per cent, and the flight of ATLAS-2 in April 1993 was timed to study its recovery. ATLAS-3 was set for November 1994 in order to observe the onset of the northern winter as well as the start of the

[2] To illustrate the scale of this problem, the SUSIM instrument on the UARS lost 90 per cent of its sensitivity during its first three years of operation.

[3] The energy delivered in the different parts of the spectrum varies over the solar cycle.

[4] The Belgian Grille Spectrometer was repackaged as the Mir InfraRed Atmospheric Spectrometer (MIRAS) and sent up to the Mir space station in 1995 to undertake an extended study.

recovery of the Antarctic ozone hole. One specific objective was to study the southern vortex as it relaxed and allowed mixing between the ozone-depleted polar and the surrounding stratospheric circulation. The Millimetre Atmospheric Sounder found that there was a significant day-to-night variation in ozone concentration in the mesopause, with less ozone on the sunlit side of the Earth. It was in this coldest region of the atmosphere that the ozone was most vulnerable. Unfortunately on ATLAS-3, this instrument suffered a power spike at an early stage and never recovered. The SSBUV verified the fall in northern-hemispheric total ozone between the first two ATLAS flights. Although the Sun was at sunspot maximum in 1991, and was still very active in the years that followed – thereby complicating the task of measuring its total output – its disk was clear of spots throughout the second half of the ATLAS-2 mission, and ACRIM and SOLCON were able to make very accurate and sustained measurements. The EURECA free-flyer, which was deployed in July 1992 and spent a year in space, carried versions of SOLCON (SOVA) and SOLSPEC (SOSP), so these two sets of data facilitated mutual confirmation. To appreciate the solar–terrestrial relationship, it was necessary to accumulate data over a long period. The initial plan was to fly ATLAS annually, but this proved to be too great an operational load. After its third flight the programme was scaled back further, partly due to budgetary pressure, but mainly to create flight opportunities for Shuttle–Mir.

The ATLAS-3 mission was combined with the first flight of a new Shuttle Pallet Applications Satellite free-flyer: the Cryogenic Infrared Spectrometer and Telescope for the Atmosphere (CRISTA) built by the University of Wuppertal in Germany. This was the first space-borne instrument with sufficient time resolution to sample with high spatial resolution. A fast-acting spectrometer was produced by chilling the detector cryogenically. Moving at 8 kilometres per second, the detector was able to take a spectrum once per second, and scan the line-of-sight across the altitude range once per minute. As the satellite pursued its orbit, it was able to extend this vertical profile around the globe and thereby chart the distribution of gases in unprecedented detail during the week in which it was operated. It yielded the first three-dimensional global survey of 15 key atmospheric trace constituents with sufficient spatial and temporal resolution to identify the small-scale structures related to the stratospheric wind phenomena. This data was to update computer models of the processes that contribute to the Earth's radiation budget. The Naval Research Laboratory's Middle-Atmosphere High-Resolution Spectrograph Instrument flew as a secondary payload on the free-flyer, and observed the limb at sunrise and sunset to measure ultraviolet absorption. As the hydroxyl free-radical has a tendency to combine with nitrogen compounds to create nitric acid – which has a voracious appetite for ozone – it was possible to make the first global three-dimensional survey of stratospheric hydroxyl and nitric oxide. And because ozone depletion is increased in the presence of stratospheric ice, thermal profiles were made in order to correlate this chemistry with the temperature at high altitudes.

Another complementary package was the Measurement of Air Pollution from a Satellite built by the Langley Research Center to measure the global distribution of carbon monoxide. This infrared radiometer detected the concentration of the gas by its characteristic absorption of the background of thermal emission from the Earth.

As it had no need to scan the limb and could observe irrespective of the position of the Sun, it was one of the first instruments capable of collecting data continuously. It was flown on four Shuttle missions over a 12-year period (its first outing being on STS-2) and on each occasion produced a 'snapshot' of the state of the atmosphere.[5] Its data has been used to track the atmosphere's response to efforts to reduce industrial emissions.

SPACEBORNE LIDAR

The Langley Research Center developed the Lidar-In-space Technology Experiment (LITE) that was carried on STS-64 in September 1994. Although a lidar is standard equipment on aircraft engaged in meteorological research, this was the first time that one had been assessed in orbit. A lidar is similar in principle to a pencil-beam radar, but uses a laser rather than a microwave system. Although a conventional weather radar detects reflections from droplets of water or crystals of ice, it does not directly sense air moisture, but the LITE could sense molecular reflectance and thus measure water vapour concentrations. It complemented instruments such as the Stratospheric Aerosol and Gas Experiment. In effect, it produced a three-dimensional map of the structure of the atmosphere in the 10- to 40-kilometre altitude range, in which its thermal profiles were correlated with the location, movement and concentration of atmospheric constituents. From its vantage point in orbit, this instrument was able to reveal the large-scale structure of clouds and storms in unprecedented detail. To calibrate the instrument in this orbital trial, simultaneous measurements were made by meteorological stations and airborne lidars at specific sites. The LITE was mounted on a Spacelab pallet together with a 1-metre-diameter telescope to detect the reflected light.[6] By the time the laser beam reached the surface it had expanded to cover a circle almost 300 metres wide. Each sample entailed emitting a brief pulse and measuring its return in the narrow-field-of-view telescope. The fact that laser light is 'coherent' simplified the task of distinguishing the reflection, but a fast-acting solid-state detector was necessary nevertheless. As the laser and its bore-sighted telescope were fixed, staring straight up from the pallet, the orbiter was placed into a dizzying 2-degree-per-second roll to scan the laser. Although it weighed 2 tonnes, the LITE instrument was sufficiently lightweight to be converted into a free-flyer at a later date, should this be deemed appropriate, or even be repackaged to operate as an independent satellite.

The Department of Defense was also interested in the results as they offered the possibility of tracking aircraft and missiles by virtue of otherwise invisible vapour trails.

[5] This instrument was repackaged for installation on Mir in 1997.

[6] The LITE telescope's mirror was actually left over from the Orbiting Astronomical Observatory programme and had been in storage for 20 years.

TERRA FIRMA

The Earth is not a sphere; its rotation makes it an oblate spheroid, which bulges at the equator and is slightly flattened at the poles. Nor is it a perfect spheroid; the continents and oceans are not arranged symmetrically and there is a 20-kilometre difference in elevation between the peak of the highest mountain and the bottom of the deepest trench on the ocean floor. Using straightforward mechanical techniques, gravity surveys in the 19th century indicated that the thin crustal layer varies in thickness and is not homogeneous, and continental rock was found to be less dense than ocean floor rock. In effect, the continents represent the 'scum' that floated to the surface when the mantle was differentiated into layers by heat escaping from the molten core. The mantle is still in a state of semi-plastic flow, where hot plumes rise and cool material sinks. It is believed that these flows cause the continents to break apart, creating new oceans that open up as the large plates split and drift apart, only to collide with one another and coalesce at some other point, millions of years later. As the various plumes have different densities, inhomogeneity extends deep into the mantle, and these variations from a uniformly dense idealised sphere manifest themselves as irregularities in the planet's gravity field.

When the US Department of Defense realised in the early 1960s that this would complicate the task of aiming an intercontinental ballistic missile, it supported the geodesy project set up by NASA. In November 1965 the first US Geodetic Satellite, Explorer 29 (GEOS-1) was launched to develop a single frame of reference based on an accurate determination of the Earth's centre of mass, by which existing geodetic surveys could be integrated to enable positions to be located within 10 metres. (When the Air Force introduced the Atlas missile, its uncertainty in Moscow's location was 10 kilometres, so this was a real improvement.) Once in orbit, the satellite extended a 20-metre boom towards the Earth so as to stabilise its radio antenna. By transmitting a spiralling beam, the satellite enabled receivers below to track its motion precisely, from which it was possible to compute the variations of the gravity field. When the first satellite fell silent in 1967, Explorer 36 was dispatched to continue the survey. Although this radio-based system initially introduced a dramatic improvement, its capacity for further improvement was soon exhausted. A laser-based system offered great potential. The resulting LAser GEOS (LAGEOS) was launched in 1976. The surface of this 60-centimetre-diameter sphere incorporated a large number of tiny corner-cube reflectors to return a laser irrespective of the angle of incidence, and such a mirror could reflect beams from different sites simultaneously. Furthermore, in its 6,000-kilometre polar orbit this satellite could be seen from sites on different continents at the same time. At such an altitude, LAGEOS was barely affected by variations in the Earth's gravity field. However, it took several years to build up a sufficiently large network of laser sites to define its orbit with sufficient accuracy to transform the satellite into a predictable reference point in the sky, against which arbitrary points on the Earth could be located to within 3 centimetres. A long series of measurements using a surveyed grid enabled the drift-rate of continents to be computed with an accuracy of 5 mm per year, which was sufficient to provide confidence in the inferred motions of 2 centimetres per year. As a surprise result, it was also possible to measure vertical motions caused by tides in the crust in response to the Moon's gravitational pull, and even ongoing crustal relaxation in the aftermath of the retreat of the great northern ice sheets thousands of years ago. In general terms, the polar radius is 21 kilometres less than the equatorial radius, and the Earth is pear-shaped with a 40-metre 'dimple' at the south pole.

In 1984 NASA and the Italian Space Agency agreed a joint programme to launch a second satellite. Two points of reference would enable positions to be determined to within a few millimetres. Although originally assigned to a Shuttle mission in 1987, LAGEOS-2 was not finally rescheduled until October 1992. Released into an orbit inclined at 28 degrees, the Italian IRIS two-stage solid rocket raised the plane to 52 degrees as it climbed to 6,000 kilometres to deliver the 400-kilogram satellite. As its orbit will not decay for millions of years, Carl Sagan seized the opportunity to etch maps inside the satellite depicting the continents as they are believed to have been 268 million years ago, as they are today and, based on observed drift-rates, as they should appear in 8 million years' time. The first objective was to identify any unsuspected influences on its polar-orbiting predecessor. In the longer term, the aim was to characterise the extent to which the axial pole wanders, and measure the rate at which the Moon – by tidal action – is sapping the Earth's angular momentum, lengthening the Earth day and increasing the size its own orbit. Taken out of context, this tiny sphere was popularly criticised as being unworthy of a Shuttle mission, but while it was the only satellite deployed, it was actually the secondary payload.

To map the variations in density in the Earth's upper mantle and crust in fine detail, it is necessary to operate a geodetic satellite at low altitude. To probe the Earth's internal structure, Germany had its GFZ satellite deployed by the Mir space station in 1995. This 22-centimetre-diameter sphere was simply ejected from the airlock. At 480 kilometres, the 20-kilogram satellite was easily perturbed by the irregular gravity field, and required intensive tracking. However, it was soon brought down by air-drag. A few medium-altitude reflectors are also in use. In 1986 Japan's Experimental Geophysical Satellite (EGS) – a comparatively large 2-metre sphere – was placed at 1,500 kilometres at an inclination of 50 degrees, and in 1993 a small French reflector called Stella was put into a 750-kilometre polar orbit. In 1989 the Soviet Union placed Cosmos 1989 and Cosmos 2024 in high orbit at an inclination of 65 degrees, but these reflectors rode into orbit with GLONASS satellites – the equivalent of the US Department of Defense's NAVSTAR Global Positioning System (GPS) – and their function was to determine the extent to which the Moon's gravity disturbed the regular orbital spacing required to operate such a network. Meanwhile, it was discovered that the GPS satellites could be used to monitor vertical crustal movements. Long-term measurements by the University of Arizona using a network of receivers distributed across the Tucson Basin revealed that the floor of the basin was sinking, indicating that it was losing its capacity to hold a water table. Accurate vertical measurements were feasible with GPS in its coarse civilian mode of operation only because the immobile receivers were able to refine their positions by taking continuous 'fixes'. A net of receivers on the slopes of a volcano might be able to detect the distortions caused by rising magma prior to a major eruption. There was seemingly no end to the assistance that satellites could contribute to geophysical research.

MAPPING FROM ORBIT

The mixed bag of instruments tested on the November 1983 Spacelab 1 mission included a camera supplied by the European Space Agency to assess the potential of high-definition photography for mapping from orbit. It was mounted in the porthole in the roof of the pressurised module and was operated directly by Robert Parker, which was fortunate as he was able to free the film-feed mechanism when it jammed at an early stage. It produced excellent results, although a microwave radar intended to take complementary imagery malfunctioned and this part of the experiment was cancelled. The camera was assigned to the first Spacelab Earth Observation Mission, but when this was transformed into ATLAS the deletion of the module meant that the European Metric Camera could not be flown. NASA flew its own Large Format Camera (LFC) on two missions in 1984. While derived from a high-altitude metric stereographic mapping camera, it was bigger and more optically and electronically advanced than an aircraft-borne camera. This 400-kilogram package was carried on an MPESS frame in the bay and remotely operated from the flight deck. Its cassette had film for 2,400 exposures, and a pair of small cameras recorded orthogonal star fields to enable the location of each site to be determined – a technique that had been first used by photoreconnaissance satellites. Its first flight was primarily for calibration purposes. In addition to nadir imagery, it took oblique and limb pictures to evaluate its capabilities. Star fields were also photographed to prepare for the flight planned for March 1986 when it was to be used to document the passage of Halley's comet. The LFC's primary terrestrial application was to support the Landsat programme in the ongoing search for natural resources. From 260 kilometres, the 30-centimetre focal length f/6 lens was able to record a 220 by 440-kilometre field on a 22 by 44-centimetre negative. As its 20-metre resolution was four times better than Landsat, it was to assist in identifying features that showed interesting spectral characteristics. However, responsibility for running the Landsat system was transferred to a newly created Earth Observation Satellite Corporation (EOSAT), and the LFC's flight to observe Halley's comet was pre-empted by the loss of Challenger.

SHUTTLE RADAR LABORATORY

On its second test flight, in November 1981, the Shuttle carried a sophisticated scientific payload on behalf of NASA's Office of Space and Terrestrial Applications. This included the Spaceborne Imaging Radar (SIR). The Landsat satellites used visual and infrared imaging to chart the Earth's natural resources, but they required clear weather. A radar, on the other hand, can see through cloud, and is therefore capable of all-weather operation. The Jet Propulsion Laboratory had adapted an L-Band side-looking Synthetic-Aperture Radar (SAR) built for airborne reconnaissance and had flown it on the SeaSat satellite in 1978. At an altitude of 800 kilometres, its 25-metre resolution imagery was sufficient to map the state of the oceans. It was particularly valuable as it was able to monitor the sea beneath tropical storms, which would otherwise be visible only as an atmospheric structure. An S-Band radar was then under consideration to map Venus through its perpetual cloud cover. The SIR was an L/C-Band dual-frequency system. From the Shuttle's altitude of 225 kilometres the radar could provide a resolution of 10 metres under optimum conditions, which was sufficient for general terrain-mapping. The 2- by 9-metre planar array system was carried on a Spacelab pallet. The Shuttle flew upside down, and the radar viewed 47 degrees off the nadir to scan a swath 50 kilometres wide. In 1979 the Soviets tested a radar on Cosmos 1076 for later use on its Okean sea survey satellites. The US Department of Defense was meanwhile developing its much higher-resolution Lacrosse reconnaissance radar.

On its STS-2 test, the SIR concentrated on locations for which comprehensive Landsat data was available in order to enable the two techniques to be compared, and when over Africa data was taken to assess the radar's ability to see through dense vegetation. It proved possible to chart jungle terrain – something that had not been possible by other means. Surprisingly, the radar could also penetrate 10 metres into the extremely fine dry sand of the eastern Sahara, the Earth's most arid region, and revealed previously unsuspected ancient dry river channels in Egypt and the Sudan. Analysis of tracks in the desert, almost undetectable at surface level because of dunes, revealed sites of human activity dating back to the Stone Age. On the SIR's flight in October 1984 on STS-41G the archaeologists had set a real challenge. Could the radar locate the lost city of Ubar, the hub of the frankincense trade some 4,000 years ago which, according to lore, was somewhere in the Omani desert? Bertram Thomas, in following ancient desert tracks on his searches in the 1930s, had criss-crossed the Empty Quarter but found nothing. Some argued that the city had never existed. The SIR revealed new tracks, and an expedition set off to look for evidence. After a fruitless search, the team made a slight detour south of the most promising track, stopping at the village of Ash Shisr where an old dry well merited further study owing to its proximity to the new track. Excavation revealed a walled enclosure that would have encompassed a city in the remote past. Furthermore, a geological study revealed that the structure had been built over a limestone cave, and when this had begun to subside in 100 AD the residents had moved on. This contribution to archaeological research was a welcome bonus as the SIR's primary utility was to complement multispectral imaging for mapping natural resources geomorphology. A natural resource in short supply in a desert is water. The radar could detect a substructure that was likely to hold a shallow water table, and water was indeed found later by drilling at seven specific sites in Saudi Arabia. To a nation that relies on costly desalination for its fresh water, satellite prospecting was an attractive option. Among the morphological features on the target list were tectonic studies of the floor of the East African Rift Valley, coastal land forms in the Netherlands, and suspected meteor impact structures in Canada. In addition, the SIR charted wave patterns beneath Hurricane Josephine and the propagation of the so-called extreme waves of ocean currents.

For its second outing, the SIR radar had been upgraded. Its planar array was hinged to enable it to peer over the starboard sill at a variety of angles. Scanning the same swath on successive passes would facilitate stereoscopic mapping. However, data sampling was severely limited on this occasion by communications problems.

First, the steerable TDRS antenna malfunctioned and had to be locked in position, preventing data relay in real time. As only one TDRS satellite was available at that time, a tape unit had been installed to record SIR data while the Shuttle was out of contact. Instead, the data from each swath was stored on tape, and the orbiter was reoriented to aim its stuck antenna at the satellite for the download. In fact, the radar mission would have been interrupted in any case, as, at an altitude of 36,000 kilometres, geostationary satellites are more exposed to the solar wind than a Shuttle in low orbit, and when the TDRS detected a sudden blast of high-energy particles it 'hibernated' to protect itself and remained off-line for 13 hours. On this flight, the radar imaging was coordinated with conventional photography by the LFC to assist with feature identification. The only problem with the radar occurred at the very end of the mission. In its original form the planar array had been placed on a fixed frame that occupied the front half of the bay. To make the array more compact in order to allow the Earth Radiation Budget Satellite to be carried further aft, it unfolded across the space made available by the satellite's deployment. Although the panels folded back, the thermal insulation blanket had expanded to such an extent that the latches would no longer engage. Undeterred, Sally Ride used the end-effector of the RMS to push the top panel flat, and it finally locked into position.

When the SIR flew again a decade later, it was alongside the European X-SAR radar, and was termed the Shuttle Radar Laboratory (SRL). In this SIR-C form, the planar array of the Jet Propulsion Laboratory's radar was 12 metres long and 3 metres wide and, for the first time, steered its beam electronically and exploited both horizontal and vertical polarisation for improved spectral discrimination. Having increased to 10 tonnes, it was the biggest payload-bay instrument ever carried by the Shuttle. The X-SAR – so-called because its SAR operated in the X-Band – was a lightweight 250-kilogram instrument with a 40-centimetre-wide mechanically tilting array using a single polarisation. It was under evaluation for a future European Space Agency remote-sensing satellite. As the two radars used different frequencies, they could run simultaneously and, by scanning the same track, their data could be combined to make three-shade false-colour imagery to highlight a wide variety of detail. On this occasion, everything worked flawlessly, and the SIR took six times more data than it had during its earlier trial. In addition to mapping 20 per cent of the Earth's land surface, it observed several test sites that had been extensively surveyed by on-site teams to evaluate its ability to discern different types of detail. A battery of video and still cameras recorded these passes to assist the analysis. In order to identify short-term changes, the orbiter manoeuvred to sample sites several days apart. The crew split into shifts to facilitate 24-hour operations, and took 133 hours of radar data on 400 selected sites. To safeguard the radar mapping, this time a high-capacity tape system was carried, and all the data was stored in addition to being periodically downlinked. A follow-up flight was made a few months later to record seasonal changes. Tom Jones flew on both in order to provide continuity in procedure. As a one-time employee of the Central Intelligence Agency, Jones is believed to have been primarily involved in airborne radar-imaging reconnaissance.

With multiple polarisation, the SIR-C was not only able to identify a crop as wheat, it could also tell the type of wheat, its condition, and measure the moisture content of the soil. These tasks were previously feasible only by multispectral optical sensors. The X-SAR could determine the type and water-equivalent content of snow, and so provide advance warning of the thaw run-off. As before, the deserts proved remarkably interesting, but this time, because cameras shadowed the radar, direct comparison confirmed that the radar could see so much more. A chain of three impact craters, each 15 kilometres in diameter, was identified near Aorounga in northern Chad. As well as following up on previously noted ancient river-beds in the Sahara, the radar spotted an unsuspected dry channel at a sweeping bend in the River Nile. Eight hours after the Shuttle reached orbit on SRL-2, a volcano on the Kamchatka peninsula, Mount Kliuchevskoi, erupted and the radar was able to see through the massive ash cloud, which was spectacular in the optical imagery, to chart the progress of the mud flows beneath. Also, when a powerful earthquake occurred off the coast of Japan a few days later, the radar looked for tsunami waves, but saw nothing. On this flight the archaeological test was Angkar Wat in Cambodia, and although most of this vast temple complex is within the jungle, the radar was able to survey its true extent. The Chinese Academy of Sciences had requested the mapping of an ancient desert section of the Great Wall, long since buried by sand, and also a tracing of the Silk Road in the north-western part of the country in which ground-surveying is very difficult. On the last few days of the SRL-2 mission, the Shuttle used GPS to refine its orbit to fly within 100 metres of the ground-track of its first mission. The resulting topographic data had sufficient resolution to reveal surface movements between the two flights to an accuracy of a few centimetres. Although the SIR could detect the expansion of a volcano due to the rise of magma prior to an eruption, and could also, in principle, yield early warning, in practice this would be useful only if the radar data could be processed in real time. In fact, the SIR issued a prodigious stream of angle, phase and reflectance data that required sophisticated processing to produce an image. It took six months to reduce the data from each SRL flight to imagery, and a further nine months to refine this to highlight the specific details required by the 52 science teams involved in the demonstration.

When the SRL programme had been proposed in 1993, it had called for one flight each in 1994, 1995 and 1996. While the second flight had been brought forward, the third had been postponed indefinitely. NASA considered a third flight in 1997, but the extension of the Shuttle–Mir programme commandeered flights. Nevertheless, in July 1996 the Defense Mapping Agency (one of the many and varied organisations in the orbit of the Department of Defense) announced that it intended to work with NASA to adapt the SRL to map the Earth with a vertical accuracy 30 times greater than previous surveys. A three-dimensional study using data secured by having the second SRL overfly the ground-track as its predecessor was spectacular, but the coverage had been very limited. For this new Shuttle Radar Topography Mission, a pair of receivers were to be used to secure such data during a *single pass*. To achieve the baseline for an interferometric analysis, the second receiver was to be mounted on a 60-metre-long mast that would extend far out over the orbiter's left wing. When Endeavour flew this mission as STS-99 in February 2000, its 57-degree orbital inclination enabled it to map 80 per cent of the planet's land surface – documenting the areas in which 95 per cent of humanity lives. It was a spectacular finale to the Shuttle's study of the Earth.

The Cryogenic Infrared Spectrometer and Telescope for the Atmosphere (CRISTA) version of the SPAS free-flyer.

10

Shuttle–Mir

COMPETITORS

The early Soviet and American space programmes were overt manifestations of the Cold War. It was competition, and space was the arena. Astronauts played out the ancient ritual of single combat, in which the opposing forces lined up and dispatched a single representative to fight to the death in order to decide the entire issue. In its modern equivalent, the nation with the most impressive space feats would be shown to have the best technology and, as a result, surely, the best society – a society that the rest of the world would wish to emulate. Nothing less than the future of civilisation was at stake. America lost the race to put a man into orbit, but it raised the stakes and developed the technology required to land on the Moon so seemingly effortlessly that the Soviets, who had suffered problems, later denied that they had ever intended to do the same.

Having established its mettle, America relaxed, took stock of the cost of its achievement, and decided that its interests would be best served by a cost-effective transportation system. Meanwhile, the Soviets launched a succession of space stations. By the time construction of the Mir complex began in 1986, the two space programmes had so little in common that there was no scope for open competition. Each nation ignored the other and pursued its own objectives. The sudden collapse of the Soviet Union in 1991 transformed the situation overnight. With the Russian economy in free-fall, and its partners in the new Commonwealth of Independent States claiming ownership of the former Soviet Union's nuclear weapons, the USA decided that it needed to support Russia, and space became the arena in which to make a public display of cooperation.

PARTNERS

At a Washington summit on 17 June 1992, President George Bush and President Boris Yeltsin agreed to coordinate their efforts in space astronomy, astrophysics, solar–terrestrial physics, and Solar System exploration. They ordered their respective space agencies "to give consideration to" a joint venture involving their human spaceflight programmes. A month later in Moscow, NASA administrator Dan Goldin and Russian Space Agency director Yuri Koptev issued a follow-up memorandum of understanding that called for a cosmonaut to fly on a Shuttle in 1994 in return for an astronaut visiting the Mir space station in 1995. This *Human Spaceflight Cooperation* protocol was formalised on 5 October 1992. The climate of cooperation blossomed after the Clinton administration took office in January 1993, and in September of that year it was decided that America and Russia would merge their respective plans for a large new space station. In December 1993, as a step in this direction, the October 1992 protocol was formally extended by Vice President Al Gore and Prime Minister Viktor Chernomyrdin to include a series of visits by the Shuttle to Mir. In June 1994 this Shuttle–Mir programme was formalised. Starting in 1995, over a period of two years, a succession of astronauts would fly tours of duty on Mir. From the Russian viewpoint, NASA's agreement to pay $100 million per annum over the years 1994–1997 (a total of $400 million) was simply another fee to lease time on its orbital laboratory, but from NASA's viewpoint it marked a notable departure from procedure because it had previously cooperated with international partners on a no-exchange-of-funds basis.

In February 1994, Sergei Krikalev flew as a mission specialist on STS-60. It was ironic that "the last Soviet citizen", as Krikalev was described by the Western media, had more experience in space than all of his colleagues combined. As a precursor for the Shuttle–Mir dockings, it had been agreed that Krikalev's backup, Vladimir Titov, would fly STS-63 to rehearse the rendezvous phase of such a mission. This had been scheduled for June, but the lack of payloads for its Spacehab module had prompted a postponement, and it did not launch until February 1995, which was just a month before Norman Thagard was to lift off on a Soyuz rocket to fulfil the original June 1992 agreement. Meanwhile, in preparation for Shuttle–Mir, STS-51 Discovery had tested the Trajectory Control System (TCS) whose laser rangefinder was to augment the orbiter's inertial navigation system in the final phase of a rendezvous with Mir. For this manoeuvring trial the ORFEUS–SPAS satellite played the part of Mir. As Atlantis retrieved the CRISTA–SPAS satellite on STS-66, it tested the procedure that was later to be used to dock with Mir. It was standard practice for a Shuttle to approach its target from directly ahead, but for Mir it had been decided to approach from below. This approach had been used by STS-61 in December 1993 in closing in to retrieve the Hubble Space Telescope because it involved the orbiter firing only its downward-pointing thrusters to overcome the gravity gradient, and so minimised the contamination of the telescope with its efflux. Atlantis also tested the deployment of the recumbent seats to be used by Thagard and his Russian colleagues during their return to Earth. Immediately upon landing, Atlantis was returned to Rockwell for a refit, during which the equipment required to dock with Mir would be installed.

CLOSE APPROACH

The most stunning of the preliminaries, however, was STS-63's rendezvous, dubbed 'Near-Mir'. In fact, the rendezvous with Mir was a secondary objective, and was to be undertaken only if doing so would not jeopardise the deployment of a free-flyer. The determinant was propellant. If the launch missed its five-minute window, then correcting the discrepancy between the two orbital planes would consume so much propellant that it would not be possible to retrieve the free-flyer. Based on previous experience, there was a one-in-three chance of success, but Discovery set off on time on 3 February 1995. However, two of the RCS thrusters began to leak soon after reaching orbit. In light of concern that the nitrogen tetroxide might coat the apparatus mounted on Mir, the final go-ahead for the final phase of the approach was made contingent on the leaks being stemmed. The plan required Discovery to halt 10 metres from the androgynous docking port that Atlantis was to use, but if the leak could not be stopped then Discovery would not be permitted to approach closer than 125 metres. Fortunately, during the three-day chase the problem was overcome by the expedient of shutting down the leaking thrusters and letting them drain dry. On 6 February, Discovery approached from below and, with its nose high and its bay facing Mir, stationed itself directly in front of the complex. This was how a Shuttle normally approached a satellite, but not how Atlantis was to make its approach. The objective on this occasion was to evaluate the orbiter's station-keeping characteristics, and the line of approach was not crucial to that test. Mir was reoriented so that the axis of its Kristall module was aligned on the velocity vector, with its docking port facing the newcomer. On Discovery, Vladimir Titov was responsible for communications with the Mir cosmonauts, who had been talking by VHF radio since the Shuttle had established line-of-sight contact.

Discovery paused at 300 metres to await permission to proceed with the final phase of the approach. From this point, to avoid blasting Mir with their efflux, the upward-firing thrusters were not to be used.[1] Mir's solar panels had been 'feathered' edge-on to the Shuttle to further protect their sensitive transducers. With permission granted, Discovery began to reduce the separation. Any manoeuvring was to be done by the orbiter and Mir was to hold its orientation. (If Mir drifted, Discovery was to pause and wait for the complex to restabilise itself.) Jim Wetherbee flew from the aft station, viewing Kristall's port through the camera in the specially fitted window in the roof of the Spacehab at a spot corresponding to where a camera would be on the Russian-built docking system, and Eileen Collins monitored the laser rangefinder and kept up a commentary of the range and closing-rate. The television networks relayed the video downlinks from the two vehicles in split-screen fashion. The closing rate was so low that it required about 40 minutes – almost half an orbit, much of which was in darkness – to reach the 10-metre limit. The video showed members of the two crews waving to each other. After 10 minutes, Wetherbee announced that he hoped their successors would be able to shake hands, and then eased Discovery back to 125 metres to enable him to make a slow fly-around to photograph the complex using the large-format IMAX movie camera in the bay. Discovery spent about three hours within 125 metres of Mir. It had demonstrated that an orbiter could approach Kristall's axial port within an 8-degree cone, and a 2-degree tolerance in orientation,

[1] This was a flight mode referred to as 'low-Z' because only thrusters that were aimed obliquely 'up' could be used to brake.

and maintain its position at a point 10 metres distant. In addition, it had exercised communications and coordination procedures between the two crews, and between Houston and the control centre at Kaliningrad near Moscow. "When all was said and done," Wetherbee reflected, "it turned out to be easy." This demonstration had cleared the way for Atlantis to attempt to close the final 10 metres.

FIRST TOUR

On 14 March 1995, Norman Thagard became the first NASA astronaut to ride a Russian rocket. His colleagues on board Soyuz-TM 21 were spacecraft commander Vladimir Dezhurov and flight engineer Gennadi Strekalov. This spacecraft, in its original form, had been introduced in the mid-1960s. On its test flight, in April 1967, a string of problems that culminated in a twisted parachute had led to the death of its pilot, but since then it had been turned into the space station programme's workhorse. The rocket on which it was launched had an even better pedigree. It was the direct descendant of the Semyorka – the world's first intercontinental-range ballistic missile developed by Sergei Korolev in the mid-1950s – and had launched Sputnik in 1957, Yuri Gagarin in 1961, and every cosmonaut since. The 7.5-tonne Soyuz spacecraft comprised three modules: the cylindrical service module at the rear contained the engines, the spheroidal orbital module on the front served as the living quarters and cargo hauler, and the descent module was sandwiched in between. At just 2.2 metres in diameter and 2 metres tall, the bell-shaped descent module was rather cramped for a crew of three pressure-suited cosmonauts. After the standard two-day rendezvous, the ferry automatically docked with the orbital complex, and the newcomers were given a traditional bread-and-salt welcome by the resident crew of Alexander Viktorenko, Yelena Kondakova and Valeri Poliakov.

As the world's first modular space station, Mir was a physical incarnation of NASA's aspirations. Its base block had been launched in early 1986, just weeks after the Shuttle had been grounded by the loss of Challenger. Derived from the 20-tonne Salyut, the first of which had been launched in 1971, the Mir base block was the first spacecraft specifically designed to be permanently occupied, and was fitted out as a habitat. Like its predecessors, it had an axial docking port at each end, but it also had a ring of four radial ports at the front. In 1987 it had been augmented at the rear by an 8-tonne astrophysical laboratory and at the front by two 20-tonne modules – one in 1989 with additional environmental systems and the other in 1990 with a set of furnaces. The assembly of the complex had taken longer than intended, and the final two modules had been indefinitely grounded, but as the only orbital facility capable of sustaining human life, Mir had found its niche as an international laboratory for microgravity research. NASA had yearned for a permanent space station since its inception. It had mounted three expeditions on Skylab, one of which had lasted three months, and had hoped to use the Shuttle to refurbish that station, but Skylab had plunged back to Earth before the Shuttle was ready. With its Shuttle restricted to a fortnight in space, NASA had seized upon the opportunity to send an astronaut for a tour on Mir.

As a physician, Thagard's primary research involved gathering biomedical data on how he adapted to the space environment. On his four Shuttle missions he had never had time to adapt to weightlessness. His data was to extend that from the Space Life Sciences missions which had, of necessity, focused on the initial phase of the process. By taking samples of blood, urine and saliva on a regular basis, and by a log of food and fluid intake, the post-flight analysis would be able to track changes in body fluid and blood chemistry in fine detail. He was also to perform tests to measure cardiovascular, regulatory, metabolic, musculatory, bone and psychological effects. Although the planned three-month tour was insignificant in comparison to the 438-day marathon that Poliakov, a Russian physician, brought to a close on his departure on 22 March, it would equal, or even exceed, the longest tour of duty on Skylab, which was NASA's only previous experience of long-duration adaptation. Previously, fee-paying international researchers had been limited to the period of a crew exchange, although in several cases this had facilitated visits of up to a month. Originally, Thagard was to have visited during an extended handover, but when it was decided to expand the cooperative programme and retrieve him when Atlantis made its first visit, his rôle had been upgraded to membership of the new resident crew. This status was symbolised by the 'Stars And Stripes' being mounted on the bulkhead alongside the Russian national flag.

The decision to integrate the national space programmes had finally released the funding required to complete the assembly of the complex. The last two modules – incomplete and in storage – were redesigned. Spektr was to be launched first, closely followed by Priroda. To overcome the fact that Mir was forever short of power, it had been decided to remove some of the remote-sensing equipment from Spektr to enable a second pair of solar panels to be fitted. Also, to enable NASA to undertake a serious programme of research, some 800 kilograms of apparatus was to be ferried up as internal cargo. Priroda was to have a substantial amount of NASA apparatus built into it. It had been hoped to fully commission Spektr prior to Thagard's arrival, but a variety of minor issues had conspired to delay its final preparation. As soon as it had become clear that the module would not be launched on time, a Progress ferry – one of the Soyuz-based spacecraft that routinely resupplied Mir – had delivered sufficient apparatus to enable Thagard to start his work. As he settled down to life on Mir, he expected the remainder of his equipment to arrive in early May, but this soon slipped to *late* May, which was a bad omen. Life on Mir was very different to that on a Shuttle. Whereas NASA mission planners broke each crew member's day into five-minute frames, arranged tasks to make the most productive use of the limited time in orbit, and then monitored progress on a continuous basis by way of the TDRS geostationary relay satellite network, the routine on Mir was considerably more relaxed. Although Mir's crew worked through a prioritised list of tasks, they worked independently and defined their own pace because the complex was typically in communication for no more than 20 minutes even on a favourable pass, and was often out of contact for up to nine hours. Furthermore, as experience had shown that it was impractical to sustain the level of activity imposed on a Shuttle crew, Mir crews worked a five-day week. Of course, a space station was not an environment conducive to the pursuance of hobbies, and most cosmonauts spent much of their spare time working on their experiments, but the fact that they had chosen to do so voluntarily made it seem less like work. Of the 200 kilograms of NASA apparatus that had been flown up by Progress, the most significant item was the Mir Interface to Payloads (MIPS) – a laptop to control NASA apparatus and send and receive data via Mir's telemetry link. Its burst mode enabled Thagard to keep in touch with his support team by e-mail, and to make optimum use of the periods during which Mir was in communication, but he could also use Mir's short-wave to relay messages via the ham radio network. Although this radio also enabled him to keep up with US news, Thagard increasingly suffered cultural isolation. When the next ship docked on 11 April, it delivered additional supplies for Thagard's programme and 48 fertilised quail eggs which he was to place in an incubator, allow to develop for various times, then freeze for post-flight analysis of the process of embryonic development.

On 12 May, Dezhurov and Strekalov made the first of a series of spacewalks to prepare the complex for the arrival of Atlantis. One of the solar panels on Kristall was to be transferred to Kvant 1, the small module at the rear of the complex. After laying cables to the motor that an earlier crew had set up they retracted the panel, and on 17 May they dismounted the 500-kilogram unit and used a Strela crane to swing it to the rear of the complex. As it took rather longer than expected to mount it on the new motor, they had no time to plug it in. On Mir, however, tasks that could not be completed could be left to another day. The cosmonauts returned on 22 May, plugged in the new motor and extended its concertina array. They then set off to retract Kristall's other panel to allow the module to be moved off the lower port to clear the way for Spektr, which had been launched on 20 May and was already on its way. Kristall was first swung up onto the axis, then over to the righthand port. After docking on the axis on 1 June, Spektr was transferred to the lower port. On 10 June, the last act in reconfiguring the complex was to return Kristall to the axial port so that, as Atlantis closed on the port at the module's far end, the orbiter would be well clear of the remainder of the complex. The irony of the androgynous docking port was that although it was about to be used in its intended rôle, the nationality of the visitor was not as had been hoped.

ATLANTIS ... NOW ARRIVING!

After several postponements due to weather, STS-71 was launched on 27 June 1995 and was flown by Robert 'Hoot' Gibson and Charles Precourt. As well as mission specialists Greg Harbaugh, Ellen Baker and Bonnie Dunbar, it was carrying Anatoli Solovyov and Nikolai Budarin – Mir's next crew – as passengers.

The Orbiter Docking System (ODS) was mounted near the front of the payload bay. Set in a twin-triangular truss, it was basically two interconnected tubes, one leading from the mid-deck hatch to the tunnel running to the Spacelab mounted further aft, the other running upwards through an airlock to the androgynous docking system. Built to enable the Soviet Shuttle Buran to dock with Mir, this system had been bought from Energiya. It had been hoped to take receipt of it in July 1994, but as permission to import the pyrotechnically fitted hardware had been delayed, it had not arrived until September and had immediately been forwarded to Rockwell to be mated with the ODS. The TCS rangefinder was on the sidewall of the bay. Of the four Shuttles, Atlantis had been selected for these Mir missions because, at the end of 1993 when the decision had been taken to undertake a series of Mir dockings, the orbiter had been in refit at the Rockwell plant in California and the modifications required to accommodate the ODS had been added to that schedule. The ODS was to become a permanent fixture in Atlantis's bay, and the orbiter was to be dedicated to the Shuttle–Mir programme. Some disgruntled scientists pointed out that this display of détente had commandeered many flight opportunities that would otherwise have serviced a variety of Spacelab missions, and that this political junket had been bought at the expense of the scientific programme. However, for this first docking, Atlantis had a Spacelab fitted as a life-sciences laboratory, and Baker and Dunbar were to study the state of adaptation to weightlessness of the Mir crew. In addition, this criticism failed to acknowledge the long-term potential for research from maintaining a succession of astronauts on Mir.

Atlantis rendezvoused with the Mir complex on 29 June. It approached from below, in the same manner as STS-63, but rather than pass by and then ascend to dock by drawing back along Mir's velocity vector (the V-bar approach), it ascended the radius vector (R-bar) that came straight up from the centre of the Earth. When Discovery had made its V-bar approach, the low-Z thrusters had consumed much more propellant than if the high-Z thrusters had been used because the low-angled thrusters directed only a small fraction of their impulse to halt the motion. On the precursor flight, this penalty in propellant had been traded against the familiarity of the line of approach. By using the R-bar approach, Atlantis would genuinely save propellant. It relied on the gravity gradient of one vehicle located below the other to brake its very slow climb during the final phase without firing its upward-directed thrusters at all. In fact, to overcome natural forces of orbital dynamics, Atlantis was required to fire its downward-directed thrusters intermittently to maintain its closure rate. The real attraction of this approach for NASA was that it was fail-safe: if the Shuttle was to become disabled, gravity would draw it away, thereby precluding the possibility of a collision. Mir was reoriented to aim its androgynous port towards the Shuttle. Dunbar, who had backed-up Thagard and was fluent in Russian, handled VHF communications with Mir.

At 14:50 Moscow Time – as kept on Mir – Atlantis was in position 300 metres directly beneath the station. After a pause, it closed to 100 metres, then paused again. At 16:23, as the two spacecraft passed north of the equator, Atlantis resumed its approach, then paused again at 10 metres. At 16:55, with Mir back over Russian territory and once again in contact with Moscow, Atlantis started the final phase of the approach. The rules required the docking to be achieved before Mir flew out of range of the tracking network, which gave Gibson about 15 minutes to complete the manoeuvre. Stationed at the rear of the flight deck, he initially peered through the overhead window, but for the docking he switched to the monitor displaying the view from the camera on the centreline of the ODS. To ensure that the two docking units were aligned, he had to remain in the centre of an ever-narrowing cone as he slowly ascended, and this was done by sighting on a stand-out bull's eye over the target in the centre of Kristall's port. Mir was put into free-drift mode to ensure that its attitude-control system would not compete with the orbiter's for control of the docked structure. Atlantis closed at a rate of 3 centimetres per second, displaced by no more than 2 centimetres from the axis of the cone, and with an angular discrepancy of less than 0.5 degree. Gibson had flown the 100-tonne orbiter with such skill that he had made the unprecedented manoeuvre appear deceptively straightforward. Exactly on time, at 17:00, while over Lake Baikal, the two sets of triple-petals meshed and the capture latches engaged. There was a momentary shudder as the two spacecraft jostled one another, but the springs within the mechanism absorbed these residual motions. After 15 minutes in this soft-docked state, Atlantis fired its thrusters to drive the guide-ring of the ODS against Kristall in order to ensure that the 12 primary latches around the rim of two collars were aligned for the hard-dock that would form a hermetically sealed tunnel between the two vehicles. An hour later, when Mir flew back into communications range, Dezhurov swung open the Kristall hatch, Gibson opened the ODS hatch, and they greeted each other in the tunnel. In Moscow Yuri Koptev and Dan Goldin savoured the moment. The first order of business on Mir was for everyone to congregate in the base block for a 'photo opportunity', which amply demonstrated that it had not been built to accommodate 10 people.

After a well-earned rest, it was time for business. In the Spacelab, Dezhurov, Strekalov and Thagard, who were well advanced with their preparations to return to Earth, were given a thorough medical examination using the life-sciences equipment that included a treadmill, a veloergometer and a Lower-Body Negative Pressure chamber. This was the first opportunity for NASA to study long-duration cases while still in space. Only a space station can provide the time for the body to adapt fully to microgravity (and Mir was the only such facility) and only by having the Shuttle visit Mir could the sophisticated biomedical technology at NASA's disposal be applied to space-adapted cases. The symbolism of this mission was strong, but this first visit to Mir produced tangible scientific results. Those who had complained that the science programme had been hijacked could hardly criticise the examinations undertaken on these three subjects, who had spent over a hundred days in space. As the retiring crew served as guinea-pigs for the biomedical studies, their successors transferred cargo. Atlantis had a great deal of water in reserve as a result of the reaction in its power-generating fuel cells. This 'waste' was normally vented, but water was a precious resource on Mir. On the whole, Mir recycled 60 per cent of its water, but only that condensed from the water vapour was potable. It had been decided to donate the excess water to top up Mir's tanks, but because there were no pipes it had to be pumped into small containers that were ferried across manually. Nevertheless, half a tonne of water was donated by this means. The formal handover between the two Mir crews occurred when Solovyov and Budarin installed their Kazbek couch liners and Sokol suits on Soyuz-TM 21, and Dezhurov, Strekalov and Thagard moved theirs to Atlantis. The cargo transferred to Atlantis included the frozen quail chicks, accumulated film and experiment materials and biomedical samples. The Russians exploited the Shuttle's payload capacity to return a number of expired items that were usually discarded with empty cargo ferries – including elements of the Salyut 5B computer – so that their service life could be reassessed. This aspect of the joint mission relied on quartermastering skills to track all the items going in each direction and ensure that they were properly loaded for

After STS-71 docked with Mir, Vladimir Dezhurov and Robert Gibson shake hands in the tunnel to the Kristall module.

A severe case of overcrowding. The designers of the Mir space station's base block could hardly have imagined that it would one day play host to ten people.

the return to Earth (a laser bar-code reader was under development to assist transshipment in the future).

Late in the preparations for the mission it had been decided that rather than keep the combined 'stack' oriented as it had been at the moment of docking with the Shuttle beneath Mir, Atlantis would reorient the stack at the beginning of each day to enable Mir's solar panels to generate maximum power. Because the differential gravity field would tend to restore the most stable position, Atlantis would need to make frequent adjustments to maintain this 'solar inertial' attitude. The stack was to re-establish the gravity-gradient attitude each evening, and be left like that overnight. When it was found that this consumed much more propellant than expected, it was realised that the Shuttle computer's mass model was unable to deal with such a large 'attached' mass, and in overcompensating using its fine-control thrusters, oscillated back and forth each side of the optimal alignment. The long-term solution was a better mass model, but on this occasion the orientation tolerance was relaxed, and the remaining minor deviations did not seriously degrade Mir's power. The inadequacy of the computer's mass model had never been suspected. With this revelation and a variety of other engineering data, NASA was racing up the learning curve of how to operate a Shuttle in conjunction with a large structure. Removing the technical risk from operations planned for the International Space Station (ISS) was the main goal of the Shuttle–Mir programme. On 1 July the hatches were closed to enable Atlantis to perform manoeuvres while docked with Mir, to test the integrity of the ODS (it did not lose its hermetic seal) and to observe the dynamics of Mir's solar panels (they wobbled alarmingly). The tunnel was then reopened. Late on 3 July, as the spacefarers bade their farewells and congregated in their respective vehicles, Atlantis pumped up the complex's atmosphere to a pressure of 15.4 psi as a parting gift, thereby obviating the need to send up any air on the next cargo ferry. Kristall's hatch was closed and the ODS hatch was closed. This must have been a sad moment for Dunbar. She had been assigned to the first docking mission to transfer to the Mir complex, along with Solovyov and Budarin to continue Thagard's programme and be retrieved by Atlantis when it returned later in the year, but this plan had had to be cancelled when it became clear that leaving an astronaut on board would conflict with the 'long' visit planned by the European Space Agency (the factor determining the size of the resident crew was the capacity of the lifeboat Soyuz). However, NASA had booked a continuous slot thereafter, and the presence of an astronaut would restrict other visitors to the duration of a handover. Only after NASA had used up its assigned time would others be able to make extended visits. The demand for access to Mir was so great that it was a problem to fit everyone in! On 4 July, Solovyov and Budarin undocked their ferry from Mir's rear port, withdrew 100 metres, then flew out to the side of the complex to photograph Atlantis's departure. When the latches released, the spring-loaded mechanism eased the vehicles apart.[2]

[2] Explosive bolts would have been fired if the latches had failed to disengage, and if the bolts failed to fully separate the two components, Atlantis had tools to enable spacewalkers to unfasten the 96 bolts that held the docking system on the ODS assembly. If this had proved necessary, of course, fouling Kristall's port in this way would have brought the Shuttle–Mir programme to a premature conclusion.

A historic picture of Atlantis with the Mir space station during the STS-71 mission, taken by the crew of a Soyuz spacecraft.

Atlantis returned to the Kennedy Space Center on 7 July. During the descent, Dezhurov, Strekalov and Thagard wore NASA pressure suits. Reclined couches had been installed on the mid-deck to preclude them from having to sit in an upright pose at the start of their readaptation to gravity – the time that they would be most at risk of orthostatic intolerance, and most likely to black out. Contrary to instructions, Thagard climbed out of his couch and walked away to the recovery van. The delay in launching Spektr, and the consequent delay in sending up Atlantis, had resulted in his planned 90-day tour being extended to 115 days. Nevertheless, he had completed his entire research programme.

Thagard's debriefing highlighted many interesting points:

- Mir was roomy and very habitable, but it definitely had the look and feel of a locker room that had been lived in for a decade.
- He had had an amiable association with Dezhurov and Strekalov, but days had passed without hearing English on the voice link, and he had suffered cultural isolation.
- He had yearned for his family.
- Being the equivalent of a laboratory rat in such a rigorous biomedical study was quite stressful.
- The requirement to log food intake was a disincentive to eating.
- He had dreaded the prospect of his tour being extended to six months.

After four Shuttle flights and a tour on Mir, Thagard retired from NASA a few months later and returned to academia.

The primary objectives of the Shuttle–Mir programme were that America and Russia should learn how to work together in space and reduce the technical risk in building and operating a joint orbital facility, but the opportunity to have astronauts serve tours on Mir prompted NASA to review the science programme for the ISS. This identified several dozen experiments that could be undertaken early, on Mir, at little or no cost-penalty to the individual project budgets, and some of these projects were brought forward. Clearly, although the first docking with Mir was primarily an engineering test flight, the ongoing Shuttle–Mir programme would enable NASA to build up science activities.

SPACE TRUCK

On 12 November 1995, STS-74 Atlantis was launched. As there were no long-stay residents to be retrieved, there was no requirement for biomedical facilities and the bay was given over to the Russian-built Docking Module (DM). At 4.2 tonnes, this 4.6-metre-long, 2.2-metre diameter compartment was really an extremely stretched Soyuz orbital module fitted with an androgynous port at each end. Two days later, while Chris Hadfield used the RMS to unstow the DM and position it directly over the ODS, the Space Vision System (SVS) processed imagery from obliquely angled video cameras to infer the relative positions of the two units from reference markings adorning their surfaces, and then calculated an animated bore-sighted viewpoint and presented it on Hadfield's control screen. The SVS was to be used to mount modules on the ports of the ISS where direct viewing would be impractical. With the DM in position, the arm was placed in 'limp' mode, and Atlantis fired its thrusters to nudge the guide ring of the ODS up against the DM to achieve a soft dock. As soon as the residual relative motions had been damped out, the ring was retracted to hard dock and then the arm was withdrawn. If there had been a problem, Jerry Ross and Bill McArthur were ready to make a spacewalk in an attempt to resolve it.

Atlantis rendezvoused with Mir on 15 November, and pursued the same R-bar approach as before. Although Kristall was now back on Mir's radial port, having the DM on the ODS provided 4 metres of extra clearance from the solar panels that projected from Kvant 2, Spektr and the base block. To provide further clearance for the DM, the base block's panels had been rotated face-on to the Kristall module. As Ken Cameron did not have a direct view of the docking system, the RMS was fixed in position above and to the side of it to give a side view of the final few metres of the approach. No orbiter had ever manoeuvred in such a confined space before, but he made it look easy. The crews met in Mir's base block for the photo-opportunity, and this time there were national flags of Russia, America, Canada (for Hadfield) and Germany (for Thomas Reiter, a European Space Agency researcher) on display – the complex had indeed become an international facility. In fact, Yuri Ķoptev had recently noted that since the Russian Space Agency had been underfunded by 180 billion roubles in the current financial year, continued Mir operations were possible only because of the money earned from leasing it to fee-paying guests.

For the first time, the Shuttle really had served as a 'space truck' by delivering a module to a space station. The cargo destined for Mir comprised 700 kilograms of food and water and 300 kilograms of items for NASA's research. The 375 kilograms of cargo to be returned to Earth included processed materials, computer disks of experiment data, biomedical samples, and a variety of expired and broken hardware that was to be returned to the manufacturers for study. The Russians had never been able to return such amounts of cargo from Mir before so, by this simple act, Atlantis was making a significant contribution to Mir operations. Overall, 275 items were taken on board Mir and 195 items were retrieved. This time it was not necessary to maintain Mir in solar-inertial attitude, because the Sun angle was better and the power level was higher. Nevertheless, Atlantis manoeuvred to test its ability to control such a large offset centre of mass. On 18 November, when Atlantis departed, the DM was left on Kristall. Shuttle manager Tommy Holloway noted delightedly that this tricky flight had "far exceeded expectation". Dan Goldin announced that Mir was "proving to be an ideal test site for vital engineering research", and that the first two dockings were "already paying back benefits" by providing "proximity and docking operations" and by "simulating an early construction flight".

COMINGS AND GOINGS

STS-76 Atlantis docked at the DM on 24 March 1996 with a Spacehab full of cargo in its payload bay. The original plan had been to carry cargo in a Spacelab, but the internal configuration of this laboratory module did not readily lend itself to bulk haulage, and Spacehab turned out to be ideal. In July 1995 a contract had been signed to fly one of the existing pair of Spacehab modules on this mission, and to mate the other with an engineering test article to construct a double-length module for use on later flights. Mounted well aft of the ODS, almost on the orbiter's centre of mass, the single-length module could carry 700 kilograms more than it could in its usual position at the front of the bay, taking its capacity to 2 tonnes. In this case, it delivered 980 kilograms of material for the Russians and 740 kilograms for NASA's programme, and off-loaded 500 kilograms. In addition, the module carried a rack of experiments, some of which were transferred to Mir and the rest run independently, and a freezer in which to return accumulated biomedical samples. The cargo transfers were routine, but the highlight of this visit was the spacewalk by Linda Godwin and Rich Clifford. As this was the first time that astronauts had worked alongside an orbiter that would not be able to chase after them if a tether snapped, they wore backpacks incorporating SAFER manoeuvring units. Also, because it was the first time that NASA astronauts had worked near Mir, with its unfamiliar surface, they were told not to venture beyond the top of the DM. After dismantling a redundant camera, they installed the Mir Environmental Effects Package (MEEP) cassettes on the short module. They then tested a new portable foot-restraint and a tether system intended for use in assembling the ISS. When the hatches were closed on 28 March, Shannon Lucid remained on Mir with Yuri Onufrienko and Yuri Usachev, who had begun their six-month tour in February. Lucid would serve on with their successors. The plan was that she would be retrieved by Atlantis on its return in early August. Her tour of duty initiated two years of continuous habitation for NASA astronauts on Mir.

Priroda – which incorporated an array of remote-sensing instruments developed collaboratively – was finally launched on 23 April 1996, docked on 26 April, and swung across to the radial port opposite Kristall the next day to complete the Mir complex. It contained a large amount of equipment for NASA's scientific research. A Space Acceleration Measurement System had been delivered prior to Thagard's tour, and Lucid's first task was to install this in Priroda to characterise its microgravity environment. She then installed video cameras and the Microgravity Isolation Mount in the glovebox and ran an experiment to determine how well the mount could damp out the ambient vibrations in order to decide whether it would

Table 10.1 Assembly of the Mir Space Station

Element	Launched
Mir	20 Feb 1986
Kvant 1	31 Mar 1987
Kvant 2	26 Nov 1989
Kristall	31 May 1990
Spektr	20 May 1995
Priroda	23 Apr 1996

Note: The Mir complex was de-orbited on 23 March 2001.

permit particularly sensitive microgravity experiments to be performed in the 'noisy' environment of an inhabited spacecraft. Over the following months, Lucid's varied programme included using several furnaces, sampling air and water quality, monitoring ambient radiation, embryological studies, and, of course, gathering data for the biomedical database. But her favourite task was geological, ecological and environmental photography, because that gave her an opportunity to gaze out of Mir's high-fidelity porthole. Each night, Houston updated a four-day task-list that Lucid worked through in her own time, ticking off the tasks as they were completed.

On 21 June, Onufrienko and Usachev were informed that financial problems had delayed the fabrication of the rocket that was to have delivered their successors in July, and that their tour would therefore be extended to late August. This was a disappointment for Lucid, as she had expected to play her part as host to the French cosmonaut accompanying the handover. On 12 July, however, she was told that she, too, would have to extend her tour by six weeks because the SRBs on STS-79 had to be replaced, and she would afterall be present at the handover. In mid-August, Lucid worked on the Candle Flame Experiment and the Forced Flow Flamespread Test. At this point, she had completed her programme, and used her extended time efficiently by installing the BioTechnology System for later use, planting wheat in Mir's Svet cultivator, ran Anticipatory Postural Activity experiments and further characterised the body in its fully adapted state by measuring muscle stimulation. Frenchwoman Claudie Andre-Deshays arrived on 19 August with Valeri Korzun and Alexander Kaleri. As a specialist in rheumatology and neurology, Andre-Deshays focused on adaptation to microgravity, then left with Onufrienko and Usachev on 2 September. Lucid spent her time with the new residents preparing to return to Earth, but on 7 September she claimed Yelena Kondakova's 169-day woman's record, and ten days later – with Atlantis already on its way – she also claimed Reiter's 179-day visitor's record. On this visit Atlantis carried the double Spacehab, but the docking on 19 September was slightly complicated by a recently deployed solar panel that projected within 3 metres of the orbiter's nose. John Blaha, who had flown with Lucid on two previous Shuttle missions and was an old friend, had come to replace her. The exchange of Kazbek couch liners and Sokol suits marked the first-ever astronaut handover. Lucid then briefed Blaha on where everything was stowed, and suggested how best to work on the complex. Tom Akers, serving as a 'loadmaster', evaluated a new method of handling the cargo. Rather than stow items in lockers and transfer them individually, everything was packed in locker-sized canvas bags, and the shipping monitored by a bar-code reader. Stowing the return cargo turned out to be more difficult, as everything had to be weighed and placed consistent with centre-of-mass requirements. Some 2,250 kilograms of miscellaneous cargo was delivered to Mir, and 1,000 kilograms retrieved. Lucid's results alone filled 20 canvas bags, and the Russians sent back an expired Orlan EVA suit to enable its state of degradation to be determined. Atlantis left on 24 September. An important test in the Spacehab involved the Active-Rack Isolation System which Boeing had built for microgravity experiments on the ISS.[3]

[3] Whereas the Microgravity Isolation Mount used magnetic levitation, the Active-Rack Isolation System relied on a mechanical suspension system to damp out micro-accelerations.

Immediately upon landing, the density of Lucid's skeleton and key musculature was imaged by an NMR scanner. To her surprise, she rapidly readapted to gravity. Shuttle–Mir programme manager Frank Culbertson opined that Lucid had "set the standard" for NASA's work on Mir, and she was subsequently welcomed into the select group of astronauts to have been awarded the Congressional Space Medal of Honor. On taking over, Blaha expressed the view that there was no better way for NASA to prepare for the ISS than to have a cadre of astronauts living and working on Mir. Everything from the exercise regime to the logistics system was new to the agency. Blaha's programme built on Lucid's, but included several new experiments. One of the packages delivered by Atlantis was a 'powered transfer' with mammalian cartilage cells for a BioTechnology System experiment involving tissue growth. The Diffusion Crystallisation Apparatus for Microgravity employed a semi-permeable membrane to grow protein crystal over a long period. The Binary Colloid Alloy Test investigated the crystallisation of alloyed colloids. Blaha also had a number of technological tests to perform in order to provide information important for fitting out the various modules of the ISS. One of these was an instrumented foot-restraint that monitored how much his body moved while he worked 'in place', and another used a push-off pad to measure the force imparted in moving around in the complex. As a further experiment to characterise the microgravity environment, the Passive Accelerometer System measured low-intensity continuous effects due to the gravity gradient and air drag. The Mir Structural Dynamics Experiment measured transient stresses on the complex during thruster firings and docking operations, and thermal effects arising from flying into and out of the Earth's shadow. Once installed, these suites of sensors were to record data whenever NASA was conducting microgravity experiments.

In early December, the wheat that Lucid had planted in the Svet in August had completed its cycle and produced grain, which Blaha harvested. This was the first time that a staple had grown to maturity, and it had important implications for the prospects of making a station self-sufficient in food production. The wheat and most of the grain was frozen for return to Earth for a comprehensive biochemical analysis. During a press conference at the end of the year, Blaha was asked if he was eager for the Shuttle to retrieve him on schedule and he replied laconically: "If it gets here, it gets here." Clearly, like Lucid, he was enjoying his tour. STS-81 lifted off on its first attempt. Despite the early doubts of the programme's critics, no launch had been scrubbed during the final moments of the countdown. Atlantis docked without incident on 15 January 1997. The double Spacehab carried the routine cargo mix. Marsha Ivins was the loadmaster, and the offloading was done primarily by John Grunsfeld and Jeff Wisoff. The assessment of a treadmill that did not transmit vibrations to the orbiter's frame was another small step towards minimising the pollution of the microgravity environment by the presence of a crew. Blaha departed on Atlantis on 20 January, and on landing in Florida two days later was "absolutely stunned" at the strength of the Earth's gravity. In contrast to his predecessors, he allowed the medics to carry him from the Shuttle.

GOOD TIMES AND BAD

On Mir, Jerry Linenger, a physician, had a programme that combined biomedical, biotechnology, fluid physics and materials-processing assignments, but the highlight of his tour was to be a spacewalk to deploy one experiment and retrieve two others that had been emplaced a year earlier.

Like Lucid, Linenger played host to a routine handover between Russian crews when Vasili Tsibliev and Alexander Lazutkin arrived with Reinhold Ewald, but this proved to be rather more eventful than expected. When more than three people were on Mir, the Vika chemical burner supplemented the oxygen output of the Elektron regenerative system. On 24 February the module at the rear of the complex suddenly filled with thick smoke. The Vika had split, released oxygen into the electronics, and started a fire. Although the cosmonauts were on the scene within a matter of seconds with portable extinguishers to smother the system with foam, the combustion was sustained for 10 minutes by the release of oxygen from the canister. The entire unit was reduced to soot-blackened scrap. For almost an hour, while the air conditioning system extracted the majority of the smoke, everyone on board wore face-masks and portable oxygen bottles, and for the next few days they wore small filter-masks to ensure that no particulates were inhaled. It was the worst fire yet. On any previous station this would have resulted in a hasty evacuation, but Mir was too valuable to be abandoned and the mess was cleared up. On 4 March, two days after Korzun, Kaleri and Ewald departed, Tsibliev tried to manually steer Progress-M 33 in to re-dock but the TORU remote-control system malfunctioned and the ferry was ordered to withdraw. The following day one of the Elektron units shut down because a bubble of air was blocking the flow of water in the electrolysis canal. This was a serious problem because the other Elektron had been out of commission for some time, and the only means of making oxygen was the one remaining Vika.[4] A lesson learned previously was that an empty station could be disabled by a fault that could easily have been fixed if a crew had been on board. Leaving Mir vacant, even for a short period, was inviting trouble. As if to emphasise this danger, on 19 March an orientation sensor failed and the attitude control system tried to correct what it took to be an unwanted motion. By the time the cosmonauts were able to intervene, the momentum wheels had built up a runaway three-axis roll. The computer-controlled gyrodynes had to be turned off, the rotation cancelled manually by thrusters, and the complex stabilised in the gravity-gradient while the fault was analysed. It is doubtful that control could have been reasserted from the ground, because problems with the radio link meant that Mir was in communication with Moscow for only 10 minutes on favourable passes. On 8 April Progress-M 34 brought a stock of 60 canisters for the surviving Vika and parts to repair the transmitter and service the balky Elektron. With the short-term future of the complex assured, Tsibliev, Lazutkin and Linenger settled down to work and, on 29 April, Tsibliev and Linenger made their spacewalk to deploy and retrieve experiments. Despite having been drummed into helping with maintenance chores, Linenger had made good progress with his science assignments.

[4] If the Vika failed, there was only a few days' supply of bottled oxygen as an emergency reserve.

Indeed, by processing 50 samples in a furnace for a liquid diffusion experiment he had not only finished his own programme but had begun to work his way through the samples intended for his successor, Michael Foale, who arrived on STS-84 on 17 May 1997. The double Spacehab had another heavy logistics load: in addition to 500 kilograms for NASA, the 1.2-tonne Russian cargo included a 120-kilogram Elektron to replace the one that had previously failed, a 165-kilogram gyrodyne and a stock of lithium hydroxide canisters. Also on board, but only as a visitor, was veteran Mir resident Yelena Kondakova. When Linenger left on 22 May, he took 400 kilograms of accumulated NASA research results with him.

Foale had been given a multifaceted programme of protein crystallisation, Earth observations, life sciences, materials processing and engineering tests, and was also to assist with maintenance tasks, help to unload the Progress scheduled for late June, control Mir while Tsibliev and Lazutkin made two spacewalks in early July, and host the August handover. The maintenance on the environmental systems got off to an excellent start, but then, completely unexpectedly, disaster struck in the shape of an errant ferry. The experimental manual re-rendezvous that had been abandoned with Progress-M 33 was to be attempted using Progress-M 34. It undocked on 24 June and returned the following day. As Tsibliev was manoeuvring the 7-tonne ferry, he lost control. Tsibliev ordered a retreat to the Soyuz at the front of the complex. As Foale crossed the cramped forward docking compartment he heard "a loud bang". Lazutkin, who was still in the base block with his feet anchored to the floor, felt a shockwave propagate through the structure. The ferry had smashed into the Spektr module, badly mangling a solar panel and puncturing a coolant radiator. The twisting of the panel's motor installation, and the force on the hull transmitted through the struts supporting the conformal radiator combined to break the hermetic seal of that module, and the air inside began to vent to space. In seconds, there was a faint hiss of air escaping. Spektr's hatch had to be closed – and quickly. If it could not be closed, the crew would have no choice but to retreat to the Soyuz and abandon the station. Cables had been laid between the modules to form an integrated power grid linking the solar panels to the storage batteries distributed throughout the complex. There was no time to unplug the cables. However, a heavy cutting tool was stored in the docking adapter for precisely such an emergency. It took several minutes to clear the aperture and fit the hatch, but by then it was evident that it was a *slow* leak. Subsequent monitoring showed that the integrity of the docking collar had not been breached and had absorbed the energy of the impact. On Earth, Linenger opined that the two worst accidents that could befall a spacecraft were fire and depressurisation. Mir had suffered both within the space of a few months.

With the station secure, it was time to tackle the immediate consequences. The quartet of solar panels on Spektr had contributed almost half of Mir's power, but, with the cables cut, this power was no longer being fed into the power distribution system. The complex had survived the decompression, but it now faced a power drain. As it was essential that the batteries should not be totally drained, each man dived into one of the other radial modules and switched off all the apparatus. Nevertheless, a crisis was in the making, because the complex needed power if it was

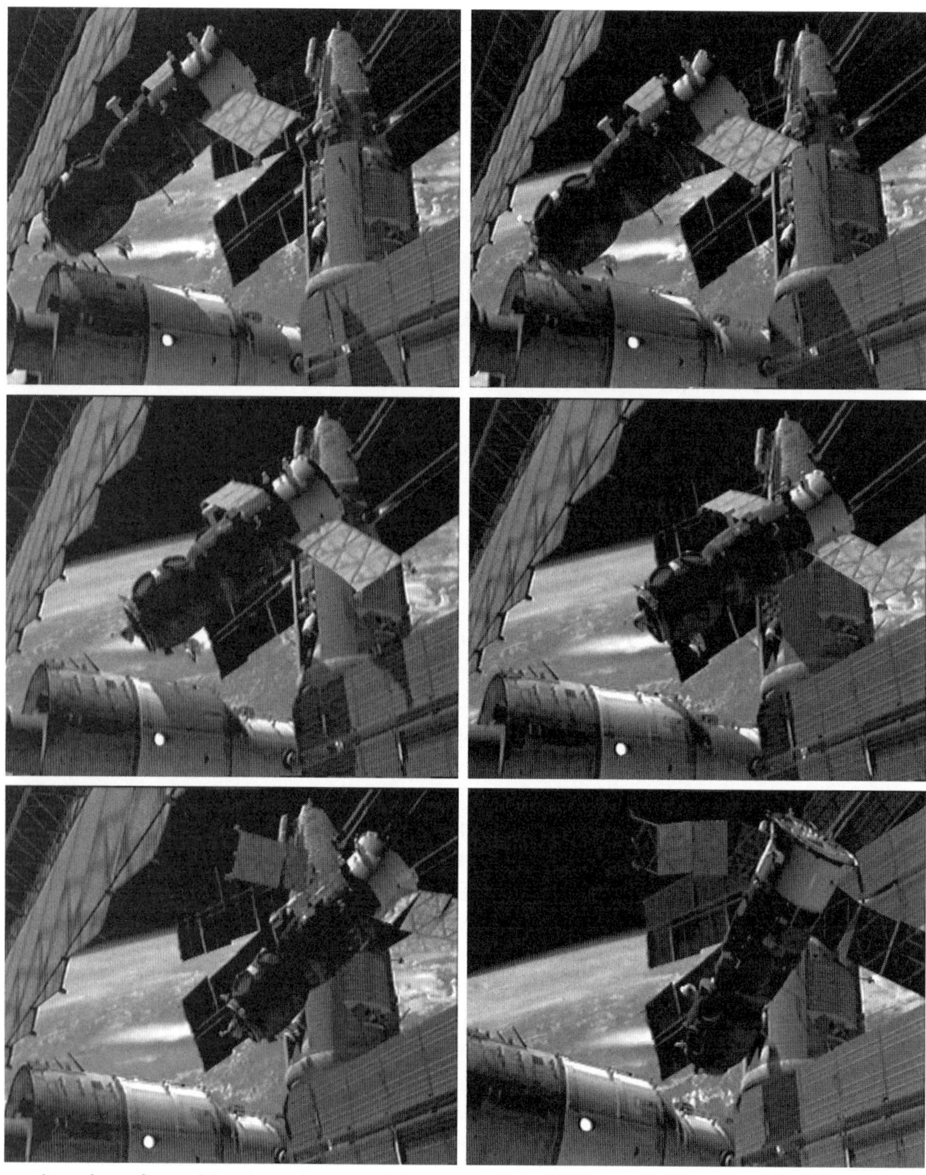

A series of graphics (running top left to bottom right) showing how Progress-M 34 damaged the Spektr module of the Mir space station. (Courtesy of Dave Woolard of Pictures on the Wall.)

to continue to generate power. The power output from a solar panel is related to the angle of insolation by a sinusoidal function, and produces its peak output only when it is face on. As Mir flew around its orbit, it generally kept the same orientation with respect to the Earth and the panels were rotated to track the Sun. It required power to run these motors, and each panel had its own motor. If the batteries were depleted, there would be insufficient power to rotate the panels, which meant, in turn, that no power would be available to recharge the batteries. It was a runaway process – a spiral to oblivion. It was therefore vital to switch off everything to conserve the batteries, and then, with half the generating capacity denied, endeavour to remain on the safe side of the power curve. The situation was complicated by the fact that the orientation of the complex was being controlled by the gyrodynes, and since they also consumed electrical power they had to be deactivated. This put the complex into free-drift. It was at this time that Mir flew back into communications range with Moscow. The first that the flight controllers knew of the problem was when the telemetry did not materialise because the transmitter had been switched off – together with everything else. In any case, there was nothing the ground team could do; the crew in space were effectively on their own.

The Soyuz had been powered up immediately, just in case it proved necessary to make a hasty escape, and now it was the only part of the complex not crippled by the power loss. Its thrusters were used to stabilise the complex and to reorient it so that the majority of its remaining solar panels faced the Sun. It took several hours of continuous adjustments to fully charge the batteries, and only then could Mir's control systems be reactivated. For the next few days, as they inspected the station, the cosmonauts worked by torchlight, and set up a sleep roster to ensure that they would not be taken by surprise by a sudden deterioration. There were still some potential problems that could, in the end, force abandonment. With power low, the cooling system was ineffective. The soaring temperature was not just uncomfortable for the crew, the overtaxed thermal regulation system made the Vozdukh carbon dioxide scrubber overheat and this had to be turned off, which forced reliance on the limited supply of lithium hydroxide canisters. The Elektron also suffered, with the result that waste water could not be electrolysed to make oxygen. As this forced reliance on the Vika, it was necessary to have masks and fire extinguishers readily available. The toilet was operational, but its reprocessing system could not be used because the tank for the Elektron was full. Although the list of small problems grew, the situation improved day by day. It was by no means clear, however, that they would not be forced out. One thing was certain: if Mir was vacated, it was unlikely to be reoccupied. The spacecraft and its crew were in a symbiotic relationship: while the crew relied on Mir for their survival, Mir relied on their presence for its continued operation. Basic survival, however, was not the objective. Mir was a laboratory, and if it was to have a long-term future, then, at the very least, the undamaged solar panels on the Spektr module would have to be brought back on line. After the re-docking Progress-M 34 was to have been discarded to clear the aft port for its successor, which was already on the pad for launch on 27 June. This was postponed to give time to work out a way to restore the power and to fabricate the necessary materials and tools. It was concluded that although it might be feasible to run external cabling from Spektr's panels to sockets outside the base block, it would be simpler to try to reconnect them internally. With Spektr exposed to vacuum, however, the hatch would have to remain closed. But how could new cables be run through a sealed hatch? An ingenious scheme was conceived: one of the Konus drogues in the docking adapter would be modified to serve as an air-tight electrical junction. There were two of these drogues, one of which was kept permanently on the axial port for dockings. The other, which was detachable, had been moved around the radial ports as necessary to facilitate the movements of the add-on modules, which could swing themselves around on their short Ljappa arms. This Konus had last been used to accept the Priroda module. It comprised the hollow drogue, which was essentially a conical guide plate, and the bulbous end cap that contained the clamp for the mechanism at the tip of the probe. The two parts were connected by a ring of bolts. The plan was to remove the end cap and the clamp, and bolt on a new unit with a set of electrical sockets on each side. If this could be mounted on Spektr's hatch, the situation should be retrievable. It would require an 'internal spacewalk', however, to redeploy this hatch and lay the necessary cables within the stricken Spektr module. Trials in the hydrotank by Anatoli Solovyov and Pavel Vinogradov, who were to be Mir's next crew, showed that the task was feasible, so the equipment was loaded on board Progress-M 35. As this ferry made its final approach on 7 July, the view from its docking camera clearly depicted the extent to which one of Spektr's solar panels had been twisted in the collision, but the other three panels appeared to be undamaged.

The Russians had hoped that Tsibliev and Lazutkin would be able to effect the repairs within a week, but when the cosmonauts reviewed the proposed procedure they requested additional time to prepare, and the internal spacewalk was pushed back a week. Fortunately, because the Konus was not already on Spektr's hatch, the preparatory work of exchanging the junction-plate for the drogue mechanism could be done in a shirt-sleeved environment. It was important that Spektr's power be restored prior to the August handover, because (a) the new crew's arrival would place an extra demand on the environmental system, and (b) power would be required for the experiments to be performed by the visitor. A spacesuited rehearsal without depressurising the multiple docking adapter was scheduled for 15 July, and the real thing was set for 18 July. With power restored to the maximum available, the next crew would be able to concentrate on the external activities required to locate and plug Spektr's leak. The module could then be repressurised and, with a little luck, much of its apparatus could be salvaged. The extent to which this proved feasible would decide NASA's future on the complex. Although this module housed 50 per cent of NASA's science apparatus, if it had to be written off it was debatable whether there would be sufficient scientific justification to continue the Shuttle–Mir programme.[5] A great deal, therefore, was riding on the outcome of this makeshift repair.

Although the preparations progressed well, the mounting stress evidently took its toll on the commander. On 12 July, Tsibliev reported sensing a heart arrhythmia.

[5] If the errant ferry had struck and depressurised Kristall, the loss of its androgynous docking system would have terminated the programme then and there. Because Atlantis would not be able to return in September to collect him, Foale would have been obliged to return with Tsibliev and Lazutkin in the Soyuz on the completion of their tour.

When a full test confirmed this the next day, he was instructed to take a combination of heart medication and tranquillisers to help him to relax. It was also decided that Foale should support Lazutkin in the difficult repair, which was put back another week to give Foale time to prepare. Because Foale had backed up Linenger in the preparations for the latter's spacewalk, he was familiar with the basic procedures for using the Orlan suit. On 16 July, while rehearsing the process of disconnecting the cabling within the docking adapter, Lazutkin unplugged the cable that distributed data from the primary orientation sensors to the gyrodynes without first switching to the backup sensors, and Mir began to drift, with the result that the power output from its solar panels dramatically fell. The crew scrambled to switch off apparatus, but the complex was rapidly overwhelmed by another power crisis. In a rerun of the previous process, they were obliged to retreat to the Soyuz and use its thrusters to reorient Mir so that the solar panels could feed power and recharge the batteries.

Undeterred, the cosmonauts argued to be allowed to proceed with the repair on 25 July, but were told that it had been decided to reassign the job to their successors. Without power to run his experiments, Leopold Eyharts, the Frenchman who was to have worked through the handover, was reassigned to the Soyuz scheduled for early 1998. Anatoli Solovyov and Pavel Vinogradov arrived on 7 August, and a week later Tsibliev and Lazutkin left Foale with the new residents and returned to Earth. The first order of business was to swap the new Soyuz from the rear to the front of the complex so that the 'lifeboat' would be available in the event of trouble during the internal spacewalk. On 22 August Foale retreated to the Soyuz's descent module. After Solovyov and Vinogradov had installed the new connector plate on Spektr's hatch, Vinogradov entered the module feet-first to hook up cables to feed the power from the three undamaged panels into the main grid. After a fruitless search for the site of the puncture in the hull, he retrieved some of Foale's personal belongings and closed the hatch. Although Spektr's surviving panels generated power, the computer could not rotate them to follow the Sun. When Solovyov and Foale went out on 6 September to inspect the damage, they manually oriented the panels at the optimum angle. As they had no time to complete the inspection, the site of the leak remained a mystery. Several spacewalks had been assigned, so it was certain that they would return to locate the hole and, hopefully, seal it.

Given the recent requirement for spacewalking, NASA decided that David Wolf should succeed Foale in September because, unlike Wendy Lawrence, who was to have had the next tour, Wolf was EVA-trained and would be better able to assist in an emergency. It was far from certain that Foale would be replaced, however. Since the fire on Linenger's tour, critics in Congress had called for NASA to cut short the Shuttle–Mir programme because "Mir was dangerous", and surely the agency had learned all that it could hope to learn from joint operations. The recent loss of power and half of the US apparatus surely meant (they argued) that there was no scientific justification for stationing more astronauts on board Mir. NASA, in contrast, maintained that it was still learning valuable lessons in operating a space station, and in any case dealing with the fire and the decompression was on-the-job training for situations that might affect the ISS. In short, NASA did not see Mir's problems as an excuse to bale out, but as a learning curve to be climbed, which is precisely why

Table 10.2 Shuttle–Mir astronauts

Astronaut	Days in space	Up	Down
Norman Thagard	115	Soyuz-TM 21	STS-71
Shannon Lucid	188	STS-76	STS-79
John Blaha	128	STS-79	STS-81
Jerry Linenger	132	STS-81	STS-84
Michael Foale	145	STS-84	STS-86
David Wolf	128	STS-86	STS-89
Andrew Thomas	143	STS-89	STS-91

it had seized upon the joint programme in the first place. Thus, by any measure, the Shuttle–Mir programme had given NASA hands-on experience of operating a space station as a long-term facility.

THE CALM AFTER THE STORM

While STS-86 was in place, Scott Parazynski and Vladimir Titov made a spacewalk to retrieve the MEEP package. Unfortunately, because the damage to Spektr was on the Priroda side of the complex, opposite Atlantis, the astronauts could not search for the site of the leak. However, they tied a Russian-supplied 'cap' to the DM for retrieval by the residents if the source of Spektr's leak was eventually traced to the motor housing of the damaged solar panel – if so, an attempt would be made to remove the motor and use the cap to seal the leak.[6]

Wolf's tour passed off uneventfully. In addition to tending to his experiments, he made a spacewalk with the Mir commander, Anatoli Solovyov, on 14 January 1998, at which time he employed a portable spectrometer to assess the condition of the outer surface of the Kvant 2 module. It had been intended to venture down onto the base block, which had been in space for over a decade, but the need to set aside time for maintenance on the airlock hatch had ruled this out. Although Mir continued to suffer computer faults, with no real drama to report it faded from the headlines, thus vindicating Goldin's decision to send Wolf to continue the American presence. The next Shuttle arrived on 24 January 1998. On this occasion it was Endeavour not Atlantis. It brought Andy Thomas who, upon being retrieved by Discovery in June, rounded out the programme.

"The Shuttle–Mir programme has been very useful in giving our astronauts good training in crisis management," wryly noted Representative James Sensenbrenner, the senior Republican on the House's space committee.

[6] In the event, the source of the leak was never identified, and Spektr was abandoned in place.

11

International Space Station

SHUTTLE AND STATION

When NASA submitted its post-Apollo plan to the Nixon administration in 1969, it proposed a reusable Shuttle to provide routine cheap access to low orbit and a space station to serve as a laboratory and a way-station for missions into deep space, but the innovative Shuttle's projected development cost was so great that NASA had to sacrifice the station. But as soon as the Shuttle was declared operational in 1982, the agency dusted off its station plan, started to lobby for support and, to its surprise, found key members of the administration receptive, particularly to the likelihood of sharing the cost with international partners. In his State of the Union address in January 1984, Ronald Reagan directed NASA to build a space station that would be permanently occupied by a crew of eight astronauts, and in an echo of Kennedy's historic challenge, ordered that this be done within a decade. A few months later, the European Space Agency announced that it would supply a laboratory to the space station, and Japan, which regretted having missed out in contributing to Shuttle operations, said that it would do likewise.

SPACE STATION FREEDOM

Although the structure shown to Reagan consisted of a compact cluster of modules with two solar panels, the design selected was Grumman's Power Tower concept. This was a 120-metre-long keel set vertically to use the gravity gradient for stability, a horizontal truss to carry the solar panels, and a cluster of modules at each end to facilitate simultaneous study of the Earth and the sky. However, as the design was reviewed, doubts surfaced. Would such a long, narrow keel be sufficiently rigid? How would the astronauts move from one cluster of modules to the other? Other issues concerned the specification of the modules themselves, which were linked together in a way that would make it difficult to isolate any compartment in an emergency. The Power Tower was therefore rejected in late 1985. One year later, having evaluated submissions, NASA accepted the Dual Keel concept of Lockheed and McDonnell–Douglas. This had two 150-metre vertical trusses linked top and bottom by 45-metre spars for increased structural strength. Like the Power Tower, it was to exploit the gravity gradient for vertical orientation and have astronomical apparatus and Earth-observational instruments top and bottom, but in this case these would be remotely operated from a single cluster of modules set half-way up and between the keels, alongside the horizontal truss that supported the solar panels. This eliminated the requirement for a frequent-use transportation system along the keel. In addition, key safety features were incorporated into the modules. Within months, however, the enormous dual keel had been deleted from the configuration, and with it went the dedicated observational sites. Stripped of its vertical structure, the module cluster on the truss with the solar panels would not be able to exploit gravity-gradient stability. The heavy keel (it was argued) would be added later, but, until then, external apparatus would have to be mounted on the truss. To give the project a sense of identity, the structure was named 'Space Station Freedom', reflecting Reagan's hostility towards what he had so recently dubbed 'the evil empire' of the Soviet Union. Hence, the space station became a symbol of the Cold War, just as Apollo had been two decades previously.

Despite having selected the design, the annual Congressional review of NASA's budget ordered a succession of redesigns, each intended to reduce the project's cost, but during much of this hiatus the Shuttle fleet was grounded after the loss of Challenger. In 1990 Congress sent NASA back to the drawing board again. In response, NASA proposed a configuration that could be 'crew-tended' early in the assembly, and then progressively upgraded over five years to facilitate permanent occupancy. To cut the number of Shuttle flights during the assembly sequence, the truss had been shortened by 30 per cent, and the modules by 40 per cent in order to ensure that they could be delivered fully outfitted. However, the fact that the capacity of the habitat had been reduced from eight to four people meant that the European and Japanese agencies – already frustrated by NASA's seemingly endless funding delays – would not be able to staff their laboratories permanently. When Dan Goldin took over from Richard Truly as NASA administrator in April 1992, he instigated yet another redesign with the aim of further slashing costs. In addition, he ordered that the station be provided with a 'lifeboat' (or an Assured Crew-Return Vehicle, as it was prosaically termed) in order to ensure that its crew would be able to return to Earth in an emergency. Following the signing of the *Human Spaceflight Cooperation* protocol with Russia in October 1992, and in an effort to eliminate the cost of developing its own vehicle, NASA proposed the purchase of a modified Soyuz spacecraft that the Shuttle would deliver and attach using the RMS. When the Clinton administration took office in January 1993, the provocative name 'Freedom' was dropped and, in March, NASA was ordered to devise an even more cost-effective design that would contribute to Clinton's promise to reduce the national deficit.

ALPHA

On 7 June 1993, NASA submitted three options. Option 'A' was to exploit proven hardware and incorporate those systems intended for Freedom that would be most cost-effective. Option 'B' was essentially what had already been proposed, and would maximally exploit the work done on Freedom. Option 'C' was a 'minimalist' facility that called for Columbia to be decommissioned and its engine unit mated to a specially modified external tank and then used to orbit a large one-piece more or less completely outfitted station, as a latter-day Skylab. On 17 June the White House selected Option 'A', and asked Congress to award the necessary funding, which was duly done on 23 June – although with a majority of only a *single* vote. NASA was given three months to 'transition' to this new configuration. Boeing was nominated as prime contractor on 17 August, and the detailed project definition for 'Alpha' was sent to the White House on 7 September. In parallel with this technical review, in which the Russian Space Agency had joined NASA's formal international partners in a 'consultancy' capacity, the White House had been working to expand the scope of the 1992 agreement with Russia, and on 2 September Vice President Al Gore and Prime Minister Viktor Chernomyrdin signed an accord that called for their space station proposals to be merged into a single facility. The fact that this announcement took NASA's international partners by surprise highlighted the rapid pace at which the programme was being transformed.

On 1 November, NASA and the Russian Space Agency submitted an addendum to the 'Alpha' plan that was issued on 7 September. This *Space Station Programme Implementation Plan* involved three phases. Phase One was a logical expansion of the October 1992 plan to fly each other's crews, and called for the Shuttle to make a series of dockings with Mir between 1995 and 1997, and for NASA to station a succession of astronauts on Mir. Phase Two called for sufficient elements of the new facility to be in orbit by 1998 to support continuous habitation. Phase Three involved flights through to 2002 to complete the assembly. This plan was submitted to the White House on 4 November, and was endorsed on 29 November. The formal invitation to Russia to participate was issued on 6 December, and the Russian Space Agency signed on 16 December. The first meeting of the international team, which included prime contractor Boeing and its subcontractors Rockwell, Lockheed and McDonnell–Douglas, was held on 24 March 1994 to review the overall systems-design. In June, Gore and Chernomyrdin signed the interim cooperation agreement to build what was being informally referred to as 'International Space Station Alpha'. Chernomyrdin was aware that critics in Congress were continuing to snipe at the proposed budget, and in response to concerns by members of Congress that the Russians would not be able to deliver on the deal, he assured – with his tongue firmly in his cheek – that Russia would push on with the project even if NASA had to withdraw!

TWO STATIONS INTO ONE

With the political agreement in place, the two agencies then met to define the detailed configuration and the assembly sequence. NASA settled on a configuration that exploited 75 per cent of the hardware intended for the 1992 version of Freedom.

However, there were significant revisions. One immediate decision was to employ a Russian manoeuvring unit – or 'tug' – built by Khrunichev, the company that had made all the Mir modules. It would not only be cheaper than the system that NASA had intended to use, but would also be able to be replenished in orbit. The simplest way forward was to order an off-the-shelf tug, and somehow mount it on the scaled-up Alpha, but to have done no more than this would have frittered away the golden opportunity offered by cooperating with Russian. On its own plan, NASA would not be able to leave a crew on the station until the habitat module was added, which was not scheduled until the station was virtually complete. During construction, it would be 'tended' only while a Shuttle was docked. The Mir 2 base block, however, was a habitat, and surely it could be docked at the other end of the tug. By starting with these two Russian 'core modules', the unproductive initial period of the assembly process could be eliminated. Although the assembly would take just as long to complete, the 'permanent habitation capability' could be advanced by two years. It was similar to fitting together the pieces of a jig-saw. The tug was given a multiple docking adapter with an androgynous axial port to mate with the first of NASA's modules and a pair of radial ports for either Soyuz ferries or Progress tankers. Its other end would be linked to the front of the base block. These modules would provide orbital manoeuvring, attitude control, power supply, logistics and crew accommodation during the build-up of the facility. Since the hybrid integrated both technologies, the Russians would be free to develop the other end as and when they desired. After substantial votes in favour of the project in both houses of Congress in July 1994, Goldin insisted that it was "no longer just a design". When Congress expressed concern that the Russian Space Agency, which was facing severe financial constraints, would not be able to deliver the two core modules, NASA agreed to fund the development of the first one, later named Zarya.[1] In January 1995, having resolved the configuration, assembly sequence, schedule and cost, NASA confirmed its contract with Boeing to build the American hardware, and the company opened negotiations with its subcontractors. As Lockheed had forged a relationship with Khrunichev to commercialise the Proton rocket, Lockheed received responsibility for the development of Zarya, and on 8 February a $200-million contract was agreed for its 'on-orbit delivery'. Khrunichev was also to build the base block (later named Zvezda) under a contract with the Russian Space Agency that would be funded by the Russian government. Although Boeing would not play a direct rôle in fabricating the European and Japanese modules, it had the responsibility of ensuring that all the various hardware would properly integrate.

On 28 September 1995, in an unprecedented move, Congress voted NASA a multi-year budget of $2.1 billion per annum through fiscal years 1996–2002, thereby guaranteeing the financial commitment required to manufacture and assemble the space station in orbit. Unfortunately, despite a pledge by Boris

[1] Although the tug was initially referred to as either the 'functional energy block' (Russian acronym FGB) or the 'Control Module', the Russians later named it 'Zarya' (for dawn or sunrise). Similarly, although the base block was for some time known as the 'Service Module', the Russians named it 'Zvezda' (star). For simplicity, therefore, these vehicles will referred to herein by their now accepted proper names.

Yeltsin, the Russian Space Agency did not have the stable cash flow required to construct Zvezda, and funding for the proposed research modules seemed ever less likely to be forthcoming. The Russian Space Agency's dilemma was that in order to work on building up the new facility, it would have to terminate Mir operations. The worst outcome, from the agency's point of view, would be to de-orbit Mir and then be denied the funding to participate in the bold new venture. In November the Russian Space Agency announced its intention to transfer Spektr and Priroda to the new station. NASA had no objection to increasing the science capability of the complex, as long as achieving it did not delay the assembly schedule or increase its cost. On 11 December, however, the Russians proposed a radical change of plan in which Zarya would dock at the front of Mir. This would make the best use of the existing asset, and would facilitate faster commissioning of the new station by virtue of supporting occupancy by larger crews. As a compromise, NASA suggested that it extend Phase One into 1998 by adding two missions to the Shuttle–Mir programme, to enable the Russians to earn some extra fees from Mir, but the assembly of the new facility must start as planned, in November 1997. In an effort to relieve pressure on the Russian Space Agency's finances by reducing the number of Proton rockets it would have to fund, NASA also offered to deliver the Russian Solar Power Platform. This agreement was ratified in July 1996, by which time NASA had dropped the name 'Alpha'. The new facility was now simply the 'International Space Station'.

With Russia's launch capability declining, it was clear that the Shuttle would be the workhorse of the assembly process. The decision to employ a 51-degree orbital inclination to enable Russia to participate, meant that the Shuttle would not be able to carry the fully outfitted NASA laboratory, and most of its equipment would have to be delivered later. NASA had originally intended to build a similarly sized logistics carrier to resupply Freedom, but this was not included in Option 'A'. Rather than try to reinstate it, NASA accepted an offer by the Italian Space Agency to supply a smaller Multi-Purpose Logistics Module (MPLM). A deal was struck whereby, in return for Italy's donation of three cargo modules, a Shuttle would deliver an Italian life-sciences module incorporating a large centrifuge. In this way NASA operated its 'no exchange of funds' principle. Assembly of the ISS would have to be borne by Atlantis, Discovery and Endeavour, because Columbia, the first of the fleet, and some four tonnes heavier, could not carry heavy payloads. Of two long-awaited upgrades, the super-lightweight aluminium–lithium External Tank was on track for introduction in 1998, but, after several stops and starts, the Advanced Solid Rocket Motor, ordered in the aftermath of the loss of Challenger, had been finally cancelled in late 1993.

By the end of 1996, it was apparent that cash flow problems were significantly delaying the fabrication of Zvezda, but Zarya, which was being paid for by NASA, was on schedule. In response to continuing concern in Congress that Zvezda was in the 'critical path', NASA decided to modify a manoeuvrable rocket stage that had been developed by the Naval Research Laboratory to ferry military satellites to their operating orbits and was compatible with the Stabilised Payload Deployment System that STS-36 had used to deploy a classified satellite. This Interim Control

Module (ICM) would provide a contingency against Zarya expending its propellant (thereby leaving the complex without attitude control) before Zvezda could be launched. In fact, the US–Russian relationship was under strain. An instruction by Chernomyrdin to release government funds to resume work on Zvezda had been undermined by the bureaucracy. "We set up a very complex programme, and the Russians have not been funding their side," said Goldin at a meeting in April 1997. Yuri Koptev, the chief of the Russian Space Agency, said he was just as frustrated as his American counterpart by the state of affairs. On 9 April it was decided to slip Zarya's launch from November 1997 to mid-1998. To launch Zarya on time, simply to adhere to the schedule, had been deemed futile. Nevertheless, NASA hoped that it would eventually be able to make up time by pursuing a more aggressive schedule in order to complete the assembly more or less on time. Reflecting that since January 1996 the Russians had promised on eight occasions to resolve their funding problem, James Sensenbrenner, the senior Republican on the House's space committee, said that the programme was "falling apart around us" and again urged NASA to "remove the Russian *government* from the critical path" by making the Russian companies direct subcontractors to Boeing. The House's space committee then expressed its frustration by telling NASA not to fund any hardware that the Russian government was meant to supply. It also ordered the agency to make monthly reports on Russia's status. "We knew from the outset", Goldin insisted, "that building an International Space Station would be tremendously challenging." It was ironic that it was the prospect of Russian participation that had saved the programme politically in 1993, yet now the inability of that government to meet its obligations had become the greatest obstacle to the station's assembly. On 11 April the Russian government arranged to release $260 million by having banks issue loans to Energiya with the proviso that the loans would be repaid with 3.5 per cent interest. With this assurance of funding flowing by the end of May, Khrunichev resumed work on Zvezda, which prompted NASA to announce that, apart from starting late, it was "cautiously optimistic" that the ISS was now back on track. However, Randy Brinkley, the programme manager in Houston, pointed out that if Zvezda could not be launched by the end of 1998 the ICM would be sent up. On 14 May the Space Station Control Board issued a new schedule on which Zarya would be launched in June 1998 and NASA's first element, the 'Unity' node, would follow in July and Zvezda in December. This new schedule had "much more flexibility" Brinkley observed. An additional Shuttle flight had been assigned to provide a "contingency opportunity" to deliver the ICM in December 1998 if the launch of Zvezda was further delayed. If Zvezda *was* launched on time, this 'extra' mission would deliver logistics. The first crew would be launched in January 1999, regardless. To provide "defence in depth" against the run down of the Semyorka rocket's production line resulting in no Progress tankers to replenish Zvezda, NASA added a mission in October 1999 to mount the ICM on its rear for a year, in order to preserve its propellant. Also, to assure a sufficient supply of power to operate its laboratory (named 'Destiny') NASA decided to advance a segment of the Integrated Truss Structure (ITS) incorporating a pair of Solar Array Wings, and to build a short truss (dubbed the Z1) on which it to install it.

Table 11.1 Shuttle manifest, c. 14 May 1997

Date	Flight	Launcher	Objective
Jun 1998	1A/R	Proton	Zarya
Jul 1998	2A	STS-88	Node 1 ('Unity') with PMA 1 and 2
Dec 1998	1R	Proton	Zvezda
Dec 1998	2A.1	STS-96	Either Zvezda logistics or the ICM
Jan 1999	3A	STS-92	Z1 truss and PMA 3
Jan 1999	2R/1S	Soyuz	ISS-1 crew
Mar 1999	4A	STS-97	P6 truss with solar power system
May 1999	5A	STS-98	US Laboratory ('Destiny')
Jun 1999	6A	STS-100	MPLM (with lab racks) and SSRMS
Aug 1999	7A	US	Joint Airlock Module ('Quest')
Oct 1999	7A.1	US	Either outfitting Destiny or the ICM

By the autumn of 1997, NASA had to admit that the ISS would not be able to be completed before December 2003, some 18 months after the previous target of June 2002. Having been asked to augment the Fiscal Year 1998 budget with $430 million to cover cost overruns, Congress was sceptical. In November, after suggesting that the agency terminate work on the ICM and reassign the cash towards the overruns, Wilbur Trafton resigned as ISS director and was superseded by Joseph Rothenberg. In February 1998, in his Capitol Hill debut, Rothenberg was "sceptical" of Zvezda being ready for launch in December. When a disbursement from Moscow which had been due that month failed to materialise, it was decided to postpone Zarya's launch to avoid 'wasting' its orbital service life. As long as Zvezda was ready by April 1999, the delay would not seriously disrupt NASA's schedule. When Yeltsin fired Chernomyrdin in March, Sensenbrenner personally flew to Moscow to assess the implications for the ISS. Meanwhile, NASA informed the House that overruns and schedule slips meant that the ISS would have cost $21 billion by completion in December 2003, which was some $4 billion more than the $17.4-billion budget. In April, NASA released the independent assessment entitled *Cost Assessment and Validation Report on the Space Station* by consultant Jay Chabrow, which reported that there was significant scope for delay during assembly (possibly amounting to three years of delay by the completion date) and warned that $130–250 million more would be required per annum than the budget provided. Chabrow warned that these two factors could easily push the total budget to $25 billion, exceeding the target by some 40 per cent.[2] "We're all concerned about the cost overruns and schedule slips," admitted Goldin, who had commissioned Chabrow's report the previous September, "and I'm not going to sugar coat them." It was clear that the $2.1-billion annual cap could not accommodate both the cost of overruns and the Russian crisis. Each

[2] Figures in the range $50–100 billion are often cited by critics as the 'true' cost of the project, but they include the $11 billion that was spent between 1984 and 1993 on the design of Space Station Freedom (whose design served as the basis of the ISS) as well as at least a decade's worth of routine operations following the assembly of the 'baseline configuration', and hence span a 30-year period.

month that Zvezda slipped was costing NASA $120 million. If Congress insisted on imposing the cap, Goldin said that the inevitable result would be to slip the completion date by "a year or two". On the new schedule drawn up by the Space Station Control Board on 30 May 1998, Zarya would be launched "no earlier than 20 November", Unity would be added by STS-88 "no earlier than 3 December", and Zvezda would be launched "no earlier than 28 March 1999". Nevertheless, NASA hoped to make up time to complete the station by January 2004.

Reminding the White House that Chabrow's report had said that the programme was underfunded by $130–250 million per year, Sensenbrenner urged Clinton to find the shortfall because it was his administration that had introduced the Russians who were now costing NASA so much concern, but Jacob Lew, the director of the Office of Management and Budget, insisted that no "additional funding" was available. In effect, the administration could not abandon its foreign policy and admit that Russia had failed in its obligations, and Congress did not wish to sanction "foreign aid" to Russia. The White House was hoping that NASA would find the money it required by raiding other programmes, and Congress was determined not to let this happen. In September, the agency proposed a two-tiered strategy. The immediate priority was to alleviate Russia's financing shortfall by "investing" to guarantee the timely production of Russian items required to put the station on track; that is, Zvezda and the Progress tankers and Soyuz lifeboats. It was equally imperative that NASA should seek long-term "independence" from Russia. It should build a replenishable propulsion module and consider the offers by the European Space Agency and Japan to provide logistics vehicles. It proposed a $1.2-billion one-time injection of funding to "fix" the programme,[3] and urged the immediate sanction of a $60-million "advance" to ensure that Zvezda was delivered on schedule. To preclude criticism that this was just another bail-out, NASA described it as "prepayment" for the two Soyuz lifeboats. It also proposed four annual $150-million instalments to guarantee three Progress tankers for orbital reboost as the station was being assembled and a pair of Soyuz spacecraft per year, each of which would serve a six-month tour. The funding would be provided on the strict understanding that it would be used to produce crucial station hardware. Senator John McCain, the chairman of the Commerce Committee, was concerned that this would simply establish a precedent that the Russian government might exploit. Sensenbrenner said that it showed that putting Russia in the station's critical path had been an expensive mistake, and that NASA ought to have seen the problem coming. As John Glenn launched on board STS-95 on 29 September 1998, former CBS anchorman Walter Cronkite asked Bill Clinton, live on CNN, if he was willing to spend "lots of money" to keep the ISS on track. "If it were required," Clinton replied, "I'd be supportive of it." He noted that NASA had had "hardly any increase in funding" since his election to office. In fact, allowing for inflation, the agency had had a 20 per cent cut under the Clinton administration. "If it were required now to help the Russians through this difficult

[3] The $1.2-billion "fix" comprised the $660 million to ensure that crucial Russian hardware was produced, $90 million to modify the Shuttles so that they could provide orbital reboost, and the rest was to develop and launch the new propulsion module in 2002 as a long-term solution to the reboost issue.

period, which will not last forever, so they could continue to participate, I'd be in favour of it. I think we're doing the right thing with this space station, and we need to stay with it." John Logsdon of the Space Policy Institute at George Washington University in Washington, DC, was encouraged by Clinton's candid support for the programme. "What Clinton said at the launch was clearly scripted," he observed. "No President goes on television and makes up commitments like that." But Clinton's support appeared to have been directed specifically towards supporting the part of the station that was sitting in the Russian critical path – the $660 million. This, however, was only half of the $1.2 billion proposed by NASA to eliminate its reliance on Russia, and a mere 28 per cent of the projected $2.4-billion shortfall by the end of the assembly period. By limiting his support to the immediate issue, Clinton had failed to address the real issue: *the station could not be built within the projected budget.*

THE GREATEST ORBITAL CONSTRUCTION PROJECT

Although there were four spacecraft in the fleet, the task of assembling the ISS fell to Atlantis, Discovery and Endeavour. For the Mir dockings, a 1.6-tonne Orbiter Docking System (ODS) had been built using a Russian-built APAS docking system, and as the ODS was connected to the outer hatch of the mid-deck, the airlock was moved to the base of the ODS to provide access to the payload bay. Mounted in its cross-bay truss, the ODS effectively became a 'permanent payload'. NASA decided to retain this configuration for the ISS. As the first orbiter to be equipped, Atlantis was assigned to the Shuttle–Mir programme, but it was due for its next service when this series was extended, and so the final few missions were flown by Discovery and Endeavour, which had been similarly outfitted in 1996 and 1997 respectively. As the first of the fleet, Columbia was significantly heavier and hence did not have the same lifting capacity, so it was not equipped with an ODS. It would routinely service the Hubble Space Telescope, deploy the Chandra X-Ray Observatory and perform such 'independent' missions as might arise from time to time. Hence, for the foreseeable future, most of the fleet would be committed to flying the sequential missions of the integrated manifest. Of course, sceptics doubted that the hardware built by various companies in different countries would fit together properly, and that their systems would function as specified. There was also lingering concern that what NASA was calling the greatest orbital construction project ever attempted was overly reliant on spacewalking astronauts. This account will therefore focus on how the ISS hardware was assembled in space, because, to a much greater extent than its predecessors, it is a station that is being *manually* assembled in space.

YEARS OF PREPARATION

Was Freedom a wasteful fiasco? Not if it is recalled that the option Clinton selected in 1993 was essentially the final Freedom configuration, and that this design – which had reached the point of awaiting the go-ahead for fabrication – enabled NASA to fabricate the hardware for the ISS more rapidly than would otherwise have been possible. In addition, over the years, many of the subsystems had been tested on Shuttle flights. A mere six months after Reagan directed that a station be built, STS-41D tested the deployment mechanism for a large solar panel, on which Lockheed had begun work a decade earlier in the expectation that it would be needed sooner or later. Although the thickness of this frame in its stowed form was a mere 75 millimetres, it unfolded concertina-fashion to a length of 32 metres. Exploiting the Shuttle–Mir programme, NASA installed the advanced gallium arsenide transducers intended for the ISS into a proven Russian deployment frame, and this Cooperative Solar Array was deployed on one of Mir's modules to verify that the high-powered transducers would withstand the harsh space environment.

When NASA started to consider possible configurations for its space station, it realised that one of the most crucial factors involved whether it would have to design a self-deploying structure for the main truss, or whether this could be assembled by spacewalkers. In November 1985 Jerry Ross and Sherwood Spring repeatedly built and stripped down several rod-and-pin structures. When a time-and-motion study of their performance confirmed that spacewalkers could indeed assemble truss frames, proposals for a self-deploying structure were shelved.

Only about 40 per cent of apparatus designed to operate in microgravity actually works first time, and the rest requires modification to rectify unforeseen flaws. The Station Heatpipe Advanced Radiator Experiment was one such case. This prototype element for the station's cooling radiators vaporised liquid ammonia, and thermally induced convection transported the vapour along a 15-metre pipe so that a fin assembly could radiate the heat to space and condense the ammonia to flow back to the evaporator. The prototype was to have been tested in 1986, but the Challenger accident delayed this to STS-29, on which occasion it became clogged by air bubbles. On STS-37 a subscale model with a transparent cover was filmed and when this, too, became clogged, the problem was identified and rectified. Although a redesigned pipe tested on STS-43 developed bubbles, it was able to clear itself. The orbiter turned to put the radiator in the shade, so that different heaters could be applied to verify that the evaporator could cope with a range of thermal loads. Finally, the heatpipe was deactivated and allowed to freeze before being restarted in order to demonstrate that it would survive an extended period off-line in shadow. While such payloads rarely made the headlines, they were vital for the long-term future of the programme. After all the rôle of testing is to reveal design flaws, and the risk of technological frustration was significant in a structure incorporating as many new systems as the ISS.

As in the case of the preparations to repair the Hubble Space Telescope, where nothing was left to chance, even apparently straightforward apparatus was verified for the ISS. This included new work stations, rigid umbilicals, foot restraints, the trays to carry pipes and cables along the truss, and even tool boxes. Tests were conducted to verify that astronauts wearing spacesuit gloves could manipulate the plugs and sockets that were to be used to hook up external payloads. Instrumented foot restraints were used to measure the forces imparted by astronauts working on typical maintenance tasks. Astronauts removed thermal insulation from a battery, unplugged it and restored it in order to identify any issues that might impede their replacement on the ISS. As time is a premium resource on a spacewalk, a 'translation aid' was deemed necessary to enable spacewalkers to move themselves and bulky apparatus rapidly along the truss, and once STS-37 had deployed its satellite and its payload bay was clear, Jay Apt and Jerry Ross tested several competing designs. Not everything went smoothly, however. When Columbia's hatch refused to open on STS-80, this forced the cancellation of a test of the ORU Transfer Device – the crane to be used to swing bulky Orbital Replacement Units such as batteries around on the space station's truss – but this device was tested a year later on STS-87.

The Shuttle–Mir programme offered another opportunity for NASA to minimise technological risk. Although Shuttles often retrieved satellites, they did so using the RMS. To dock with Mir, the Shuttle was required to manoeuvre precisely down an ever-narrowing path leading to the docking unit, and had to stay within an 8-degree cone and maintain a 2-degree tolerance in orientation. When Atlantis made its first approach, it closed at a rate of 3 centimetres per second – offset by no more than 2 centimetres from the cone's axis – and had an angular error of less than half a degree, verifying that the Shuttle is indeed a remarkable flying machine. It later performed the even more difficult task of docking 'in the blind', because the pilot's line-of-sight was blocked by the body of the module that was to be attached to Mir. As the Shuttle resupplied the Russian station with an assortment of cargo, NASA learned many lessons concerning loadmastering. By 1998, therefore, NASA was in a much stronger position to assemble and operate a space station than it had been when Reagan assigned it the task.

ZARYA AND UNITY

At 01:40 Eastern Standard Time[4] on 20 November 1998 a Proton launch vehicle lifted off from Tyuratam in Kazakhstan and placed Zarya in orbit. The mood at the post-launch press conference was jubilant because, at long last, the assembly of the ISS was underway. "Now, we only have 44 launches to go, about 1,000 hours of spacewalks and countless problems, ... but because of the trust and mutual respect ... the ISS is going to be a reality," assured Goldin. Koptev took advantage of the occasion to say that he was actively "working with external investors" to secure the funding to continue Mir operations into the new millennium. Goldin, however, was insistent: "We expect the Russian government to live up to its ISS commitments." In fact, the basis of a compromise was becoming evident. The Russian government could satisfy NASA by denying funds to Mir and releasing Energiya to continue to run Mir as a commercial venture with fee-paying clients. However, running both programmes would create competition for the limited supply of Soyuz and Progress spacecraft and Semyorka rockets. The production lines would have to triple their current low-level deliveries to keep Mir operating, which seemed unlikely. Viktor

[4] Unless otherwise stated, all times are US Eastern (see the Mission Log on pp. 387–411 for more details).

Blagov, deputy director of the control centre at Kaliningrad near Moscow, opined that Mir's physical condition was "fully satisfactory", and insisted that continued operations would not draw resources from the ISS: "Enough Soyuz and Progress are already built to ensure Mir's use through 1999." In fact, Mir had enough propellant to operate through 2000. Fee-paying international missions had recently contributed $120 million per year, Blagov noted. However, this income was mostly due to the Shuttle–Mir contract. On the other hand, the fact that NASA had booked Mir's 'research seat' for so long had precluded others from using its facilities. As soon as it was clear that Mir was not to be immediately de-orbited, France booked a 100-day mission starting early in 1999. To NASA's chagrin, therefore, Mir would continue to distract the Russians from their obligations to to ISS.

The basic design of the 12.5-metre-long 20-tonne Zarya vehicle derives from the Almaz reconnaissance platform that was designed by Vladimir Chelomei in the late 1960s. Its configuration is similar to Mir's Kristall module, with a small spheroidal compartment incorporating multiple docking ports. It has a pair of solar arrays, each 10.6 metres in length, which, between them, generate an average of 3 kilowatts. In addition to two large engines for orbital manoeuvring and reboost, it has 24 large and 12 small thrusters for attitude control. Its externally mounted tanks can hold a total of some 6 tonnes of propellant. The port at the 'rear' is a hybrid combining a probe with an APAS collar to dock with Zvezda. The forward axial APAS port is to mate with the first US element. The radial ports are of the standard type to accept Soyuz and Progress vehicles, with the nadir one incorporating plumbing for the propellant tanks to be replenished.[5] Early in the assembly, Zarya would provide the passive US elements with electrical power, orbital reboost, attitude control, and communications, but its functions would be progressively subsumed.

Within minutes of being released, Zarya successfully deployed its solar arrays and antennae. Three hours later, the Russian flight controllers radioed a command to set it slowly rolling on its primary axis in a passive thermal roll which would ensure that energy from the Sun was distributed evenly across its exterior, to prevent fluids from freezing and bursting pipes.

The following day, one of the main engines was fired for the first of a series of manoeuvres to refine its orbit in preparation for the first Shuttle, and the two black-and-white television cameras to be used while docking with Zvezda were tested. On 22 November, tests suggested that both the 1.2-metre antennae of the TORU manual docking system might have failed to deploy. They should have unrolled when the bolts that held them in place fired. Although the telemetry failed to verify that they had fully deployed, the engineers thought it likely that they had done so and that the sensors had failed. On 23 November it was discovered that Battery 1 indicated a higher charge than its five companions. Further tests over the next two weeks established that it was functioning correctly and the fault was in the associated Storage Battery Current Regulator System (known by its Russian initials PTAB). It

[5] Zarya's zenith port will remain unavailable while access to it is obstructed by the solar panels on NASA's P6 truss.

was decided that the STS-88 crew should inspect it. By 26 November, it had been concluded that one of the TORU antennae was at least partially deployed. It was then decided to have the spacewalkers on this flight deploy the antennae manually, as these would be needed if a Kurs system failure meant that the docking with Zvezda had to be undertaken manually. By now, Zarya was essentially ready for NASA to add its first element.[6]

Because NASA's modules were to be delivered and installed by the Shuttle, they were not obliged to incorporate their own systems for propulsion, flight control or power generation, and their entire mass could be devoted to the function they would serve when part of the ISS. An early decision, dating back to the design of Space Station Freedom, was that pressurised modules would exploit the full width of the Shuttle's bay. In contrast to the multi-segmented Russian core modules, NASA's modules are cylinders with an external diameter of 4.42 metres and an internal diameter of 4.25 metres, and are connected by 2-metre-diameter Common Berthing Mechanisms (CBM) which have a 1.3- by 1.3-metre hatch with a centrally mounted 20-centimetre-diameter window. As an interconnecting 'resource node', the 5.5-metre-long Unity module incorporates six CBMs – one in each shallow conical end-cap and a set of four radial ones that are offset towards one end to house racks at the other with command-and-control and power-distribution apparatus. In addition to serving as the primary structural element for the 'US core', Unity's 10 kilometres of power, data and video cables, and 200 lines for fluids and gases, route resources to the other pressurised modules and attached elements. Fabricated by Boeing in Huntsville, it was delivered to the Space Station Processing Facility at the Kennedy Space Center in June 1997.

The Russian involvement introduced a complication. The Space Station Freedom design, with which the Canadian, European and Japanese modules were to have been compatible, was to have run on 120 volts DC, and this aspect of the design had been retained, but the Russians had carried over the 28 volts DC used on Mir. Although power converters were installed to feed power in each direction, there is limited inter-operability between the two 'zones'.

The architecture of the Command and Data Handling System uses standardised Local Area Network technology for its computer systems. Within any given module the connections are by shielded cables using standard Ethernet, but communications between modules employ radio-frequency Ethernet because, in light of experience on Mir, it was deemed unwise to string cables through open hatches.[7] The astronauts interact with the systems via IBM Thinkpad laptops, most of which run the Solaris operating system, obviating the requirement for vast panels of dedicated switches and displays. Nevertheless, the Russians were wryly amused that while the lights on Mir could be operated by throwing a switch, a laptop has to be booted up in the ISS to issue the command.

[6] In the arcane nomenclature of the programme, Zarya's launch was SSAF-1A/R, indicating that the first ISS assembly flight was an American-owned payload carried on a Russian rocket; the flight of STS-88 was SSAF-2A, which was all-American.

[7] Mir was criticised for stringing power and data cables through open hatches, and this became an issue when it proved necessary to seal its hatches rapidly, both during a fire and a depressurisation following a collision.

As the CBM hatches are much wider than the APAS docking system used by the Russians (and, hence, by the ODS) NASA developed an adapter in the form of a truncated conical shell some 2.5 metres in length that had a CBM at the wide end, an APAS at the narrow end, and incorporated a 0.6-metre axial offset between the end rings to provide additional clearance for the cabin of a Shuttle manoeuvring to dock. This structure was fabricated by machining a stack of welded aluminium alloy ring forgings of progressively diminishing diameters. Three of these Pressurised Mating Adapters (PMA) were built at Boeing's Huntington Beach factory near Los Angeles in California. A pair of PMAs were placed on Unity's axis in March 1998 and, after final tests, the hatches were sealed on 20 September.

For STS-88, Endeavour was rolled out from the VAB to Launch Complex 39's Pad B on 21 October, and preparations for launch begun promptly. Unity was transferred to the payload changeout room on the pad on 27 October and installed in Endeavour on 13 November. Ten days later, the Flight Readiness Review confirmed the intention to launch in the 10-minute window on 3 December, and the countdown commenced on 30 November. The Air Force meteorologists predicted a 60 per cent likelihood of a postponement because of low cloud and showers, but by the time the propellants had been loaded this had been reduced to 30 per cent. Bob Cabana's crew boarded Endeavour on schedule and ran through the final hours of the count. At T–4 minutes and 24 seconds, the Master Alarm sounded on the flight deck because the pressure in Hydraulic System 1 had temporarily dipped as the Auxiliary Power Units that operate the orbiter's hydraulics were switched from low to high power. Unfortunately, by the time that the clock could be restarted it was evident that the launch window would expire before the count could reach zero, and so a 'scrub' was ordered. After STS-88 had been made safe, the disappointed crew disembarked, and the launch was rescheduled for the following day. Critics who doubted that the Shuttle could be routinely launched in such narrow windows, proclaimed 'I told you so'. In making his communications check after re-entering Endeavour the next night, Cabana told the conductor in the Launch Control Center, "Let's go do this tonight". This time, the count was flawless and the Shuttle lifted off at 03:35 on 4 December. As the rendezvous with Zarya began, the pressure in Endeavour's cabin was reduced from nominal 'sea level' at 14.7 psi to 10.2 psi to significantly shorten the time that the astronauts who were to undertake the EVAs would need to spend breathing pure oxygen to purge bubbles of nitrogen from their bloodstream preparatory to donning their EMU suits, which operated at 4 psi – a protocol designed to prevent a bout of the painful affliction known to divers as 'the bends'.

On the 5th, Nancy Currie powered up the RMS, engaged the grapple fixture on PMA 2, and lifted the 11.6-tonne stack 4 metres straight out of the bay, doing so very slowly and carefully because there was only a 2.54-centimetre gap on each side between the node and the wall of the bay. She then rotated the 11-metre-long stack and positioned PMA 2 a few centimetres above the extended ring of the ODS. With the arm in 'limp' mode, Cabana fired Endeavour's thrusters to drive the collars together. The docking ring on PMA 1 was cycled several times to verify its function in advance of the rendezvous with Zarya. The small 'vestibule' between the ODS and

PMA 2 was then pressurised, and the hatches opened to enable Cabana and Jerry Ross to cap vent valves in preparation for entering Unity later that week.

The following day, as its orbit reduced the range to 15 kilometres, Endeavour initiated the terminal phase which, over the next hour and a half, brought the orbiter to a point 180 metres below Zarya. Up to this point, the profile was as used in an R-bar approach to Mir. However, because the Russians did not want the Shuttle to block the line of sight of Zarya's antennae to their ground station, Cabana, working at the aft flight deck, pursued a half-loop 100 metres in front of Zarya to a point 75 metres above it, where he halted with PMA 1 pointed towards Zarya. Since Unity blocked his view of the Russian vehicle, the RMS had been stationed out to the side to provide an oblique line of sight. After capturing Zarya using the RMS, Currie positioned it directly over the Unity stack. With the arm once again in limp mode, Endeavour drove PMA 1's extended ring against the APAS on the multiple docking compartment for a 'soft' docking. However, when the mechanism was retracted the latches in the collars failed to mate. It was belatedly realised that the RMS, although limp, was tipping the module over slightly, so the process was repeated after the arm had been withdrawn and, some 40 minutes behind schedule, the collars engaged to establish a hermetic seal. Currie then swung the RMS further up the nascent ISS projecting 23 metres above Endeavour's bay and confirmed that the pyrotechnic pins that had held Zarya's TORU antennae in place for launch had indeed fired, but either stiff cabling or interference from thermal blankets had prevented the antennae from fully unrolling from their spools to form long thin rods.

On the 7th, on their first EVA, Jerry Ross and James Newman connected power and data cables between Unity, its PMAs and Zarya – the cables linking Unity to its adapters could not have been connected prior to launch because of the payload bay constraints. Newman retrieved the cables from where they had been stowed for launch, and Ross, on the RMS operated by Currie, plugged them in. In all, some 40 cables were strung from PMA 2 to Unity, from Unity to PMA 1, and from PMA 1 to Zarya. As he worked, Ross put a thermal cover on each connector to protect it from the space environment. When the Russian flight controllers activated two Russian–American Power Converters to enable Unity to draw power, telemetry began to flow to Houston. With Unity's heaters operating, Ross removed the thermal covers from two computers (known as multiplexer-demultiplexers) on the exterior of PMA 1. Finally, they installed a number of handrails that, again, could not have been fitted for launch due to payload bay constraints. With their primary tasks accomplished, Currie hoisted Newman on the RMS to enable him to inspect Zarya's stuck antennae visually.

The following day, commands were relayed to pressurise PMA 1 between Unity and Zarya and to activate the fans and air filters in Unity, and on the 9th Ross and Newman went out again. Their first task was to affix a pair of box-like antennae on Unity, one on the starboard CBM and the other on the port-side CBM. These were part of the S-Band Early Communications System that was to increase the number of telemetry channels and provide a voice-link for crew working in the module. They then hooked up an external cable to this system to support video-conferencing from Zarya. Working together, with Newman on the RMS, they put a sunshade on one of

The Zarya–Unity stack on the ODS in Endeavour's bay on STS-88, with James Newman, assisted by Jerry Ross (on the arm), stringing cables.

the multiplexer-demultiplexers, released the launch restraints on Unity's four radial CBMs, and put thermal covers on the trunnion pins on which the module had rested in the bay. At this point, Newman, still on the arm, was positioned so that he could prod one of Zarya's fouled antennae, causing it to unfurl to its proper length. The following day, Endeavour's cabin was pumped up to full pressure and the hatches opened. Cabana and Sergei Krikalev entered Unity simultaneously to symbolise the cooperative nature of the venture. When they were joined by the rest of the crew, the apparatus that had been bolted down for launch was released and installed – one item being the electronics for the S-Band Early Communications System. To assist astronauts who might become disoriented in the module, the rims of the CBMs had signs providing directions. A video-conference was held in which Krikalev spoke to Bill Shepherd and Yuri Gidzenko, who were to accompany him as the first resident crew. Cabana and Krikalev then repeated their 'double act' by pushing on through PMA 1, first into Zarya's docking compartment and then into its corridor-like main compartment. While Krikalev and Currie replaced the faulty device that controlled the discharging of energy from one of its six batteries, the others unstowed hardware and logistical supplies that had been stored behind Zarya's panels for launch and positioned them ready for Shepherd's crew. In addition, tools and various supplies were left in Unity for the next visiting Shuttle crew. At 19:12 on the 11th, the hatch to Unity was shut to enable Endeavour to lower its cabin pressure preparatory to the EVA the next day.

As video monitoring by engineers in Houston had led to the suspicion that 'ties' on four cables strung during an earlier EVA would deny the cables enough 'play' to accommodate the cyclic thermal stress of the space environment, Ross and Newman reduced the tension on the cables on their final EVA. Then they affixed to PMA 1 a bag of wrenches, power grip tools, ratchets and portable foot restraints where it would be conveniently located for spacewalkers on the next Shuttle – the tools could not be left *inside* Unity because to do so would require the next crew to negate their pre-breathing protocol during the rendezvous by opening up the ISS. Moving up the stack, Ross and Newman put handrails on Zarya that would have interfered with the aerodynamic shroud if they had been installed prior to launch. Ross was to have placed a cover on the pin with which the RMS had manoeuvred Zarya, in order to minimise the thermal stress on the underlying structure, but this became detached and drifted away. Currie then manoeuvred Ross on the arm to enable him to release the antenna on the far side of Zarya. The final task was to disconnect and stow the cable by which the APAS docking system on PMA 1 had engaged Zarya. The cable had been run from PMA 2 around Unity prior to launch, and it would have to be removed before PMA 2 could be disconnected from Unity's axis – as it was to be, later in the assembly process. A test on STS-86 had found a faulty valve in the SAFER unit that prevented its nitrogen gas jets from firing. With their 'hard hat' work complete, Ross and Newman verified the resulting modification.

The PMA 2 docking latches were released at 15:25 on 13 December. With their attitude control systems in standby, both vehicles were passive. After the springs in the ODS had opened a separation of about half a metre, Fred Sturckow increased the opening rate using Endeavour's thrusters and Moscow reactivated Zarya's flight control system. Endeavour halted 135 metres away and flew a slow loop around the

ISS to document it. Once the Shuttle had departed, Zarya was ordered into a vertical orientation, Unity downwards, and set rolling every 30 minutes to minimise thermal stress. From time to time, Zarya's control system was told to cancel this roll and invert the stack to place Unity on top to confirm that it could set up for the arrival of the next Shuttle.

At this point, the docked combination could be monitored and controlled from either America or Russia. Houston had 'primary oversight' – meaning that it had final approval on all plans – and Moscow directed operations in real-time via Zarya, with the NASA engineers who were based there supervising commands sent to the US hardware. However, Houston would assume both the primary rôle in command and control functions and direct real-time operations when the addition of Destiny provided upgraded communications.

BACK TO WAITING

In January 1999 NASA realised that Zvezda's launch would have to be postponed for six months, into the autumn. The spacecraft was essentially complete, but post-manufacture testing was taking more time than expected, which in turn was delaying its shipment. On 1 February, Congress rejected NASA's proposal of a $600-million "investment" in the Russian Space Agency. On 25 February Joseph Rothenberg admitted that while the official launch date for Zvezda was September, "we have our reservations", and opined that it might not be ready until November or December. A meeting with the Russians was set for 15 April to review the situation, but NASA said that it would not commit to a launch date until June or July. Even so, astronaut Frank Culbertson, the deputy programme manager for ISS operations, was upbeat: "It will launch, it is just a matter of when." In early March, some of the senior staff changed hats. Randy Brinkley resigned and was superseded as the ISS programme manager in Houston by Tommy Holloway, and the veteran flight director Ron Dittemore filled the vacated slot of Shuttle programme manager. Meanwhile, the Russians were assessing their options in the event of Zvezda being unable to dock. Although Zarya's TORU could be operated from the ground, it was judged to be preferable to launch a crew – Gennadi Padalka and Nikolai Budarin – to dock at the rear of Zvezda to steer in the Zarya–Unity stack manually. This contingency option was added to the agenda for the management meeting in April. The US media argued that the Russians were simply seeking an excuse to occupy the ISS ahead of the 'official' crew, which would be commanded by Bill Shepherd, but the Russians insisted that after spending 10 days checking Zvezda's systems the 'interim' crew would return to Earth. At the meeting, the Russians confirmed that Zvezda would be sent to Tyuratam in May, and launched on 20 November. Even if the docking was successful, the option of sending up Padalka and Budarin in order to make an early start on commissioning it was discussed but rejected. "As far as I can tell," reported Goldin when Zvezda was rolled out on 26 April, "we're on track." Nevertheless, NASA could not pick up the pace of the assembly process until Zvezda was in place. By the summer, former astronaut Brewster Shaw, now heading the Boeing space

Table 11.2 Shuttle manifest, c. April 1999

Date	Flight	Launcher	Objective
24 May 1999	2A.1	STS-96	ISS logistics
20 Nov 1999	1R	Proton	Zvezda
3 Dec 1999	1P	Progress	Tanker and cargo
12 Dec 1999	2A.2	STS-101	–
22 Dec 1999	2P	Progress	Tanker and cargo
25 Jan 2000	2R/1S	Soyuz	ISS-1 crew
24 Feb 2000	3A	STS-92	–
6 Mar 2000	3P	Progress	Tanker and cargo
23 Mar 2000	4A	STS-97	–
20 Apr 2000	5A	STS-98	–
5 May 2000	2S	Soyuz	'Taxi' to replace ISS 'lifeboat'
19 May 2000	4P	Progress	Tanker and cargo
29 Jun 2000	5A.1	STS-102	–

station programme, pointed out that the Space Station Processing Facility at the Kennedy Space Center was "chock full" of hardware awaiting launch.

Recognising the progress in fabricating Zvezda, NASA decided not to pursue the option of fitting the ICM. This released STS-96. The plan had been to launch this as soon as Zvezda was in place, in order to outfit the module with all the systems that could not be installed on the ground. However, the ongoing delay was consuming the service life of some of Zarya's vital systems. If the ISS crew had been in space, this would not have been an issue. It was therefore decided to have STS-96 perform this maintenance on Zarya, and introduce an 'extra' flight (STS-101) to the manifest for the commissioning of Zvezda.[8]

MAINTAINING ZARYA

Since January 1999, barely three months after Zarya had been launched, its batteries had been experiencing a slight loss in charging capacity in the regular recharge cycle. A trio of 'charge controllers' (known by the Russian acronym MIRTS) indicated the level of charge of each battery, and it was suspected that these were faulty. Because the entire project would be jeopardised if Zarya were to lose power, they were to be replaced at the first opportunity. Another task that would have to be undertaken before the first crew took up residence, was to install customised acoustic insulation (dubbed 'mufflers') over fans in Zarya's ventilation system in an effort to reduce the background noise in that module to a level consistent with US regulations relating to health and safety at work.

[8] This made STS-96 ISS-02A.1 and STS-101 ISS-02A.2. However, STS-101 was itself split to fly prior to the launch of Zvezda as ISS-02A.2a with STS-106 flying as either ISS-02A.2b in the event that Zvezda docked or ISS-02A-2c if Zvezda had been lost and the ICM had to be fitted instead.

On 27 May 1999, after a 174-day 'dry spell' in Shuttle operations, STS-96 was launched with a Spacehab Double Module and an Integrated Cargo Carrier in its bay to "preposition" almost 2 tonnes of internal logistics and emplace external apparatus for later use. It was a mixed bag of items, including six IBM Thinkpad laptops (there was already one on Unity) and associated peripherals, replacement parts for Zarya's environmental system, tools for undertaking maintenance, medical supplies, clothing for the resident crew, scientific experiments, the kits with which to service Zarya's batteries and acoustic damping to be installed in Zarya's ventilation system – some 750 items in all. The 'dry' cargo was carried in the Spacehab, which also hosted the Volatile Removal Assembly to test apparatus designed to recycle water on the ISS. Spanning the bay above the Spacehab's tunnel, the ICC carried the operator's post for a Russian Strela crane, the ORU Transfer Device and the Spacehab Oceaneering Space System Box (an enclosure in which to deliver up to 180 kilograms of tools and equipment for use during EVAs). Also in the bay was the Shuttle Vibration Forces Experiment, with sensors to characterise the stresses on payloads during launch.[9] Its data would define criteria for testing ISS equipment, which must be subjected to vibration tests as part of its certification for flight.

Shortly before Discovery initiated the terminal phase of the rendezvous, Zarya was commanded to orient the ISS upright with Unity on top, and Zarya 'feathered' its solar panels to minimise contamination by thruster efflux. After coming up from below, Kent Rominger flew a half loop in front of the ISS and assumed a station-keeping position about 50 metres directly above it, where he waited to fly into range of the Russian communications network. This configuration was designed so that the Shuttle would not block the line of sight of Zarya's radio antennae, because immediately prior to the docking its attitude control system would have to be ordered off, lest it 'fight' manoeuvres imparted by the Shuttle. Meanwhile, Tamara Jernigan prepared the ODS systems. Given the go-ahead, Rominger, closed to 10 metres, paused to verify the situation, then made the final approach at 2 metres per minute to dock with PMA 2 at 00:24 on 29 May. Once Ellen Ochoa and Valeri Tokarev had stowed the docking target and verified the tunnel's hermetic seal, Rominger and Rick Husband removed four boxes of electronics that supplied power to the CBM to clear the way for transferring the bulky items of cargo through the narrow adapter.[10]

The mission's EVA was made by Jernigan and Daniel Barry on 29 May. Ochoa swung Jernigan on the RMS and Barry moved around using a tether. Julie Payette coordinated activities. Their first task was to retrieve the two cranes from the ICC. As part of the US crane for manipulating heavy Orbital Replacement Units, the 95-kilogram ORU Transfer Device had a 1.5-metre-long boom that could be extended out to 5 metres. Jernigan placed it on a temporary mount on PMA 2. Although it would not be needed until later, it had been assigned early in the manifest simply because this *ad hoc* logistics mission provided an opportunity to deliver it. By the

[9] This was the second flight for this package, which was first flown on STS-90.
[10] The boxes were stowed for reinstatement prior to the PMA being removed to enable Destiny to be installed.

time that Jernigan returned, Barry had released the 90-kilogram operator's post for the Strela. This was mated with a 15-kilogram adapter plate and mounted on a grapple fixture on PMA 1. Two new portable foot restraints designed to accept both American and Russian space boots, and three bags of tools and handrails for use in future assembly operations were transferred from the Spacehab Oceaneering Space System Box onto the ISS. With their main work achieved, Jernigan and Barry put a thermal cover on a trunnion pin on Unity (the cover that Ross had intended to install had floated off), inspected one of S-Band Early Communications System antennae on Unity (the system was operating at a low signal strength, but they saw nothing obviously amiss with the antenna) and then photo-documented the painted surfaces on both Unity and Zarya for evidence of flaking or discoloration in the harsh environment. Upon returning to the airlock, Jernigan experienced difficulty in reconnecting her suit to the ship's umbilical, and it took almost two hours to resolve the issue, with the result that at 7 hours 55 minutes this became the second-longest of the 45 spacewalks from Shuttles.[11] Later analysis of the pictures would reveal that, due to misleading documentation, the ORU Transfer Device had been incorrectly installed and was loose on its interface socket.

The hatches to the ISS were opened the next day. Barry and Husband replaced a power distributor and transceiver in the S-Band Early Communications System – in doing so restoring it to full capability. Payette and Tokarev went through into Zarya to make a start on the time-consuming task of replacing the recharge controllers on the six batteries under its 'floor'. Because they had to coordinate with the Russian flight controllers, who commanded the battery systems on and off, the working pace was driven by the gaps in communications coverage. Nevertheless, they managed to replace 12 MIRTS, each the size of a mobile phone, thereby restoring four batteries, leaving the remainder for the next day. Meanwhile, having measured the background noise level in Zarya at 70 decibels, Barry wrapped a number of mufflers around fans and ducts in an effort to reduce it. Concerned that these blankets of insulation were constricting some of the ducting and causing the humidity in the module to increase, he downloaded video for the engineers to examine; they promptly suggested that he remove the insulation from ducting that was deformed, and the humidity returned to normal. NASA's limit for ambient noise in its own modules was 55 decibels, which was equivalent to an office environment. A waiver had been issued to permit 60 decibels in the Russian modules. When STS-88 had measured the noise to be as high as 75 decibels at some places in Zarya the mufflers had been devised, but they only reduced the level to 65 decibels. In the comparatively tranquil Unity, Jernigan and Husband fitted shelving into racks that had been launched empty, to take some of the apparatus being transferred from the Spacehab. Serving as loadmaster, Ochoa documented the removal of each item from the module and recorded its placement on the ISS, because Shepherd's crew would need to know where each item was stored – in all, some 1,620 kilograms of miscellaneous cargo was loaded on board, including

[11] Recall that NASA defines the duration of an EVA as the time an astronaut draws on the suit's life-support system, but the Russians measure the duration from the moment the hatch is opened until it is sealed.

312 kilograms of water, in bags, drawn from Discovery's fuel cells. By the time they were finished, Zarya's corridor-like compartment was stacked so high that there was barely room for an astronaut to squeeze through. After the hatches were closed on 3 June, Discovery boosted the ISS's orbit (in an effort to save Zarya's propellant, this would become a routine operation for visiting Shuttles) and then undocked. Husband backed off 120 metres and flew a loop in order to document where things had been stowed for future EVA crews. For Tokarev, this had been an interesting first mission – having trained to fly the Buran shuttle he had found himself riding its American counterpart in order to service a Russian module on an international orbital complex.

After Discovery departed, flight controllers ordered Zarya to rotate the stack to a vertical orientation with Unity at the bottom, and reinstate the thermal control roll. A week or so later, when Battery 1 failed to fully discharge for its six-monthly 'deep cycle' maintenance, it was commanded off-line and the Russians requested that it be replaced by the STS-101 crew. In July, a problem developed with the S-Band Early Communications System when using the port-side antenna on Unity, and STS-101 spacewalkers were given the task of replacing it. Zarya continued to perform well, despite being low on power. During the second phase of 'deep cycle' maintenance in November, Battery 2 failed to discharge and had to be taken off-line. In its semi-dormant state Zarya could run on four batteries, but some of Unity's heaters were switched off to minimise power usage.

AWAITING ZVEZDA

The short-circuits that almost knocked out the SSMEs during the first few seconds of STS-93's ascent in July 1999 were traced to damaged cables. When inspections discovered that all the orbiters were similarly affected, the fleet was grounded.[12] In response, NASA asked Russia to postpone Zvezda's launch from 12 November to early 2000. Meanwhile, a string of failures led to the Proton rocket being grounded. Consequently, *both* the Russian and American launch vehicles were out of action in late 1999 and early 2000. On 27 January NASA decided to "protect the option" for STS-101 to fly prior to Zvezda to replace the faulty batteries and their ancillary equipment in Zarya. This would also be an opportunity to deliver logistics for the first resident crew. After the successful flight of a Proton on 12 February, it was decided to proceed with STS-101 "not earlier than 13 April" and launch Zvezda in mid-July. Another new mission (STS-106) was introduced that would fly a month after Zvezda docked in order to commission it or, if it was lost, to install the ICM to relieve Zarya of responsibility for manoeuvring. In the reshuffling of the crews, Yuri Malenchenko, Ed Lu and Boris Morukov – the STS-101 mission specialists who were to work on Zvezda – were reassigned to STS-106, and Yuri Usachev, James Voss and Susan Helms (the ISS-2 crew) were added to STS-101 in order to give them

[12] In the worst case – Endeavour – there were 20 strips of damaged cables, some of which were so eroded as to expose bare wire.

Table 11.3 Shuttle manifest, c. March 2000

Date	Flight	Launcher	Objective
17 Apr 2000	2A.2a	STS-101	Service Zarya
12 Jul 2000	1R	Proton	Zvezda
19 Aug 2000	2A.Ab	STS-106	Commission Zvezda
21 Sep 2000	3A	STS-92	Z1 truss and PMA 3
30 Oct 2000	2R/1S	Soyuz	ISS-1 crew
30 Nov 2000	4A	STS-97	P6
18 Jan 2001	5A	STS-98	Destiny
15 Feb 2001	5A.1	STS-102	MPLM Leonardo to outfit Destiny
19 Apr 2001	6A	STS-100	MPLM Raffaello and SSRMS
30 Apr 2001	2S	Soyuz	'Taxi'
17 May 2001	7A	STS-104	Joint Airlock Module (Quest)
21 Jun 2001	7A.1	STS-105	MPLM Donnatello

an early look at the nascent facility. The wisdom of the decision to service Zarya as soon as possible was reinforced a few days later, when Battery 5 showed signs of distress.

In March 2000, the General Accounting Office expressed concern to NASA that Zvezda, like Zarya, would exceed US regulations for noise in the work environment – an issue that would be more serious in Zvezda because it was the living quarters for the resident crew. NASA insisted that it would "resolve the problem in orbit", by which it meant that the crew would wear earplugs until sufficient acoustic insulation could be fitted to reduce the hum from the ventilation system to an acceptable level. However, the mufflers that had been installed in Zarya by STS-96 had not proved as effective as hoped, and this module would require further attention. The GAO also said that the Russian modules had insufficient protection against micrometeoroids, and NASA said that spacewalkers would install additional shielding on Zarya. However, to fit additional shielding to Zvezda prior to its launch would place it beyond the Proton's payload capacity, so, once again, the shielding would have to be added in orbit. The GAO's concern was that should one of the Russian modules be punctured and lose pressure (even if this occurred sufficiently slowly for the crew to evacuate) the electronic systems, which were air-cooled, were not capable of reliably operating in a vacuum. But Khrunichev, which had a long history of supplying pressurised shells for space station modules, countered that Mir's prolonged life demonstrated that the micrometeoroid protection was adequate. The rationale for the GAO's investigation of this issue was that the Russian modules were vital to the ISS in the early phase of its assembly, and a serious failure would jeopardise the entire project.

Back from its 10-month refit during which (among other upgrades) it received the new 'glass' cockpit utilising colour multi-function displays, Atlantis was dispatched on 19 May 2000 as STS-101. Much of the 2,225 kilograms of cargo in its Spacehab Double Module – batteries, fans, air filters, fire detection and suppression kits – was to service Zarya. It was not that this equipment was faulty, but simply that it was due for routine replacement in order to extend Zarya's warranty. In addition, NASA was sending up clothing, sewing kits, hygiene items, heath care supplies, food, video equipment, fans, computers and exercise equipment. The ICC in the bay carried the boom of the Russian Strela crane and the Spacehab Oceaneering Space System Box with a miscellaneous cargo for external activities.

As previously, upon arriving at a point 200 metres beneath the ISS, which was oriented vertically with Unity on top, the Shuttle flew a half loop around in front to take up position 76 metres above it. After waiting at 50 metres until it was in range of the Russian communications network, Atlantis docked at 00:31 on 21 May. The crew then retired with the cabin pressure reduced in preparation for the mission's EVA by Jeff Williams and James Voss later that evening. Their first assignment was to reinstall the ORU Transfer Device that was incorrectly fitted by STS-96. Mary Ellen Weber, operating the RMS, then swung the spacewalkers down to the ICC, and once they had retrieved the Strela's boom she swung them up to enable it to be mounted on its operator's post. After installing eight handrails and a TV camera on Unity's exterior, they replaced the malfunctioning S-Band Early Communications System antenna on its port-side CBM. It was routine maintenance. The hatches were opened the following day. The first task was to modify the ducting to improve the air circulation. After landing, the STS-96 crew had admitted that they had suffered headaches while working *in situ* for extended periods, particularly while replacing the MIRTS on Zarya's batteries. As this was deemed to have been caused by the local build-up of exhaled carbon dioxide, some of the air ducts in the ventilation system were re-routed to improve circulation. In addition, portable fans were set up whenever a task required someone to remain in one place. Meanwhile, Weber, serving as loadmaster, prepared the Double Module's cargo for transfer. On the first day, Yuri Usachev and Susan Helms replaced the first two batteries in Zarya and the first battery was charged overnight. The following day it was realised that the third battery could not be replaced while the second was being charged, and the recharging had to be interrupted. Furthermore, this maintenance had to be phased to ensure that there was always sufficient power on-line to run the ISS, whose electrical load was increased while a crew was on board. In all, four batteries were successfully replaced; STS-106 would replace the other two. Meanwhile, an expiring memory card in Zarya's radio-telemetry system was replaced, and a cable strung to ensure that Zarya's flight control system would be able to be switched off from the backup communications system to provide redundancy in case the primary system failed when the system was to be switched off following Zvezda's docking. The hatches were closed on 26 May, Atlantis undocked, and Scott Horowitz few a half-loop to document the station's exterior. Half an hour later he performed the separation burn, so there was no lingering around for this crew. Tests conducted over the next few days verified that all six of Zarya's batteries were performing satisfactorily.

ZVEZDA, AT LAST!

As a direct descendant of Mir's base block, which was itself derived from the Salyut space stations, Zvezda's systems are considered to be mature. It comprises a trio of pressurised compartments. The main compartment forms a single long 'room'. At its rear is a narrow axial tunnel to the aft docking port. A forward hatch provides access to a spheroidal compartment with an axial port and a pair of radial ports (one on the zenith and the other on the nadir). In fact, it has two different types of docking port. The rear port is a standard Russian drogue, but those at the front have a hybrid design combining a drogue with an APAS interface.[13] A pair of solar panels with a tip-to-tip span of 30 metres provide power. The 13-metre-long 20-tonne module is to provide early power generation and distribution, data management, propulsion and attitude control (taking over this rôle from Zarya), living facilities and life support for the resident crew.

After a Joint Programme Review and a General Designers' Review in late June 2000, Zvezda's launch was confirmed for 12 July. It was fuelled on 5 July, mated to its Proton launch vehicle on 6 July, and rolled to the pad the next day. Twenty-six months behind the original ISS schedule, Zvezda finally lifted off 'on time' at 00:56 on 12 July, and no sooner had it been placed in orbit than it initiated its rendezvous. The next day, the flight controllers confirmed that Zvezda's solar panel drive motors were maintaining its photovoltaic arrays facing the Sun, and that power was being routed to the five batteries that had been installed for launch. Three other batteries, together with the regenerative elements of the life-support system – namely the Elektron that would liberate oxygen from water and the Vozdukh that would scrub carbon dioxide from the air – had been omitted in order to lighten the craft for launch; these were all to be installed by the crew of STS-106 as part of the commissioning process prior to the arrival of the station's first residents. On 25 July, after Zvezda had completed its slow and deliberate rendezvous, Zarya closed in using the Kurs automatic system and docked at the first attempt. A few days later, after Zvezda's computers were integrated with those of Zarya and Unity, Zarya's flight control system was switched off and Zvezda took responsibility for attitude control and orbital reboost. At this time, Zvezda was linked to Unity's S-Band Early Communications System to enable the Russian controllers to relay commands to Zvezda via Houston.

As Zvezda had used a fair proportion of its propellant in the rendezvous, the immediate task was to replenish its tanks. After the standard two-day rendezvous, a Progress's Kurs system steered it to dock at Zvezda's rear on 8 August. In addition to 1,560 kilograms of propellant specifically for Zvezda, which was pumped on board through pipes incorporated in the docking collars, the Progress could use 260 kilograms of its own propellant for docked manoeuvres. The 615 kilograms of 'dry' cargo mounted in racks in the orbital module – which included an Elektron, a

[13] In fact, there is only one drogue in the forward compartment. It was initially on the axis, but once Zarya had docked it was removed. It will be installed in place of the 'flat plates' that seal the radial ports as required to enable each new module to dock.

Table 11.4 Shuttle manifest, c. autumn 2000

Date	Flight	Launcher	Objective
6 Aug 2000	–	Progress	Tanker and cargo
8 Sep 2000	2A.2b	STS-106	Spacehab DM
5 Oct 2000	3A	STS-92	Z1 and PMA 3
30 Oct 2000	–	Soyuz	First ISS crew
30 Nov 2000	4A	STS-97	P6

Vozdukh, food and computers – was to be unloaded by the STS-106 crew. When an irregularity was detected in Battery 4 in Zvezda on 21 August it was decided to have STS-106 replace it, and when the fault was traced to the PTAB that monitored the recharging, it was decided to replace this too. Battery 6 on Zarya was showing signs of ageing, but it was already on the list of items to be replaced by STS-106. In the weeks prior to this mission's launch, one of Zvezda's flight control computers failed. There were three computers for multiple redundancy – any two could 'vote down' a third if it misbehaved, and any one could run the systems on its own. In fact, if all three computers went off-line the systems could be remotely commanded from the ground, but only while in radio communication.

With Zvezda integrated into the ISS, the station now had its main control centre and living quarters so, after two years of frustrating delays,[14] the programme could finally kick into high gear. The pressure was therefore now on NASA to perform. In truth, the protracted delay with Zvezda had 'shielded' Destiny, which had suffered software problems and, to NASA's embarrassment, would have become the 'pacing item' if Zvezda had been launched when initially planned.

COMMISSIONING ZVEZDA

Atlantis was launched for STS-106 on 8 September 2000. On arriving beneath the ISS, which was oriented with Unity on top, Terry Wilcutt flew a half-loop around in front and paused 76 metres above it. Once in range of the Russian communications network, he docked. The Spacehab Double Module had 3,000 kilograms of supplies, and the ICC carried apparatus for use during the EVA. When Yuri Malenchenko and Ed Lu ventured out on the 11th, Rick Mastracchio used the RMS to emplace them as far up the stack as the arm could reach. Then, with Dan Burbank directing them around the external apparatus, the spacewalkers used tethers and handrails to make their way along Zarya onto Zvezda, where they installed a 'Yakor' foot-restraint on its main compartment. Standing on this foot-restraint, they retrieved a 1.8-metre boom and a magnetometer that had been stored on the exterior, inserted the boom into a receptacle on the surface of the craft, and placed the magnetometer

[14] On the original plan that involved Russia in the ISS, the launch of Zvezda had been scheduled for April 1998, so it was two years late.

on the tip. The magnetometer was to enable Zvezda's attitude control system to determine the orientation of the ISS with respect to the Earth's magnetic field – this was a backup sensor just in case the primary system failed. To be effective the magnetometer had to be positioned away from the vehicle's hull. Rather than develop a self-deploying boom, it had been decided to stow it under a cover for launch, and have spacewalkers install it. Located some 34 metres above Atlantis's bay, this marked the furthest that astronauts had ventured from a Shuttle without using the MMU 'flying backpack' (which was no longer in service). On their way back, they ran nine power, data and communications cables between Zvezda and Zarya. When STS-97 installed the solar power system, four of these cables would route power into the Russian modules – particularly when the Russian solar panels were shaded by the much larger American arrays. Two cables were to enable Zvezda to command Zarya's solar panels to rotate to face the Sun in concert with its own (as Zarya's control system had been deactivated when Zvezda docked). Two cables were for a closed-circuit video to enable the crew in Zvezda to monitor dockings at Zarya's nadir port. As a 'get ahead' task, Malenchenko and Lu strung a cable from Zvezda ready for attachment to the Joint Airlock (to be installed on Unity by STS-104) to relay Orlan suit telemetry to Zvezda, for transmission to Earth.[15] With the EVA over, Atlantis's cabin pressure was restored and, after a well-earned sleep, the hatches to Unity were opened.

Zvezda had to be transformed from its launch to its operational configuration – a miscellany of tasks involving unpacking items that had been stowed for launch and installing them ready for use. The fire extinguishers, for example, had been bolted for launch. (One of the problems discovered when a fire broke out on Mir was that some of the fire extinguishers had never been unbolted.) Other tasks, such as installing six ground repressurisation inlet caps, were simply chores. Most of the crew's time was devoted to transferring cargo. In fact, cargo was being loaded on board the station from *both* ends, because the Progress had to be unloaded to enable it to clear the rear port for the Soyuz that would deliver Shepherd's crew. Although the profession of astronaut conjures up the image of piloting spaceships, for mission specialists it is an engineering vocation. While Malenchenko and Lu installed Zvezda's final three batteries, Dan Burbank and Boris Morukov set to work on the batteries that needed servicing, using a hammer and chisel on rivets to remove a floor bracket that blocked access to the PTAB on Battery 4. Two of Zarya's batteries were replaced (the first four of its original set had been replaced by STS-101, and that task was now complete). When one of the new batteries in Zvezda showed problems, and an inspection established that one of the connections was faulty, the battery was disconnected. (Seven good batteries were more than sufficient to support the first resident crew.) After fitting power converters to enable the Russian modules to draw power from the exterior cables that had been hooked up, Malenchenko and Morukov installed Zvezda's toilet, the Elektron, the Vozdukh, and a 'ham' radio system. They then detached and stowed the docking drogue to improve passage

[15] Until the Joint Airlock was installed, EVAs in the absence of a docked Shuttle would be feasible only from Zvezda's docking compartment.

The ISS after STS-106. Top to bottom, Unity, Zarya, the Zvezda base block and a Progress ferry.

through the forward docking compartment. Meanwhile, Scott Altman and Lu installed a Treadmill with Vibration Isolation and Stabilisation in Zvezda as the resident crew's primary tool for physical exercise. Its complicated mounting (which had been tested on Shuttle missions) was designed not to transmit vibrations to the module's structure. The failed computer in Zvezda was replaced. The Orlan EVA suits were unstowed and checked. Since it was no longer required, Zarya's docking probe and TORU electronics unit were retrieved to be returned to Earth for reuse. In PMA 2, Mastracchio reinstalled the four CBM controllers that had been removed by STS-96.[16] The docked mission was extended by one day to provide time to finish the long list of tasks. Asked about the noise level in Zvezda, Wilcutt said that it was "just fine". When STS-106 undocked on 17 September, Zvezda had been made habitable. All the discarded packing material had been loaded into the Progress, to be taken away when that ship departed. Soon after STS-106 left, Battery 5 in Zvezda began to show signs of distress by failing to hold its charge, possibly due to a faulty wiring connection on either the charger or the voltage controller. The launch of STS-92 was only a two weeks away, but as its crew were not scheduled to enter Zvezda the task of inspecting and servicing this battery would have to be left to Shepherd's crew.

THE Z1 TRUSS

The Z1 truss, so-named because it is berthed on Unity's zenith, was not inherited from the Space Station Freedom configuration, it was introduced to the ISS to house miscellaneous essential systems required early in the revised assembly sequence and to act as a temporary mount for a segment of the Integrated Truss Structure (dubbed P6) with a pair of Solar Array Wings. The requirement for it became evident when it was decided to advance Destiny in the sequence to enable the crew to make an early start on scientific research, and since the power generation capacity of the Russian modules could not support the laboratory's systems this required advancing one of NASA's solar power modules. Because Unity offered the only mount, the squat 9-tonne *ad hoc* section of truss was provided with a CBM collar. It eventually came to house such a variety of external apparatus that it was became known as "the junk box". Its assemblies include:

- *Control Moment Gyroscopes (CMG)* – A set of four electrically driven momentum wheels can be spun up or down to transfer angular momentum on specific axes to either induce or cancel out rotations for non-propulsive attitude control.[17] Apart from providing finer control over the ISS's orientation, use of the CMGs will save propellant, which is always a premium item to be conserved for emergencies. The control system for the CMGs would be delivered in Destiny.

[16] These electronics units were reinstalled because the next two missions, which would not carry a Spacehab Double Module, would not transfer much cargo through the narrow adapter. They would have to be in place to enable STS-98 to remove this PMA prior to installing Destiny in its place.

[17] CMGs had been used on Skylab and on Mir (the Russians referred to them as gyrodynes).

- *S-Band and Ku-Band antennae* – The S-Band provides telemetry and voice communication between the ISS and the ground. The S-Band Antenna Support Assembly was to be mounted on the P6, at which time the antennae on Unity for the S-Band Early Communications System would become obsolete (they were later to be deleted). The large dish of the Ku-Band Space-to-Ground Antenna is on a long boom in order to track the geostationary satellites of the Tracking and Data Relay System. As it had in excess of 250 times the bandwidth of the S-Band, the Ku-Band's bulk data capacity was to enable the ground to monitor and remotely operate scientific apparatus in Destiny. The electronics for the Ku-Band would be delivered in Destiny.
- *Plasma Contactor Units* – To preclude the build-up of static whose discharge might damage apparatus or 'zap' spacewalkers, the ISS is 'grounded' to the electrically charged ionospheric plasma through which it travels. Of the two Plasma Contactor Units provided, one would be continuously active and the other would be on standby in case a fault disabled the primary. Because its power generation was likely to prompt electrostatic effects, the contactors would not be activated until the P6 was in place.
- *Early External Active Thermal Control System (EEATCS)* – Four accumulators were fitted in the Z1 to charge the EEATCS with ammonia, accommodate thermal expansion of the fluid, and maintain the proper operating pressure. The associated plumbing incorporated a dozen Quick Disconnects to facilitate the connection of ammonia transfer lines directly between the P6 and the Z1, and between the Z1 and Destiny by way of a hinged 'umbilical tray' (to be folded down by spacewalkers).
- *DC-to-DC Converter Units (DDCU)* – A pair of DDCUs were to be installed on the outside of the Z1 by spacewalkers (physical constraints prevented these being fitted prior to launch) in order to regulate the electrical power from the Z1 to Unity once the P6 was providing power.
- *Manual Berthing Mechanism (MBM)* – A CBM was incorporated into the 'forward' face of the truss as a temporary mount for PMA 2 during the installation of Destiny. This mechanism is 'manual' in the sense that its latches are operated by spacewalkers rather than by computer command.
- *Micrometeoroid Shield* – A solid plate was installed behind the MBM in order to protect the internal components from damage by micrometeoroids and orbital débris.

The Z1's structure was designed to support load paths at launch and in orbit and to maximise the packaging of its various black boxes. "It's like a well-packed trunk," noted Brian Duffy, commander of the Shuttle that would deliver it. "It's as if you're going on vacation and you strategically pack the trunk of your car so that you don't waste a square inch – it is packed very smartly." The plan drawn up in July 1994 called for it to be launched on assembly flight 3A, with the power module following on 4A and the laboratory on 5A. While the timeline had slipped considerably, with logistics missions being inserted, this sequence was imposed by the integrated nature of the manifest. To pick up the pace after STS-106 commissioned Zvezda, the launch of STS-92 was scheduled as soon as possible after its predecessor's return and it lifted off on 11 October 2000. As previously, the ISS was oriented vertically with Unity on top. On arriving 180 metres beneath the ISS on the 13th, Brian Duffy flew a half-loop around in front and halted some 76 metres above it. When in range of the Russian communications network, he docked Discovery so that the view from its aft flight deck looked out across Unity's nadir. In the bay, PMA 3 was mated with an MBM that was bolted to a Spacelab pallet, and the Z1 was further aft supported on trunnion pins. The Z1 was carried with its CBM facing aft, its MBM and Umbilical Tray facing the bay floor, and its large antenna boom stowed and protected by a thermal blanket.

The next day, as Koichi Wakata was preparing to retrieve the Z1 using the RMS, a short-circuit cut power to the Orbiter Interface Unit that would send commands to the ISS systems, the SVS that would provide cues for operating the RMS, and a TV camera located at the keel of the bay that faced upwards to provide supplementary visual cues while lifting the Z1 from the bay. However, a plan was devised to route power to reinstate the SVS. After hoisting the Z1 out of the bay, Wakata rotated it through 180 degrees to let its CBM face the module. Relying on the SVS cues, because Unity's bulk was blocking his view, he then mated it with the node's zenith, and as this was the first time that a CBM had been operated in space there was great relief when it activated without incident. Once the Z1 was mounted, Pamela Melroy used a laptop to tighten the 16 bolts in a four-stage process. As they were running two hours behind schedule, flight controllers opted to defer the transfer of some apparatus from Unity into Zarya until later in the mission, but the hatches were opened to enable Melroy and Jeff Wisoff to open Unity's upper hatch to access the pressurised dome on the base of the Z1, to electrically 'ground' it to the ISS. They then retreated and closed the hatches to allow Discovery's pressure to be lowered in preparation for the spacewalk the following day by Leroy Chiao and Bill McArthur. After preparing the bay for an hour, McArthur mounted the RMS and Wakata manoeuvred him so that he could hook up six power cables between the Z1 and Unity. This work had to be coordinated with the ground, which switched off each Russian–American Power Converter in turn to ensure that each socket he worked on was not carrying current, and also that power was not denied to Unity from Zarya. They then turned their attention to the communications systems incorporated into the Z1. After stripping off a thermal blanket, they unpacked the S-Band Antenna Support Assembly from its stowage site on the side of the Z1 and placed it on a temporary fixture on the truss. (It would not be used until affixed to the P6 truss by STS-97.) Once McArthur had run four more cables between Unity and the Z1, the heaters in the truss could be activated. Changing places, Chiao mounted the RMS and held the dish of the Space-to-Ground Antenna while McArthur unbolted it from its stowage location; the dish was fixed to its boom, and the boom deployed. Finally, as a 'get ahead' task, they fetched an EVA Tool Support Device from the Spacelab pallet and fitted it on the side of the truss, where it would be conveniently positioned for the team that was to install the P6 truss.

On the 16th, on the mission's second excursion, Jeff Wisoff and Michael Lopez-Alegria released the latches on the MBM on the Spacelab pallet to enable Wakata, at the RMS, to retrieve PMA 3. Mounting it on Unity was straightforward because he had a direct view of the nadir CBM, but first he had to swing it over the sill of the bay and around the node, and rotate it so that it faced the CBM. After giving visual cues to assist Wakata in slipping PMA 3 over the sill, the EVA team made their way around Unity and up to the top of the Z1 in order to uncover the latches that would hold the P6 truss. By now, Wakata had manoeuvred PMA 3 into a position for berthing, and they returned to assist him with its emplacement. When Melroy had engaged the CBM latches, they ran power and data cables from it to Unity. The following day, Chiao and McArthur reconfigured cables to route power from PMA 2 to PMA 3, to make it ready for STS-97.[18] Next, they retrieved from the bay the two 58-kilogram DDCU that would regulate the power from the P6 and installed them on the Z1 in the compartment from which they had previously removed the S-Band system. After relocating the Z1's keel pin assembly, they installed the second EVA Tool Support Device on the truss, collected the bag of tools stowed on Unity by the STS-96 crew and neatly stocked the two tool boxes. Making the fourth EVA in four days, Wisoff and Lopez-Alegria removed the Z1's grapple fixture (the one by which the RMS had manipulated the truss). Once Lopez-Alegria was on the arm, he held the Umbilical Tray (which overlapped the lower rim of the MBM in its stowed position on the front of the Z1) so that Wisoff could remove its four launch restraint pins, and then he swung it down across Unity's rim and held it while Wisoff locked it in position. When the P6 and Destiny were installed, the Umbilical Tray was to feed power and coolant to the laboratory. Next, they made their way to the top of the Z1 where Wisoff used a pistol-grip tool to open, close and finally leave open the latches that would engage the P6 truss. In addition, the 'claw' that would capture the P6 and hold it in place until the latches could be engaged was left open.[19] The latches on the MBM on the front of the Z1 were then cycled to verify their function in advance of PMA 2 being temporarily berthed there during STS-98's installation of Destiny. Although all the assembly tasks had been completed, Wisoff and Lopez-Alegria were running late, so they made only one of the two SAFER demonstrations by taking turns to test the ability of their thruster packages to move 15 metres in a straight line, simulating an astronaut with a loose tether returning to the safety of the nearest part of the ISS structure; the second demonstration, which was to assess manoeuvring with a disabled colleague, would be reassigned to a future crew.

On 19 October, the hatches were reopened. While McArthur and Chiao made the final internal connections to the Z1 and use a laptop to command the CMGs to test their heaters, and to spin their flywheels up to 100 revolutions per minute to verify their operation (when on line, they would spin at up to 6,600 rpm). Meanwhile, the other astronauts transferred cargo to Unity and Zarya – mostly computer equipment and an IMAX camera with which the residents were to document life on board. At Shepherd's request, a small bell was also installed – holding the rank of captain in the

[18] PMA 3 *had* to be successfully fitted by STS-92 because assembly flights 4A and 5A (on which axial dockings would not be possible) would require it to be on Unity's nadir. Its placement there was temporary, as it would later be moved to the port-side CBM, where it would remain until the second node was added.

[19] The 'claw' is the slang term for the Rocketdyne Truss Attachment System.

US Navy, he wished to continue the tradition of having a ship's bell. A variety of material was offloaded for return to Earth, including a Protein Crystal Growth experiment left by STS-106 in early September – it was the first microgravity science experiment to be conducted on the ISS. The CMGs would not be used until Destiny was in place, as it contained the controlling electronics. Until then, the ISS would rely on Zvezda for attitude control which, by consuming propellant, would increase the load placed on the limited fleet of Progress tankers. With the cargo transferred, the hatches were closed and Discovery undocked the next day. To satisfy thermal requirements, the ISS was left with PMA 3 pointed towards the Earth, but on 29 October it was rolled through 180 degrees to point PMA 3 at open space and the commands for the forthcoming Soyuz docking were transmitted to verify Zvezda's functionality. One of the three flight control computers tripped off-line, but the other two continued to function and controllers in Moscow set about determining what had prompted the third unit to shut down. After this rehearsal, the remaining propellant in the Progress's tanks was pumped on board Zvezda. Meanwhile, Shepherd's crew – after protracted delays, the most-trained crew in the entire history of spaceflight – was making its final preparations.

FIRST RESIDENTS

Soyuz-TM 31 lifted off from Tyuratam on 31 October 2000. Later in the day, the Progress undocked. The Soyuz automatically docked at the recently vacated port on 2 November. After powering down the ferry, Shepherd, the ISS commander, invited Yuri Gidzenko and Sergei Krikalev to enter Zvezda first. Having flown to the ISS on STS-88, Krikalev became the first person to visit it twice – although on his first trip Zvezda had not been present. Their initial tasks included verifying communications, charging the batteries for power tools, and activating the food warmers, water processors and toilet systems.[20] They soon slipped behind schedule, however. The installation of the food warmer, for example, was expected to take 30 minutes, but took several hours. Upon finally taking up residence, Shepherd asked the flight controllers to use the call sign 'Space Station Alpha'. The Russians complained that this would imply that the ISS was the *first* space station, whereas it followed seven Russian Salyuts, Skylab and Mir. Nor was Dan Goldin very happy, pointing out that the ISS was not Shepherd's to name, but agreed to using it as their radio call sign. If the mission unfolded as planned, their tour of duty would run well into the new year. Initially, they were to confine themselves to the Russian modules; although Unity was drawing power to run its heaters, its other systems were off. In any case, they would be busy setting up Zvezda and sorting through the cargo that was so stacked up in Zarya that it resembled a warehouse.

A routine was soon established. Shepherd and Krikalev used the cabins recessed

[20] Zvezda's toilet was based on the design developed for Mir. In fact, when Shuttles visited Mir, the astronauts often preferred to avail themselves of its facilities rather than that in the Shuttle – which had a tendency to become blocked, and whoever blocked it was charged with unblocking it.

into the walls of Zvezda's main compartment and Gidzenko put his sleeping bag in a convenient corner of the module. After the daily wake-up call, some 40 minutes was reserved for dressing and ablutions using the shared facilities followed by breakfast. In the early days, each member of the crew in turn was to have an hour's exercise on the treadmill prior to lunch, and a second hour before dinner. They ate their meals at the fold-down wardroom table. One Solid Fuel Oxygen Generator (a canister similar to the Vika used on Mir) was 'burned' per man per day. Most of the first week was spent configuring the environmental control systems. The first day's assignment for Gidzenko and Krikalev was to set up the Vozdukh regenerative air-scrubbing unit in Zvezda, but the procedure took longer in weightlessness than it had in training. The carbon dioxide extracted from the air was to be vented overboard. The non-regenerative system, which involved canisters of lithium hydroxide, would now serve in a backup rôle. Later in the day, Gidzenko and Krikalev set out to troubleshoot the battery in Zvezda that had failed soon after its installation by the STS-106 crew. They found a bent or broken pin in one of the connectors, and Moscow decided to send up parts to effect a repair. The controllers in Moscow urged them to re-establish the flight plan, which frayed tempers somewhat until it was accepted that the ISS crew would have to be allowed to work at their own pace until the start-up activities were complete, at which time, hopefully, the schedules based on experience on Mir would become practicable. In part, the slippage was because the ISS had only intermittent communication with Moscow – each orbit provided at most 15 minutes of radio link with the network of Russian ground stations, with the result that queries concerning the whereabouts of cargo had to be saved and read down as soon as they flew into range, and had often to wait another orbit for the answers. To overcome this, Shepherd spent part of his first day establishing a link from his laptop to the S-Band Early Communications System in Unity to talk to Houston through the Tracking and Data Relay Satellites.

Normally, the ISS residents were to operate on a standard American Civil Service week involving an 8-hour day from Monday to Friday, and have the weekend off. However, they elected to work through Saturday 4 November in an effort to catch up by completing work on the Vozdukh and making a start on the Elektron, the regenerative unit designed to liberate oxygen from water. When the compressor was installed in the air-conditioning system, this drew the moisture from the humid cabin air. The reclaimed water could be used either for personal hygiene or be pumped into the Elektron. One of the two batteries in Zvezda had not been holding its charge, and when it was repaired the Elektron was activated to enable the use-once canisters to be stockpiled in case the unit malfunctioned, although until the final battery came on line the Elektron would have to be turned off when power ran low. When the computer link was established and it became possible to e-mail the control centres in Moscow and Houston, Shepherd revealed that they were having trouble locating the proper cables for the different Russian and American laptops. Interestingly, the first use of English was heard at this time. The crew also experienced difficulty in starting up some of their various computers. Gidzenko and Krikalev set up the monitor and hand controls for the TORU backup rendezvous and docking system in Zvezda. It would be used if the Kurs system of a cargo ship malfunctioned. The next Progress was to be launched on 16 November, dock at Zarya's nadir port two days later, and be unloaded as a priority so that it could depart to provide manoeuvring room for STS-97, which was to dock at PMA 3 in the first week of December. Even when it was in perfect condition, the ISS required ongoing maintenance. For example, when the air-conditioner tripped off because its water tank was full, Krikalev emptied it (on this occasion the water was used for personal hygiene) and restarted it. On the 9th, with increased solar activity expected over the next 48 hours, the astronauts were told to install a radiation detector inside Zvezda to monitor their exposure. Although the increase in activity was not expected to present a serious threat, they spent their next two sleep periods in the aft section, close to the aft tunnel where the module had additional shielding for this purpose. The following day they configured cables for Zarya's nadir port in readiness for the Progress's arrival. The first week had been so hectic that they were told to take Saturday, Sunday and Monday off. The only task assigned to the 11th was to start wiring up Zvezda's 'ham' radio station: astronauts loved to chat with schools and enthusiasts as the planet rotated below. Rested after their extended weekend off, they began their second working week with a rehearsal of the Progress docking in which the ISS was manoeuvred to the correct attitude and Gidzenko and Krikalev tested the TORU remote-control system. The next job, which occupied all three men for a considerable time, was using a computer that read barcodes to inventory the cargo already on board, before another 1,815 kilograms of stores arrived on the ferry.

After rendezvousing on the 17th, the Progress flew around the ISS to align itself with Zarya's nadir for a docking at 22:08, but as it closed within 100 metres its Kurs system failed to lock onto Zarya's transponder. The ship's docking camera showed the ISS moving in and out of the field of view as the vehicle slewed back and forth in search of the transponder. Deciding to intervene, Gidzenko assumed control with the TORU at 22:02, stabilised the ship facing the ISS, withdrew it somewhat, and then slowly began to fly it in, but the view on Gidzenko's screen progressively degraded as the Sun shone into the lens. Abandoning the approach when the range was down to 10 metres because the poor lighting rendered the docking target indistinguishable, he halted the ship 5 metres out. A visual check through the porthole showed that the vehicle was misaligned, so he eased it out another 30 metres. Moscow told Gidzenko to hold the ferry in that position until sunset, at about 22:45. Once in the Earth's shadow, the spotlit view was clear and Gidzenko made a second approach, this time without incident, achieving the docking at 22:48, some 40 minutes behind schedule. Examining the vehicle's television camera through a pair of binoculars, he saw a thin film of ice on the lens. This had exacerbated the 'flare' caused by the unfavourable illumination. Although this incident underlined the utility of the TORU system, it also highlighted the impracticability of having a ferry approach when the station was silhouetted by the Sun; it was a significant procedural error on the part of the flight planners. The crew spent several hours powering down the Progress's systems and then retired. After their second weekend off, the astronauts began to unload the new cargo. Each item was unbolted from its rack, logged into the computerised inventory, and transferred to the appropriate position on board the ISS. The cargo included a resupply of oxygen and lithium hydroxide canisters. The unloading was done by the end of the week. Meanwhile, the engineers had determined that the ferry's Kurs had been unable to differentiate between the transponders on Zarya and Zvezda. Such a situation did not arise when approaching from the rear. (This was the first time that a docking from the side had been attempted.) Having developed a software patch to overcome the problem, they requested that after the Progress had undocked it should attempt to return. It undocked on 1 December and was 'parked' in a parallel orbit awaiting a decision on running this test after STS-97 had been and gone.

ADDING POWER

When Endeavour was launched on 30 November 2000 as STS-97, its bay contained the P6 segment of the Integrated Truss Structure that was to be mounted on the Z1 truss. This 15.875-tonne, 14-metre-long pre-integrated segment is a Long Spacer and a Photovoltaic Array Assembly with its Integrated Equipment Assembly. The 8.5-metre-long spacer would mate with the top of the Z1 and support a pair of radiators for the Early External Active Thermal Control System.[21] The Photovoltaic Array Assembly – which comprises a Solar Array Assembly and its associated controls – generates, regulates, stores and distributes electrical power, and has its own radiator. The Solar Array Assembly has a pair of 1-tonne, 34-metre-long Solar Array Wings with a total span of 73 metres, each of which has two blankets of photovoltaic cells (one on each side of an extensible framework mast) with an 11.6-metre overall span. Nevertheless, for launch each blanket was carried in a 4.6-metre-wide box that was no more than 50 centimetres in depth. Each mast is on a Beta Gimbal Assembly that can rotate its wing to face the Sun for maximum power generation. The 7.6-tonne, 4.8-metre-long Integrated Equipment Assembly incorporates the energy management system and thermal control system, and serves as the primary structure supporting the storage batteries and power conditioning equipment; it has 12 batteries (six for each Solar Array Wing), each containing 76 hydrogen–nickel cells. A 'transmission line' technology is used to minimise power losses in the cables from the batteries to the cluster of pressurised modules. The 26 kilowatts of power from the P6 was to run the Destiny laboratory.[22]

When Endeavour arrived on 2 December, the ISS was set 'horizontally', with the module chain aligned along the velocity vector and with Unity in the lead. Pausing directly beneath, Brent Jett turned the orbiter tail-first and then manoeuvred to line up the ODS with PMA 3. Once Zvezda and Zarya had 'feathered' their solar panels to minimise the contamination by thruster efflux, Jett closed in using the profile that was developed to dock with Mir – repeatedly firing the orbiter's downward-facing thrusters in order to overcome the tendency of the gravity gradient to draw the lower

[21] The ammonia-coolant 'heat pipes' for the radiators had been tested on Shuttle missions. This was fortunate because it had taken several refinements to make them work properly. If it had not been for this flight testing, the ISS would have been in serious trouble.

[22] Mounting the P6 on the Z1 was only an expedient during the assembly, once the rest of the ITS is in place the P6 is to be relocated. With four such units operating, the ITS will deliver 104 kilowatts.

vehicle down. The two space crews were on different diurnal cycles. After watching the docking from Zvezda, Shepherd's crew retired. Meanwhile, Joseph Tanner and Carlos Noriega opened the ODS to deposit some urgent supplies in the vestibule to PMA 3, which the residents would retrieve when they awoke. An hour or so later, Marc Garneau used the RMS to hoist the P6 truss from the bay, tilted it to an angle of about 30 degrees, and locked the arm to enable the structure to attain thermal stability 'overnight'. While Jett's crew slept, the ISS crew made their first foray into Unity to retrieve the items left in PMA 3 – a laptop, video-conferencing equipment, a hard drive for one of the computers in Zvezda, and some eagerly awaited coffee and fresh food. After closing up Unity, they retreated to Zvezda.

The next day, Tanner and Noriega made their first EVA. While they prepared in the bay, Garneau, using SVS cues, oriented the P6 directly above the Z1 truss. When the 'claw' had captured the P6, the spacewalkers made their way up and locked it in place utilising a pistol-grip tool to engage the bolt at each corner. Only then did Garneau withdraw the arm. When Michael Bloomfield took over the RMS, Noriega stood on it and was positioned to link up nine cables for power, commands and data. Meanwhile, Tanner scrambled up the truss and placed the Solar Array Blanket boxes – two on each Solar Array Wing – in the ready-to-deploy position. However, when computer commands were issued to release the securing pins, they failed to release. Although the starboard boxes released when the procedure was repeated, the others remained in the stowed configuration. It was decided to proceed and deploy the starboard wing. As the central mast slowly extended over a 13-minute interval, dragging the blankets from their boxes, the stop-and-go fashion in which the blankets unfolded concertina-style sent 'waves' along the delicate structures. Although one of the blankets was drawn taut, the other was not because the waves had prompted the tensioning cables to jump off their reels. It was decided to study the situation before tackling the other wing the following day. Moving on, Tanner and Noriega released the launch restraints on one of the three radiators. Shortly after the astronauts ended their excursion, the pins in the blanket boxes on the port-side wing were commanded to release, this time successfully. Two hours later the P6 radiator was unfolded. This was to dissipate the heat that would be generated by the electronics contained within the Integrated Equipment Assembly once the power system was operational. The next day, the blankets on the port-side wing were successfully deployed. In an effort to eliminate the disruptive 'waves', the mast was extended in a series of short steps over a period of two hours. The only incident occurred when two rows of cells on one of the blankets emerged from the box stuck together, but these were separated by slightly retracting and then re-extending the mast. This time, to everyone's satisfaction, the tensioning system was not disrupted. With both solar wings generating power, all 12 batteries in the Integrated Equipment Assembly were soon fully charged.

On their second EVA, on 5 December, Tanner and Noriega hooked up the power and data cables and the ammonia-coolant umbilicals from the P6 to the Z1, and removed the thermal covers from a power conditioner and a signal processor. After collecting the S-Band Antenna Support Assembly from its temporary mount on the Z1, they rode the RMS as far up the P6 as the arm could reach, then set off using tethers, passing the antenna by hand as they leapfrogged each other. The antenna was placed high on the Integrated Equipment Assembly to give it a clear line of sight. On their way back down, they released the launch restraints on the radiator that was to cool Destiny's systems while the ISS was being assembled. It was deployed shortly after the conclusion of the EVA. When they awoke, Shepherd's crew re-entered Unity to configure cables to enable the ISS to draw power from the P6. While Shepherd reset the S-Band to utilise the newly installed antenna rather than the 'early' antennae that had been affixed to Unity's port and starboard CBMs (and which were soon to be deleted) his colleagues configured the American–Russian Power Converters to feed 3 kilowatts of power from Unity to the Russian modules to supplement their own limited capacity (particularly at times when the shadow of the arrays on the P6 fell on the much smaller arrays on Zarya and Zvezda and rendered them ineffective).

Meanwhile, the engineers had worked out a procedure for retensioning the wires on the slack starboard blanket, and a video of David Wolf rehearsing this procedure in the hydrotank was uplinked. When Tanner and Noriega went out on the 7th, they made their way to the top of the P6. After the mast had been retracted by 1 metre, Noriega reseated the slack tensioning cables on the dispenser reels and Tanner rewound the spring-loaded reels to recover some of the 'spring'. As the mast was re-extended, they ensured that the cables did not dislocate from the reels. The next task was to fit the Floating Potential Probe on top of the truss so as to measure the electrical potential of the ionospheric plasma through which the ISS was passing at a speed of 8 kilometres per second – this data was to provide a basis for assessing the effectiveness of the Plasma Contactor Unit on the Z1 which, now that the P6 was generating power, had been activated to eliminate electrostatic arcing. On the 8th, with Endeavour's cabin pressure restored and with Unity now open for business, the two crews were finally able to meet. Shepherd marked their arrival in naval style by ringing the ship's bell. The cargo transferred from the mid-deck was stored in Unity awaiting an inventory and stowage. The following day, after unwanted items had been offloaded from the ISS, Endeavour departed, leaving the station with more than sufficient power for the remainder of the assembly process. Although STS-97 had tested the Floating Potential Probe, Shepherd discovered that *his* laptop could not communicate with it after Endeavour left – it turned out that his laptop (unlike the one on the Shuttle) did not have the latest software, but when this was uploaded a few weeks later he was able to start to monitor the probe's data.

Five days into the STS-97 mission, when the recovered SRBs were inspected, it was found that the primary pyrotechnic separation system on the lefthand booster had partially failed to fire – the booster had been jettisoned by the backup system. While the problem was investigated, it was decided to delay STS-98, which had been due on 19 January 2001. The knock-on effect would delay STS-102, which was to make the first ISS crew exchange in February, so on 13 December Shepherd's crew were told that their tour of duty would have to be extended to mid-March. This news was not a cause for concern, however, because, with the ISS functioning smoothly, the crew were in excellent spirits. By mid-December, most of the bags of stores that had cluttered up Zarya for so long had been placed out of sight in the vestibule at the base of the Z1 truss. NASA accepted the Russian plan for the Progress to return on the 26th to test its updated software, and although the approach was flawless, once it had successfully locked onto Zarya's transponder Gidzenko took control for the final approach and re-docking, and Krikalev promptly uninstalled the electronics for the ship's Kurs, for return to Earth and reuse.[23] The new year was marked by the replacement of the faulty converter on Zvezda's eighth battery, which finally brought the power supply of the Russian modules to full capacity.

NASA'S MICROGRAVITY LABORATORY

As NASA's budget provided only for the fabrication of the hardware that would fly, it was operating a 'success oriented' strategy. If a module was damaged during final preparations, the assembly of the ISS would grind to a halt. There was, therefore, a great sigh of relief when the 14.5-tonne Destiny laboratory was safely installed in Atlantis's bay. Destiny is a 4.42-metre-diameter shell created by welding together three annular rings and adding domical end caps (each of which has a CBM) to create an 8.5-metre-long cylindrical module. It was manufactured by Boeing in Huntsville and was delivered to the Space Station Processing Facility in November 1998 to be outfitted. The aluminium pressurised shell is protected by a thermal blanket and an aluminium débris shield. Destiny's cylindrical internal volume can readily accommodate either the International Standard Payload Racks (ISPR) developed for Spacelab or the new EXPRESS[24] racks, both of which are comparable in size to a US domestic refrigerator, are curved around the rear, and present flat faces. Internally, the laboratory is a square cross-section 'room' with 6 racks on each 'wall', 6 on the 'ceiling' and 5 on the 'floor'; that is, 23 in all. While the hope is that the ISS will eventually pursue materials-processing on a commercial basis, it will start by serving the Centers for Commercial Development of Space that NASA established with industrial and academic co-sponsorship, and which, by use of the Spacehab mid-deck augmentation module, have finished their development work. As the ISS must support an integrated research programme on an ongoing basis over a lifetime of several decades, a modular strategy is essential. Plug-in compatibility will enable a variety of apparatus to be combined in a 'pick-and-mix' manner for specific experiments. This approach will enable apparatus to maintain pace with the state of the art, and enable the focus of the scientific research to evolve over the years.

Even with an EDO of cryogenic fluids in its bay, a Shuttle on a solo mission is restricted to a fortnight in space, which, for example, limits the size to which tissue cultures and slow-growing protein crystals can be grown. The ISS will transform the way that experiments are undertaken. Scientists working in a terrestrial laboratory often repeat their experiment, tweaking its configuration in order to produce a crucial measurement, then they vary it to establish the sensitivity of the data to parameters.

[23] This was a cost-saving measure, as the Ukrainian manufacturer of the Kurs system had increased the price.
[24] The acronym stands for EXpedite PRocessing of Experiments to Space Station.

This activity takes time, and *time* on a Shuttle is a precious resource that must be shared between the many payloads that have to be carried to justify the cost of the mission. On the ISS, however, the economics are different. The cost of the science is decoupled from the Shuttle's operating cost. Once an experiment has been delivered, it will be able to be run for as long as required, and be repeated, *at no extra cost* to the Shuttle. At long last, therefore, the Shuttle, *in combination with* the ISS, will cut the cost of performing scientific research in space, and materials-processing should become financially justifiable for the markets that can support the premium imposed by the unique production costs. To date, it has not been just the operating overhead that has deterred commercial processing in orbit, it has been the reliance of any activity on the vagaries of the Shuttle manifest – that is, infrequent flights on an indeterminate schedule. It is difficult to justify commercial investment in a process that can be invoked for only a fortnight each time, and twice a year at most. Spacehab Incorporated faced this dilemma from the start – development funding and initial customers were forthcoming only after NASA had reserved slots in its manifest for the company's module. The ongoing nature of operations on board the ISS will remove both of these obstacles, and it will be possible to run a process on a semi-continuous basis. If the raw material is delivered in bulk, 'mass production' will be feasible, and the output will be collected whenever convenient. If necessary, the process can be run on a free-flyer that operates either autonomously or by remote control, and from time to time returns to the ISS to be serviced. Production will at last have been decoupled from the constraints of flight operations.

Having 'sold' the ISS as an orbital laboratory for microgravity science, in early 2001 NASA found itself in the unfortunate position of having to cut this effort back due to the fact that the projected overrun on the budget for the assembly phase had sky-rocketed to $4 billion.[25] It cancelled the US Habitat module and the Propulsion Module (the one that was to be based on a node test-article) and halted work on the X-38 Crew Return Vehicle. This effectively limited the resident crew to three people for the foreseeable future. Although it would be possible to double this occupancy by docking two Soyuz craft, to do so would in turn increase the logistics requirements. This would essentially double the rate of production of Soyuz and Progress ships and rockets, but the Russians were in no position to fund such an increase. As the scale of the financial crisis sank in, NASA cut the science budget by 40 per cent, axing the Materials Science rack for Destiny and placing other microgravity disciplines under review. One proposal was to equip Shuttles so that they could remain docked for a month, but turning more assembly flights into 'utilisation' flights would impact the turnaround flow, which would in turn slow the flight rate and, thereby, increase the cost of the project by extending the duration of the assembly process. A few months later, NASA signed an agreement by which the Italian Space Agency would build a habitat using a pressure shell derived from its MPLM design, and donate this to the ISS in return for a 'free' launch of Italian scientific apparatus and greater access for Italian astronauts.

[25] This overrun is in relation to the $17.4 billion allocated for the assembly phase. Later in the year, a $25-billion cap was imposed.

Following the problem with the separation charges on STS-97's lefthand SRB, STS-98's roll out had been delayed from the scheduled 11 December date for checks. On the 15th, a wire in the lower strut supporting the faulty SRB failed a resistance test. When STS-98's SRB separation charges were X-rayed, the shielding on several cables in one booster were discovered to be eroded. On the 19th, Atlantis's roll out was postponed to 3 January 2001. Ten days later, with STS-98 now on the pad, the thousands of reusable electrical cables in the SRBs were tested and a number of data cables that ran the length of the boosters were found to transmit the test signals only intermittently. As potential 'single point' failures that could threaten a mission, they had to be rectified immediately. On the 14th, NASA rescheduled the launch for "no sooner than 6 February". On 17 January, after Destiny had been removed and placed in protective storage in a pad facility, the crawler transporter retrieved the Shuttle and returned it to the Vehicle Assembly Building; on the 26th it was returned to the pad and Destiny reinstalled.

STS-98 finally launched on 7 February 2001. The Progress departed the next day, taking away trash. As Atlantis made the rendezvous, Shepherd's crew marked their 100th day in space. This time both crews were more or less on the same sleep cycle (with the Shuttle's crew awakening about two hours after the ISS crew) so they could undertake joint activities. As on the previous mission, Atlantis adopted a tail-first orientation and rose from directly beneath the ISS to dock at PMA 3 on the 9th. The hatches were opened immediately to enable a variety of supplies – including a spare computer for Zvezda, cables to be used in powering up Destiny, and three 12-gallon bags of water (a by-product of the Shuttle's fuel cells) – to be transferred, and then sealed again to enable Atlantis to reduce its cabin pressure in preparation for the first EVA. However, as Tom Jones and Robert Curbeam prepared their apparatus, they realised that some of the cables they were to mount externally had inadvertently been transferred to the ISS, and were retrieved utilising PMA 3's vestibule as an airlock. On the 10th, while the spacewalkers were in Atlantis's airlock, Marsha Ivins used the RMS to remove PMA 2 from Unity's axis. Using SVS cues (because Unity blocked her view) she swung the adapter up in front of the MBM on the Z1, and Jones, who was by now in position, engaged the mechanism. Meanwhile, in the bay Curbeam disconnected the Assembly Power Converter Unit umbilical that fed power to Destiny's heaters so that Ivins could lift the module. Once it was clear of the bay floor, she rotated it through 180 degrees end-over-end and positioned it directly in front of Unity. The spacewalkers then went to the Umbilical Tray (which a previous crew had folded down from the Z1 truss over the front of Unity) to hook up power and data cables and coolant lines to Destiny. Curbeam accidentally let a small amount of fluid leak out. As the coolant loop used 99.9 per cent pure ammonia (in marked contrast to 1 per cent solution used domestically) this spillage prompted a decontamination protocol on their return to the airlock. Curbeam was obliged to remain in the glare of the Sun for 30 minutes to vaporise the residue on his suit. After entering the airlock he was 'brushed down' to dislodge any ice crystals, and the airlock was partially pressurised and flushed out prior to finally being repressurised. As a further precaution, after doffing their suits, they wore oxygen masks for 20 minutes while the environmental system removed any remaining contaminants. Later that day, PMA 3 was reopened and Shepherd and Ken Cockrell, Atlantis's commander, accessed the vestibule between Unity and Destiny to plug in a laptop with which to activate the new module's basic environmental systems. This activation process was continued by flight controllers as the two crews slept. Early on the 11th, Destiny's hatch was opened and Shepherd made a ceremony of signing the chit that officially transferred its ownership from Boeing to NASA. Its addition was a milestone, giving the ISS a greater pressurised volume than Mir in its completed state.

Even using the super-lightweight ET, the Shuttle could not lift the fully outfitted laboratory into the steeply inclined orbit that was required for the Russians to be able to reach the ISS. It was therefore launched with only the systems racks in its floor. The Environmental Control and Life-Support System, which was to provide temperature and humidity control and 'revitalise' the air in the US modules, offered a degree of redundancy to back up Zvezda's systems. The Thermal Control System operated the internal water loop of the cooling system and transferred heat to the external loops that circulated ammonia through radiators to shed excess heat to space (initially, it would use the radiators on the P6, but later it would use those on the ITS). The Guidance, Control and Navigation System contained the electronics to operate the CMGs in the Z1 truss. This initial configuration would be augmented by later Shuttles.

Jones and Curbeam ventured out again on the 12th. After Ivins had grappled PMA 2, Jones disengaged the MBM to enable her to retrieve it from its temporary mount. Meanwhile, Curbeam made his way to the end of Destiny, stripped off a thermal cover to expose the CBM at its end, and provided visual cues as Ivins mated the adapter there.[26] The spacewalkers then moved rapidly through a variety of tasks, including putting thermal covers on the trunnion pins on which Destiny had rested while in the payload bay, installing a non-propulsive vent for the laboratory's pressure regulation system, installing a Power and Data Grapple Fixture to serve as an anchor for the robotic manipulator that would be delivered by a later mission, fitting handrails on the module's exterior, and pre-positioning sockets and cables for future use. Finding themselves ahead of schedule, they made a start on tasks planned for their final excursion by attaching an exterior shutter for the large window on the laboratory's side, and running power and data cables between Destiny and PMA 2. As the two crews worked independently on the 13th, flight controllers first tested the CMGs in the Z1 truss at their maximum speed of 6,600 revolutions per minute and then verified that control of the station's orientation could be transferred from Zvezda to the CMG control system in Destiny. If ever the CMGs dropped off line, the computers were to coordinate so that Zvezda took over and used its thrusters to stabilise the station. While the CMGs were in operation, however, the rate at which propellant was consumed would be greatly diminished, and this in turn would reduce the requirement for resupply tankers. On their final EVA, on the 14th, Curbeam and Jones tackled a miscellany of tasks, including double-checking connections between Destiny and PMA 2, storing a spare S-Band antenna on the Z1, releasing the launch

[26] Note that whereas the asymmetric cone of PMA 2 had been offset 'upwards' (towards the Z1) on Unity, when it was reinstalled on Destiny the offset was in the opposite sense in order to enable a Shuttle to dock in a 'nose up' orientation.

restraints of the third radiator, inspecting the power connections from the P6 truss, and finally the long-delayed SAFER test of a spacewalker's ability to manipulate an incapacitated colleague. With the external work complete, the hatches were reopened for 36 hours of joint activities. After 1,360 kilograms of cargo had been transferred to the ISS, and 386 kilograms of unwanted items – including old Russian batteries – offloaded, Atlantis departed on the 16th. Meanwhile, the preparations for STS-102 were well advanced.

On 24 February, in an activity reminiscent of Mir, Soyuz-TM 31 was relocated from one port to another. This involved partially mothballing the ISS by deactivating the Vozdukh, water supply, toilet, ventilation and thermal control system, removing the air ducting in Zarya, closing the internal hatches in the Russian modules so as to seal off their docking compartments and to isolate the US segment, reconfiguring the communications system, and switching off the many laptops; in short, preparing the ISS for untended operations in the event that the re-docking failed and the crew had to return to Earth. This took several hours. After sealing themselves into the Soyuz, Gidzenko undocked at 05:06 from Zvezda's rear, withdrew 30 metres, then flew around and lined up to dock at Zarya's nadir at 05:37. While testing the integrity of the docking system's hermetic seal, they had their lunch. On re-entering the ISS, they noted that the temperature had fallen somewhat while there was less equipment in operation. Reactivating the station then took several hours. Hence, while the act of swapping docking ports involved only about half an hour, it consumed most of the working day. Nevertheless, this was necessary to clear the way for the next Progress resupply ship on the 28th, which delivered some 730 kilograms of propellant and a variety of dry stores.

CANADA'S CONTRIBUTION

Given its expertise in developing robotic manipulators for use in space, it was logical that Canada's main contribution to the ISS should be the Mobile Servicing System (MSS) consisting of:

- *Space Station Remote Manipulator System (SSRMS)* – This is a development of the manipulator used by the Shuttle, having enhanced functionality. At 17.6 metres in length, it is a few metres longer than the RMS. It has a Latching End-Effector at each end and its three booms incorporate seven joints for versatile and precise movements. It is designed to be repaired in space in the event of a fault.
- *Special Purpose Dexterous Manipulator (SPDM)* – This is a two-armed robotic adjunct for the SSRMS to provide dexterity for a range of delicate assembly and maintenance tasks that previously only a human spacewalker could undertake.
- *Robotic Work Station (RWS)* – This is a cluster of hand controllers and video monitors located inside the ISS, from which an astronaut will operate the various systems of the MSS.

The ISS after STS-98 installed the Destiny laboratory.

- *Space Vision System (SVS)* – This system enables astronauts operating a robotic manipulator to work 'in the blind'. The exteriors of the elements of the ISS are decorated with small black-and-white reference markers, and the SVS infers the position and orientation of a payload on an arm by integrating the perspectives from widely separated TV cameras. It is crucial to the trickier parts of assembling the ISS.
- *Power and Data Grapple Fixture (PDGF)* – This is an 'anchor' to provide power to the SSRMS and its payloads, and to transfer commands and video signals to and from the RWS. Such fixtures will be installed at strategic points around the ISS (including on the MBS) and the SSRMS will be able to 'walk' from one to the other, making its way across the structure in the manner of an inchworm.
- *Mobile Base System (MBS)* – This is a work platform for the SSRMS, with fixtures for gripping payloads and storing tools for spacewalkers. It has four PDGFs for the SSRMS and/or the SPDM.
- *Payload ORU Accommodation (POA)* – This attachment on the MBS will hold Orbital Replacement Units (ORU) for the externally mounted systems (most notably the batteries in the Integrated Equipment Assemblies) that will require replacement from time to time during the operation of the ISS. The MBS has a removable Camera/Light Pan and Tilt Assembly. Situated on a mast behind the POA, this camera will have a view of the upper surface of the MBS and unobstructed views of all four of the anchor points.
- *Common Attach System (CAS)* – This attachment on the MBS will carry pallets containing work tools and small scientific experiments.
- *Mobile Transporter (MT)* – This is an 886-kilogram box-like aluminium 'flat car' some 2.75 metres long, 2.4 metres wide and 1 metre thick designed to run along the 'railway tracks' on the leading edge of the Integrated Truss Structure, carrying the MBS on its back. It has three pairs of wheels, with one pair being driven by individual electric motors, and the others set in spring-loaded units in which rollers grip both sides of the track to prevent the transporter leaving the tracks (which must maintain their gauge in the extreme variation in temperature from sunlight to shade). It can drive at speeds ranging from a few millimetres to a few centimetres per second, and halt at 10 specific 'work sites' and be locked down to provide a firm base for the SSRMS to operate.[27] The carriage capacity (including the MBS and its facilities) is 22 tonnes.

The MSS was designed by Spar Aerospace under contract to the Canadian Space Agency, with the MT, which had to interface to the ITS, being built in California by the subsidiary Astro Aerospace under direct contract to Boeing. The MT and MBS were united on 12 June 1988, at which time the mechanical and electrical interfaces were tested by NASA astronaut Paul Richards and Canadian astronaut Julie Payette employing standard EVA tools. The SSRMS was delivered to the Space Station Processing Facility on 16 May 1999.

[27] The 'rail track' does not extend onto the trusses carrying the Solar Array Wings.

When Discovery was launched on 8 March 2001 for STS-102, its bay contained the first MPLM, named Leonardo. The module's 5-tonne payload included the first science rack (Human Research Facility-1) and six more systems racks for Destiny: two with power distribution equipment, two containing the Portable Robotic Work Station that would be used to control the SSRMS (which would be operated from within Destiny pending the installation of the first Cupola), one with the electronics for the Ku-Band Communications System, and one with a variety of health-care equipment. The payload also included three resupply racks and four resupply stowage platforms of apparatus to augment existing systems, including spare parts. The ICC in the bay had a pair of Assembly Power Converter Units and the 136-kilogram Lab Cradle Assembly. On arriving 180 metres beneath the ISS on the 10th, Jim Wetherbee flew a quarter-loop to position Discovery directly in front of the ISS in a nose-up attitude, and then slowly closed in to dock at PMA 2. The hatches were briefly opened a few hours later to transfer some items of cargo and to retrieve the cable connectors that would be required on the forthcoming EVA, then closed to enable the Shuttle to reduce its cabin pressure. The following day, Susan Helms and James Voss went out and Paul Richards coordinated their activities from the aft flight deck. When a 4.75-kilogram foot-restraint floated away, a replacement had to be retrieved from the stowage box mounted on the ISS before the planned activities could start. Their first task was to electrically isolate PMA 3 to enable it to be unberthed. They then disconnected the 45-kilogram antenna of the S-Band Early Communications System on Unity's port-side CBM, which had been surplus since the installation of the S-Band Antenna Support Assembly on the P6. Next, they retrieved the Lab Cradle Assembly from the ICC and affixed it to a trunnion pin on Destiny's 'roof'.[28] After fitting a Rigid Umbilical Tray on Destiny's exterior in preparation for the installation of the SSRMS, they retreated to the bay while Andy Thomas used the RMS to unberth PMA 3 and move it to Unity's port-side in order to clear the nadir port for Leonardo. As soon as the CBM latches were confirmed to have engaged, Helms and Voss re-entered the airlock. At a few minutes short of 9 hours, it had been the longest EVA in Shuttle history. After the two crews had slept, the hatches were reopened and Thomas lifted the 10-tonne MPLM from the bay. Its berthing on Unity was delayed while Shepherd routed a cable to transfer the video from a TV camera in the small window in the centre of the CBM to one of the screens on Discovery's flight deck to give Thomas a view of the module as he manoeuvred it in. After some confusion concerning the location of the power cable that was to route power from Unity to Leonardo, Discovery's hatch was sealed to enable its cabin pressure to be reduced again.

The mission's second excursion was made by Richards and Thomas on the 13th, with James Kelly operating the RMS and Helms coordinating activities. They began by installing a stowage platform on the ISS for spare parts, on which they placed an ammonia coolant pump. After finishing the task of stringing cables on the exterior of

[28] The 'claw' on the Lab Cradle Assembly was to be used on the next mission to temporarily hold the pallet on which the SSRMS would be delivered, and on a later mission to grasp the central segment of the ITS.

Destiny to provide power and control of the SSRMS, they scaled the P6 in order to inspect the faulty latch on the port-side array's structural brace. One of four, this brace had not latched into place properly when the arrays were installed. Although the other three braces were secure and the array's stability was not a concern, it was a frustration, so the news that they had managed to 'tap' it into its latched position was very welcome. While on the truss, they inspected the Floating Potential Probe, which had been operating only intermittently, and reported that they could not see the instrument's status lights. Meanwhile, Shepherd, Usachev and Voss unloaded cargo from Leonardo and, on Discovery, Wetherbee verified the function of the claw on the Lab Cradle Assembly and Gidzenko and Krikalev exercised to prepare their bodies for the return to Earth after four and a half months in space. After the hatches were reopened the following day, the unloading resumed. Once the racks had been transferred, most of the rest of the cargo was contained in soft-sided canvas bags. On the 15th, Thomas supervised the task of loading into Leonardo the 900 kilograms of items that were to be returned to Earth. The RWS was installed in Destiny. This comprised a trio of flat-screen display units, a laptop, and a small console with two hand-controllers, altogether providing a facility equivalent to that on the Shuttle's aft flight deck for controlling the RMS, but with additional screens because its operator has restricted external vision. When a Shuttle was docked, a cable would supply TV from the payload bay cameras to provide the SSRMS operator with a variety of perspectives. (Later, this 'situational awareness' would be provided by TV cameras mounted on the exterior of the ISS.) After a 24-hour extension of docked operations to give more time to load Leonardo, its retrieval on the 18th was delayed by several hours due to a leaky hose used to depressurise its vestibule. Monitoring the pressure in the vestibule 15 minutes after it should have been a vacuum, flight controllers noted that it was not. After making an inspection, Voss tightened a loose fitting on one of the hoses, and once the seals to Unity and Leonardo were verified and the vestibule depressurised, Thomas retrieved the MPLM and returned it to the bay.

After a sleep period, Shepherd lined up the two ISS crews facing each other in Unity and rang the ship's bell as he formally transferred command to Usachev. "We are on a true space 'ship' now, making her way above any Earthly boundary," he said. Wetherbee added insightfully: "This ship was not built in a safe harbor but on the high seas." When the hatches were finally closed an hour later, the 136-day first 'increment' of ISS habitation ended.[29] In fact, the exchange had been staggered, with the retiring residents transferring their individually contoured Kazbek couch liners and lightweight Sokol pressure suits from the Soyuz to Discovery (where they were stowed away) and the newcomers (not necessarily direct successors in terms of crew assignments) installing theirs in the Soyuz. If an emergency had developed that required the Shuttle to undock, those whose couch liners were in the lifeboat would remain on the ISS, irrespective of the crew to which they belonged. Shepherd summed up his crew's tour of duty: "We basically put the space station in

[29] The 'increment' marks the crew's time on board the ISS. Their time in space is longer, because it includes the flights up and down.

commission. We have taken something that was an uninhabited outpost and we now have a fully functional station where the next crew can do research. I think that is the substance of our mission." Although content to be returning home, he was "not that anxious" to return to full gravity. For the descent, Shepherd, Gidzenko and Krikalev used recumbent seats on the mid-deck, as had their predecessors returning from Mir.

Even before STS-102 departed, the new crew members had started the Hoffman Reflex neurological experiment to monitor how their nervous systems adapted to the space environment. On 21 March, Helms configured the Ku-Band to relay data from the racks in Destiny, but a software error prevented the dish antenna from tracking the geostationary satellite. (After the engineers had identified the fault, they devised a software 'patch', but this did not become available until early April.) Over the next few days, they started both the Bonner Ball Neutron Detector and the Dosimetric Mapping experiment, storing the data on the computer in the HRF-1 rack. One of this crew's primary tasks was to characterise the radiation environment inside the ISS and in early April a major solar flare provided a useful point of reference. Although Usachev and Voss slept in Zvezda's cabins, Helms set up home in Destiny, in a Temporary Sleep Station slotted into the space where a science rack would later be installed. This box was lined with dense polyethylene for protection against such 'hard' radiation as might penetrate the module's walls, had a door for privacy, and incorporated all the comforts of home.

On 4 April it became possible for flight controllers to use Zvezda's computer to fire the engine of a docked Progress, so that it could boost the orbit of the ISS using the reserve of 250 kilograms of propellant that it carried for this purpose. This new command path was achieved by installing a computer in the ferry and linking it to Zvezda's computer. The command path would be retrieved before the craft left, and would be installed in each successive Progress. Whenever practicable, the rear port would be reserved for Progress ships in order to save the 'service life' of Zvezda's engines. After being loaded with trash, the Progress departed on 16 April. Two days later, the ISS was placed in semi-hibernation while the crew relocated Soyuz-TM 31 from Zarya's nadir to the rear of Zvezda in order to clear the way for a replacement 'lifeboat' – the 'taxi' crew would take away the 'expired' Soyuz and thereby free the rear port once again.

When STS-100 docked on 21 April, it brought MPLM Raffaello with 3.4 tonnes of cargo which included two more science racks for Destiny,[30] and a Spacelab pallet with the 1.75-tonne SSRMS and a UHF antenna. The hatches could not be opened because Endeavour's cabin pressure had been reduced, but the ISS crew opened the vestibule to PMA 2 to pass a battery-operated hand-tool that would be required on the EVA, and in return the visitors passed back some computers, water containers, and a cable that was to bypass an overly sensitive circuit breaker in the RWS. The next day, Scott Parazynski and Chris Hadfield went outside. Umberto Guidoni and Jeffrey Ashby operated the RMS, on which Hadfield rode for much of the time, and

[30] These were EXPRESS racks 1 and 2, loaded with CGBA, PCG-High Density, PCG with a Single Thermal Enclosure System, Advanced Astroculture, MAMS, SAMS, ARIS and the Experiment on Physics of Colloids in Physics.

John Phillips coordinated activities. While they were still in the airlock, Ashby lifted the 1.36-tonne Spacelab pallet and positioned it to engaged the 'claw' on the Lab Cradle Assembly, then withdrew the arm. One of the Latching End-Effectors of the SSRMS had been pre-engaged with the Flight Support Grapple Fixture on the pallet. Parazynski strung power, command and video cables from Destiny to this fixture to enable the SSRMS to be tested by the RWS in Destiny. Next they collected the UHF antenna from the pallet and mounted it on Destiny. This was the first of a number of such antennae that were to be installed to facilitate communications during EVAs from the Joint Airlock (once this was installed) when no Shuttle was present. On returning to the pallet, Parazynski and Hadfield stripped off the thermal blanket that had protected the SSRMS. After verifying the power supply, the launch restraints were released, the main segment of the arm was unfolded and a pistol-grip tool (set in 'manual' mode to gain the correct degree of torque) was used to secure several expandable fasteners in order to lock it rigid. Once Voss and Helms at the RWS had issued a command to the SSRMS to verify its response, the spacewalkers retreated to the airlock. Early on the 23rd, the hatches were opened. Helms made a methodical test of the arm's seven joints, testing both the primary and backup modes for each, and putting the arm through a variety of contortions prior to finally grasping the Power and Data Grapple Fixture on Destiny (which had been installed by STS-98), in which configuration it was left. Meanwhile, Parazynski had berthed Raffaello on Unity's nadir to let the others make a start on unloading its cargo. Finally, the hatches were again closed to enable Endeavour to reduce its cabin pressure. On the 24th Hadfield and Parazynski went out again. The first task was to delete the S-Band Early Communications System from Unity's starboard CBM, to clear this for the Joint Airlock. When Helms reported that a backup power circuit in the SSRMS had failed to respond to a command, the spacewalkers opened a panel on the Power and Data Grapple Fixture on Destiny and rewired it to complete the redundant power path. With the arm operational, the 'claw' was commanded to release the pallet, which Helms manoeuvred into position preparatory to 'handing it over' to the Shuttle's arm the next day. At the conclusion of the EVA, the hatches were reopened so that everyone could participate in the unloading of Raffaello.

As the two crews slept after an exhausting day, Houston saw a loss of signal on Command and Control Computer 1 (C&C-1), one of three identical units that ran in parallel.[31] This resulted in an inability to transfer bulk data between the computers and their mass storage devices. Suspecting a software fault, Houston turned C&C-2 from 'standby' to 'primary'. When diagnosing the fault the following day, Houston commanded C&C-1 to illuminate a light in Destiny, which it did, and turn it off again, but when Helms tried to use C&C-1 to transfer a file to the RWS it failed. The plan to work with the SSRMS had to be shelved. Meanwhile, crew members not involved in this diagnostic effort unloaded Raffaello. By the start of the next day, C&C-1 and C&C-3 were *both* down with what appeared to be software faults, and C&C2 was running without a level of redundancy. Overnight, two fault-protection

[31] The C&C computers act as the interface between the ISS systems and hard drives that store bulk data and executive programs.

computers in Unity, whose rôle was to respond to malfunctions in other computers by rebooting them, also failed. But early on the 26th, C&C-1's command and data functions were restored, and it was commanded to dump its state to the ground to permit the engineers to study it. NASA decided to extend docked operations by up to two days to give time to complete the planned activities. This introduced a problem. According to the 'rule book', the Russians could not manoeuvre one of their spacecraft while a Shuttle was docked. If the first Soyuz 'taxi' lifted off on schedule on the 28th, Endeavour would still be present when it arrived on the 30th. To dock at Zarya's nadir port the Soyuz would have to pass close to the Shuttle's vertical stabiliser, which was a risk that NASA had no wish to take. NASA requested a postponement of one day, but the Russians continued the countdown and promised that if Endeavour was there when the Soyuz arrived, their spacecraft would stand off and await its departure. While the astronauts slept, Houston worked on in the hope of recovering at least the fault-protection computers, and succeeded in resolving most of the problems. C&C-1 was found to have suffered a hard drive failure. The failed unit was removed for return to Earth, and replaced by Payload Computer 2 (PC-2). A software patch was uploaded to C&C-3, but this was found to contain an error. However, when this was rectified the computer operated perfectly. Meanwhile, Parazynski retrieved Raffaello, which had been loaded with 726 kilograms of unwanted items. On the 28th, with the computers finally back on line, Helms powered up the SSRMS, which was still holding the pallet high above Endeavour's bay. Once Hadfield had grappled the other fixture on the pallet using the RMS, the SSRMS withdrew, thereby completing its 'step' from the pallet onto Destiny. Hadfield put the pallet back in the bay. This handover from one arm to the other was crucial, because the pallet could not have been left fouling the claw, because it would be required when installing the central segment of the ITS. Fortunately, it was not necessary to run into a second day's extension, so the hatches were closed on the 29th and Endeavour departed. The Soyuz, which had been launched on time, was only 14 hours away from completing its rendezvous. Once the Shuttle had departed, Houston reformatted the hard drive in PC-2 (now serving as C&C-1) and copied the hard drive of C&C-2 to C&C-3, thereby restoring the ISS to a fully redundant system. On the ISS, Usachev, Helms and Voss had a well-deserved afternoon off.

TAXI

When Soyuz-TM 32 docked at Zarya's nadir on 30 April 2001, it delivered Talgat Musabayev, Yuri Baturin and Dennis Tito – a 60-year-old American who had made his fortune in the insurance business and paid $20 million to become the first 'space tourist'. Originally, he was to have flown to Mir, but this station was de-orbited on 23 March and the Russians honoured the deal by flying him to the ISS. This greatly irritated NASA – before the agency would consent to Tito's visit, it insisted that he sign a declaration that he would be personally liable for any damage that he caused, and that he would not sue in case of personal injury. Unless supervised by an ISS crew member, Tito was to remain in the Russian modules. In practice, he spent most of his time in Zvezda admiring the Earth slowly passing below, and made himself useful by doing mundane chores for his hosts. When Tito held a press conference in Zvezda, NASA dissuaded Voss and Helms from participating, did not relay it via its Ku-Band link (it had to be transmitted directly to Moscow while within range of the Russian ground network) and NASA-TV did not relay it. The rôle of this flight was to exchange the station's 'lifeboat', whose service life was six months. Musabayev, Baturin and Tito duly departed in Soyuz-TM 31 on 5 May and landed just after dawn on the Kazakh steppe the following day.

TWO AIRLOCKS

After a hectic few weeks of playing host and dealing with the computer problems, Usachev's crew settled into a routine. A regular task – which was undertaken every Thursday for six weeks – was to put the SSRMS through a full dress rehearsal of how it was to retrieve the Joint Airlock from the next Shuttle's bay and manoeuvre it for mating with Unity's starboard CBM, something that the RMS would have difficulty doing because it was shorter and inconveniently located for operating on that side of Unity. On 16 May, NASA announced that both Solar Array Wings on the P6 had been locked into position, preventing them from turning to face the Sun to maximise power generation while engineers assessed abnormally high temperatures in the Beta Gimbal Assemblies. The next day, while Helms was testing a backup mode on the SSRMS, the arm malfunctioned, but it was not immediately evident whether this was a software or a hardware issue. The operating rules said that the Shuttle with the Joint Airlock could not be launched unless both the primary and backup systems on the SSRMS were functional. The next Progress, launched on 20 May, rode the uprated 'Soyuz FG' launch vehicle, which enabled it to carry an additional 250 kilograms of payload – in this case 1.2 tonnes of propellant and 1.4 tonnes of dry cargo, the latter including Pizza Hut products with a variety of toppings that the crew were to be filmed eating in return for a $1 million commercial contract. When the ferry docked at the rear of Zvezda, vibration sensors in Destiny monitored the transient vibrations that were transmitted through the structure in order to determine the extent to which dockings would influence microgravity experiments. The Kurs package was removed from the Progress for return to Earth on the next Shuttle, for reuse. Helms's efforts to diagnose the fault with the SSRMS continued through the end of the month, but as it was proving unresponsive to software patches the launch of STS-104 was postponed to "no earlier than 7 July" and planning was initiated for a later Shuttle to replace the affected hardware. On 8 June, Usachev and Voss donned Orlan suits, sealed themselves in Zvezda's forward docking compartment, unbolted the 'flat plate' on the nadir port and replaced it with the 68-kilogram docking cone to prepare for the arrival of a new Russian module. During this 'intravehicular activity', Helms had retreated to Zarya to ensure that she would be able to reach the 'lifeboat' in the event that Zvezda's compartment could not be repressurised. The following day, the Orlan suits were serviced by recharging the batteries, replenishing the oxygen tanks, cleansing the carbon dioxide scrubber, and drying out the sublimator that used water to shed heat to the vacuum of space. By 20 June, an analysis of a series of software patches had finally determined that the SSRMS was suffering an intermittent fault in either a specific cable or an associated chip (located inside the arm) that generated a communications error between the computer and the pitch joint on its 'shoulder'. To ensure that the backup mode would be available during the installation of the Joint Airlock, the computer was instructed to ignore this spurious error. Meanwhile, it had also been found that the motors in the Beta Gimbal Assemblies were suffering thermal stress, and decided that a Shuttle crew should install insulation blankets.

STS-104 lifted off on 12 July with the Joint Airlock Module, named 'Quest',[32] and a Spacelab pallet carrying the High-Pressure Gas Assembly. Quest comprises two cylindrical compartments. The Equipment Lock has a CBM to mate with Unity and the long narrow coaxial Crew Lock is derived from the internal airlock developed for the Shuttle's mid-deck. The Crew Lock is the section that can be depressurised. In contrast to the airlock on the Shuttle, which vents its air to space, the pumps in the Equipment Lock can reclaim 75 per cent of the air in the Crew Lock in order to minimise losses. As on Shuttle spacewalks, there will be no solo excursions, and the Crew Lock can readily accommodate two suited astronauts oriented head-to-foot. The inward-opening exterior hatch faces the nadir. In addition to providing storage for suits to prevent them from cluttering up the ISS, the Equipment Lock has facilities for servicing suits. Quest is 'joint' in the sense that it includes systems to support both American and Russian suits.[33] Four High-Pressure Gas Tanks were to be installed around the exterior of the Equipment Lock with nitrogen and oxygen to replenish the losses from operating the lock,[34] and spacewalkers will also be able to 'top up' from the oxygen tanks. Two tool boxes are on the exterior of the Crew Lock for convenient access. The airlock and its gas tanks were developed by Boeing in Huntsville.

Atlantis's docking at PMA 2 late on the 13th was monitored by the sensors in Destiny. As the two crews were on overlapping shifts, the hatches were opened to review the plan for the first EVA, Helms and Janet Kavandi rehearsed how they would coordinate the transfer of Quest, then the hatches were closed again to enable Atlantis to reduce its cabin pressure. After sleeping, Michael Gernhardt and James Reilly made their first excursion. Helms operated the SSRMS, Kavandi operated the RMS and Charles Hobaugh directed activities from the aft flight deck. The first task was to prepare Quest for removal from the bay. While Gernhardt removed thermal covers, Reilly installed the attachment points (dubbed 'towel bars' because that was what they resembled) for the tanks. (The bars could not have been fitted for launch because they would have impinged upon the bay walls.) Finally ready, Gernhardt unhooked the umbilical that had powered Quest's heaters and Helms lifted it out of

[32] It may have been named *Quest* after one of the ships used on Ernest Shackleton's 1921 Antarctic expedition (the ship on the 1914 expedition was *Endurance*).
[33] Astronauts wearing EMU suits cannot pass through the narrower hatches of the Russian-designed airlocks.
[34] As well as servicing the airlock, the High-Pressure Gas Assembly represented several months of 'spare' air for the ISS crew in an emergency.

the bay. After an intricate series of manoeuvres lasting some 2.5 hours – with the benefit of a perspective view from the TV camera on the RMS that was fed to the RWS in Destiny – she placed it close against Unity's starboard CBM, which she could not see directly, and berthed it using the visual cues of the spacewalkers. After connecting cables to power Quest's heaters, Gernhardt and Reilly retreated to Atlantis. As soon as the CBM vestibule had been pressurised, Voss and Helms opened the hatches to inspect Quest's systems. When the two crews were reunited, they held a 'ribbon cutting' ceremony. Over the next two days, Quest was commissioned. This involved hooking up the water coolant loop of the Thermal Control System, fitting valves to connect Quest to the common Environmental Control and Life-Support System, testing the equipment for servicing spacesuits, testing the lines that would be used by future Shuttles to replenish the external gas tanks, testing the spacesuit communication systems, and testing the evacuation and repressurisation of the Crew Lock. Several problems were encountered. Firstly, air bubbles in a coolant line resulted in a water spillage which had to be mopped up. The depressurisation of the Crew Lock was frustrated by air leaking into it. Although Kavandi and Voss could not pinpoint the problem, they traced it to a valve in the Inter-Module Ventilation Assembly, and the leak stopped when this was capped. Because the third and final EVA of the visit was to be made from Quest, there was a sense of urgency. It was decided to replace the valve with one taken from Destiny that would not have been used until the second node was mated (there would be plenty of time to fit another valve in *that* position). Voss and Steven Lindsey completed this transfer on the 18th, and a check confirmed that the leak had been eliminated. A dress rehearsal was then undertaken, including depressurising the Crew Lock to 10.2 psi. As this maintenance had slipped the flight plan by about half a working day, it was decided to extend docked operations and postpone the final EVA by one day to ensure that Quest was fully operational.

The EVA on the 17th began a little later than intended due to the need to reboot one of the C&C computers. After Helms had grappled one of the gas tanks using the SSRMS, and Gernhardt had released the launch locks, Helms lifted the 545-kilogram 'dog house'-shaped tank off the pallet and positioned it alongside one of the 'towel bars' on the Equipment Lock. Gernhardt mounted the RMS and Kavandi placed him next to the tank. Meanwhile, Reilly, working tethered, had fitted PFRs and guides to align the tank for attachment. Standing on the newly installed PFRs, they latched the tank into place and hooked up hoses to Quest. A micrometeoroid débris shield and a multi-layer insulation blanket protected the tank from the harsh space environment. While Gernhardt rode the RMS back to the pallet, Reilly moved the PFRs and guides for the second tank, which was installed with equal ease. After installing insulating covers on several airlock fixtures, including the grapple fixture and the four trunnion pins, they found themselves so far ahead of schedule that they decided to 'get ahead' and transfer the third tank. Late on the 20th, Gernhardt and Reilly prepared to make the first outing from Quest. Although the Shuttle's cabin pressure could readily be reduced overnight to help to bleed nitrogen bubbles from the bloodstream, lowering the pressure of the ISS would not be feasible. When using Quest, the pre-breathing protocol involved 10 minutes of vigorous exercise followed by 80 minutes on oxygen masks prior to suiting up in the Crew Lock. On this occasion, a problem with the equalisation valve in the hatch between the Equipment and Crew Locks meant that the depressurisation process took 40 minutes instead of the expected 7 minutes, and the exterior hatch could not be opened until after midnight. After the fourth gas tank had been installed, they made their way up the P6 and reported that there was nothing obviously wrong with the Beta Gimbal Assemblies, which were still locked. The next day, Atlantis departed with just over 1 tonne of unwanted items.

STS-105's mission was to retrieve Usachev's crew and deliver their successors: Frank Culbertson, Vladimir Dezhurov and Mikhail Tyurin. Discovery's bay held Leonardo with EXPRESS racks 4 and 5 for Destiny[35] and the ICC carrying the Early Ammonia Servicer, which was a tank of coolant. The hatches were opened soon after docking at PMA 2 on 12 August. The next day, Leonardo was berthed on Unity's nadir. This time, the three crews made a 'human chain' to pass the smaller items of cargo from hand to hand – "like buckets of water to a fire" – in order to evaluate whether this was more efficient than having people float back and forth with individual items.[36]

Although Quest was operational, the two EVAs of this mission were to be made from Discovery. The hatches were closed on the 15th to enable the cabin pressure to be reduced. The next day, Scott Horowitz grasped the Early Ammonia Servicer using the RMS. Once Patrick Forrester and Dan Barry had released the launch restraints, they held onto the handrails on the tank and hitched a ride as the RMS positioned it alongside the P6. Barry affixed a PFR on the truss, stood on it, and held the EAS to allow the RMS to be withdrawn. Forrester then manoeuvred the tank to engage a hook-like fixture, and when this was engaged he inserted a 'pip pin' to secure it. The rôle of the hook was to ensure that the tank did not float off while it was being bolted to the truss. This done, the spacewalkers attached cables to power the heater in the tank. In effect, the 'servicer' was an in-orbit reserve in case it became necessary to top up the external coolant loop of the Thermal Control System. After riding the RMS back into the bay, the Materials ISS Experiment was retrieved from the ICC, and when this was clamped onto a handrail on Quest it became the first experiment to be deployed outside the ISS.[37] With the hatches reopened, the formal handover ceremony was performed on the 17th. On the second EVA the next day, Barry and Forrester retrieved bags of handrails and cables, Horowitz positioned them beside Destiny, where they relocated two of its handrails, added a dozen more, then used these to string cables from the Umbilical Tray at the base of the Z1 along the

[35] The EXPRESS racks were not being delivered in numerical order: racks 1 and 2 had been ferried up by STS-100 in April 2001 and rack 3 was assigned to STS-111, which was due in mid-2002. Taken together with the HRF, these two additions increased to five the number of science racks in Destiny.

[36] In addition to two EXPRESS racks, Leonardo contained six resupply stowage racks and four stowage resupply platforms. Some of the stowage racks were temporarily transferred to the ISS, but many hardware items and clothing, food, miscellaneous supplies for the crew were transferred individually in canvas bags.

[37] The Materials ISS Experiment (MISSE) comprised two pallets of materials that were simply to be exposed to the space environment for 18 months in order to assess their degradation. In this case the experiment involved 750 samples ranging from lubricants to solar transducer cells.

module, leaving them conveniently positioned to be plugged in to power the heaters of the central segment of the ITS when this was delivered by STS-110. After Leonardo had been loaded up with 1,360 kilograms of miscellaneous items for return to Earth, it was retrieved on the 19th and Discovery left on the 20th, taking with it the retiring crew and their completed experiments.

The next Progress lifted off on the 21st. Its predecessor (having been on Zvezda since May) left on the 22nd, taking with it the accumulated trash, and the new ship took its place on the 23rd. As Jerry Linenger had remarked while serving on Mir, the Russians operated the Progress logistics ships with the precision of a well-organised railway timetable.

By mid-2001, NASA had relaxed its antipathy towards 'tourists' accompanying taxi crews, and in September it agreed standards that 'spaceflight participants' must satisfy: they had to be fluent in both English and Russian, pass a psychological test and be in excellent physical condition. Mark Shuttleworth, an internet millionaire from South Africa, was already in training.

In early September, Dezhurov set up the TORU in preparation for the arrival of the first module that was to expand the Russian section of the ISS. This lifted off on 14 September and arrived on schedule two days later. It was an *ad hoc* configuration in which the propulsion module of a Progress ship was carrying an airlock module named 'Pirs' (meaning 'pier'). At orbital insertion, the vehicle was 7,130 kilograms. Half of this was the module and the 800 kilograms of cargo with which it had been packed. It had a hybrid docking system on its nose in order to dock at Zvezda's nadir, where the drogue had been fitted in June. Dezhurov, ready at the controls of the TORU in Zvezda, had a side view of the ship's final approach from a TV camera placed in Soyuz-TM 32's window. The Kurs system's approach was flawless, but the shock at the moment of contact was sufficiently hard to prompt Culbertson to report, "We really felt that!" Pirs is essentially cylindrical in form, being 2.55 metres in diameter at its widest, and 4.9 metres in length – very similar to the Docking Module that was installed on Mir. It was mated to the Progress propulsion module by an annular skirt that housed the supplementary avionics.[38] The airlock has two outer hatches (set 180 degrees apart) that face to port and starboard of the ISS, with inward-opening hatches (in contrast to the troublesome outward-opening hatch on Mir) of 1 metre diameter. Its rôle is to provide ready access to the Russian end of the facility. Although the compartment is cramped, apparatus intended to be taken outside can be placed in Zvezda's forward docking compartment, which will serve as an alcove. Unlike Quest, Pirs's systems are specific to Orlan suits; it is not a 'joint' facility. When not in use, the airlock will serve as a store room for the Orlan suits. On the 20th, Zvezda's flight control system was updated to communicate with Pirs, and on the 26th explosive bolts were fired to jettison the propulsion module. Its departure exposed a standard docking system to provide a third site for Soyuz or Progress ships. In fact, the Soyuz taxi flights would now utilise the nadir ports on Zarya and Pirs, and the rear port would be reserved for Progress ships, from which position they could boost the ISS's orbit.

[38] This *ad hoc* configuration was designated Progress-M-SO 1.

A good view by STS-110 of the Quest airlock, with its external gas tanks, a Soyuz docked at the nadir port of Zarya and, in the background, the Pirs airlock with its Strela cranes in place.

On 8 October Dezhurov and Tyurin made the first spacewalk from Pirs. After Culbertson had retreated to Zarya, the apparatus that they were to take outside was placed in Zvezda's forward docking compartment, whose axial hatches were sealed. With the airlock depressurised, Tyurin opened one of the outer hatches and led the way out. Dezhurov passed out several handrails, which Tyurin affixed around the rim of the aperture (the handrails could not have been fitted prior to launch as their presence would have obstructed the launch vehicle's aerodynamic shroud). A ladder was then installed to provide ready access to Zvezda, and power and data cables run between Pirs and Zvezda. After removing and jettisoning thermal blankets, the cosmonauts began the 'heavy' work. One of the items of cargo delivered by Pirs was the second Strela crane. The stubby operator's post was eased out of the hatch and mounted on a nearby fixture, and the extensible boom was added. They paced their work by resting during orbital darkness. After orienting the SSRMS to allow the ground to monitor activities via the arm's TV camera, Culbertson observed through one of the windows of Soyuz-TM 32. Having slipped behind schedule, it was decided to go ahead and install two antennae and an optical target for a Kurs docking system and postpone the final task, which was to have seen Dezhurov swing Tyurin around on the end of the crane to evaluate its rigidity. They went out again on the 15th. Once across the ladder, they made their way along the length of Zvezda using handrails and tethers. The Kromka package was emplaced to sample the residue from the thrusters arranged around the module's rear, and it was to be returned to Earth later to enable the chemicals that stained the vehicle's surface to be identified. Moving to a nearby site, they assembled a small truss structure on which they mounted three suitcase-sized packages for the Japanese Space Agency. Normally, micrometeoroids and particles of orbital débris are vaporised on impact with a shield, but one experiment used a foamy substance to slow and 'capture' them for return to Earth. Another experiment exposed materials such as paint, insulation and lubricants to the harsh space environment. On their way back, they set up another exposure experiment, this one bearing the logo of a company that was to supply the cameras to document future external activity.

On 17 October 2001, after almost a decade at the helm, during which the space station had been transformed from a paper project into an orbital reality, Dan Goldin resigned. He was succeeded a month later by Sean O'Keefe. As the deputy director of the Office of Management and Budget, O'Keefe had criticised NASA for ignoring costs and concentrating primarily on technical and safety issues. He announced that his mandate was to bring "firm accountability" to the agency. At his swearing in, he warned that the ISS should not be pursued "at the expense of everything else this organisation does". He said that he would decide in June 2003 "whether it is prudent to continue" with the *final* construction of the ISS as designed – the key factor in his decision would be the spiralling costs. Meanwhile, Chris Kraft, the former lead flight director and chief of the Johnson Space Center, wryly noted: "NASA today is asked for 100 per cent reliability. If we'd had the same oversight in the 1960s we wouldn't have left the ground." It would be possible to "decrease costs dramatically" if NASA were to make trade-offs against the "no risk" philosophy, by allowing Shuttles to lift off with some redundant systems 'out', rather than insisting on the maximum of redundancy. He also opined that the $4 billion recently assigned to the Space Launch Initiative that was to investigate technologies for a replacement for the Shuttle should be devoted to upgrading the Shuttle, in much the same manner as Boeing continually upgrades its airliners.

With Soyuz-TM 32 nearing the end of its life, Culbertson's crew transferred it from Zarya to Pirs on 19 October to enable Soyuz-TM 33 to dock at Zarya's nadir on the 23rd with Viktor Afanaseyev, Konstantin Kozeev and Claudie Haigneré – a researcher who had previously visited Mir as Ms Andre-Deshays – and was flying as part of a long-term contract between Russia and France. After the ceremonies, Dezhurov exchanged Kazbek couch liners and Sokol suits between the two vehicles. On the 30th, having conducted her scientific work in Zvezda, Haigneré left with her colleagues in Soyuz-TM 32. Stationing the 'lifeboat' on Zarya meant that the third member of the crew would not be obliged to retreat to it during an EVA from Pirs as there was little that a person in *that* position could do to assist in the event of a problem. No sooner had the visitors departed than preparations got underway for an EVA scheduled for 5 November, but at NASA's request this was slipped by a week to give the crew a rest after playing host. On the 12th Dezhurov opened Pirs's hatch and fitted more handrails, and when Culbertson followed him out they installed more cables between Pirs and Zvezda. After inspecting a small section of one of Zvezda's solar panels that had not deployed properly, they tested the boom's rigidity – a task left over from the previous excursion. One man tethered himself to the Strela, which was then extended to a length of 10 metres and hand-cranked in pitch and yaw. The test was monitored by Tyurin via the TV camera on the SSRMS which, because he was not qualified to operate it, had been previously set by Culbertson. The Progress departed on the 22nd. Its successor had to be in place before STS-108 arrived for the crew exchange. Although the Kurs steered the replacement in flawlessly on the 28th and made a 'soft' docking, the 12 latches in the two collars failed to engage when the probe was retracted. A replay of the ferry's docking camera footage showed that something had fouled Zvezda's drogue. With the ferry held only by the small latches at the tip of its probe, manoeuvring of the ISS would have to be severely restricted. It was decided to conduct a spacewalk to extricate whatever had fouled the docking mechanism. NASA immediately postponed STS-108's launch (set for the next day) to await a resolution of the situation. On the 29th, the ISS crew began to service the Orlan suits. On 3 December Dezhurov and Tyurin exited Pirs, crossed the ladder to Zvezda, and made their way to its rear where, on peering over the rim, they saw a rubber ring-seal trapped in the narrow gap between the two docking collars. Moscow commanded the probe to extend in order to open the gap to 40 centimetres, and Dezhurov cut the ring to enable it to be withdrawn. As they watched, the probe was commanded to retract, which it did to form a hermetic seal.[39] As they made their way back to Pirs, Dezhurov and Tyurin had the satisfaction of knowing that a human presence had once again saved the day.

[39] A similar operation had been performed aboard Mir in 1987, when a bag of trash had obstructed the approach of the Kvant 1 module.

NASA promptly reinstated STS-108's count, aiming for launch on 4 December, but poor weather ruled this out and it finally lifted off on the 5th. Endeavour carried Raffaello in its bay, with 3 tonnes of food, supplies, experiments and miscellaneous apparatus. The hatches were opened three hours after docking at PMA 2 on the 7th, and Yuri Onufrienko, Carl Walz and Daniel Bursch exchanged Kazbek couch liners and Sokol suits with the retiring crew. The next day, Mark Kelly and Linda Godwin used the RMS to place Raffaello on Unity's nadir to enable the three crews to make an early start on offloading its cargo. On the 9th, Endeavour was sealed to reduce its cabin pressure in preparation for the next day's EVA. Kelly operated the RMS and Dominic Gorie acted as the coordinator. After Godwin and Daniel Tani had retrieved two thermal insulation blankets from the bay, they rode the RMS as far up the P6 as this was able to reach, and then they completed the ascent in hand-over-hand manner using tethers. Once at the top, they wrapped the blankets around the Beta Gimbal Assemblies – the engineers believed that reducing the thermal stress on their motors would eliminate the intermittent 'spikes' in their current consumption. On their way down, a cover that had been stripped off the S-Band Antenna Support Assembly by an earlier crew, and left in place, was retrieved as it had been decided to return it to Earth for reuse. Finally, they undertook a few 'get ahead' tasks by positioning cables and two Circuit Interrupt Devices on the exterior of the Z1 in preparation for the STS-110 crew's installation of the central segment of the ITS, and retrieved several tools from the EVA Tool Support Devices on the Z1 that were no longer required. By the 12th, some 2,700 kilograms had been offloaded from Raffaello, together with 300 kilograms from Endeavour's mid-deck. On the 14th, after 900 kilograms of miscellaneous items had been stowed in Raffaello, Kelly retrieved the MPLM and Endeavour left the next day. There was no rest for Onufrienko's crew, however, because they had to unload the Progress ship. On the 18th, the engineers were surprised when the insulated motor of the port-side Beta Gimbal Assembly stalled, but once it was commanded to restart it displayed no further sign of distress.

At the start of an EVA from Pirs on 14 January 2002, Onufrienko extended the Strela crane's boom to its full 15-metre length, cranked it around and, taking care not to damage any apparatus, lowered it alongside Zarya. Walz made his way along the boom in hand-over-hand manner and unlatched the crane that had been stored on PMA 1.[40] After Onufrienko had retracted the boom to return Walz and the payload, they installed it on Pirs. (There was now a Strela on each side of the complex.) This done, they used the newly fitted crane to make their way along Zvezda to put a ham radio antenna on the module's rear periphery, as the first of four whip-like antennae that would eventually project out at 90-degree intervals. The electronics, currently in Zarya, was to be relocated to Zvezda to provide the operator with greater comfort. On the 25th, Onufrienko went out with Bursch. After Onufrienko had inspected a flaw in one of Zvezda's windows, which he said was reminiscent of "an exploded air bubble" in the glass, they installed six plates to deflect the plumes from the thrusters around the periphery at the rear of the module,

[40] The operator's post had been installed by STS-96 and the boom added by STS-101.

to protect its surface from the toxic residue. The Kromka sample collector was bagged for return to Earth to provide a point of reference for evaluating the effectiveness of the deflectors, and another Kromka was fitted. After erecting the second ham antenna, they set up three more materials cassettes and an experiment to capture the low-energy heavy nuclei in cosmic rays. As they returned, they attached a series of 'fairleads' to the handrails to prevent their tethers from fouling exterior apparatus on future excursions. Their progress was monitored by Walz through the SSRMS camera, who coordinated their activities.

The next EVA was the first from Quest without a Shuttle being present, and was conducted using NASA suits. On 20 February Bursch and Walz undertook tasks in preparation for the installation of the central segment of the ITS. The activities were coordinated by Joseph Tanner in Houston, who monitored the TDRS-relayed video feed from the TV camera on the SSRMS – which was operated by Onufrienko. The Circuit Interrupt Devices left by STS-108 were hooked up to the recently emplaced power cables on Destiny in order to verify their function. As the telemetry showed spurious readings, the circuit was left 'on' for evaluation. While Walz removed four thermal blankets from the Z1 and stowed them inside its trusswork, Bursch retrieved tools from the EVA Tool Support Devices that would be required by the STS-110 spacewalkers and stored them in the 'handy' tool boxes on Quest. They inserted ties to secure the latches on the tanks of the High-Pressure Gas Assembly, retrieved the adapter plate which had temporarily held the Strela on a trunnion pin, stowed it on Zarya, and visually inspected and photographed the radiator on the P6 for signs of micrometeoroid strikes that might have punctured the ammonia coolant tubes. On noticing that some of the samples in the Materials ISS Experiment package were already peeling off, this was also photographed. While outside, Bursch and Walz had worn radiation sensors for an experiment to monitor radiation doses endured by spacewalkers. As the suits were being serviced following the EVA, using the system in Quest's Equipment Lock that was designed to 'bake' the metal oxide canisters that had removed carbon dioxide from the EMUs, an unpleasant odour was emitted. After switching off the system, they closed the valves of the Inter-Module Ventilation Assembly to isolate the contamination, retreated to Zvezda and closed Unity's hatch to leave the Trace Contaminant Control System in Destiny to cleanse the air overnight. It was decided that the apparatus was outgassing and that, with a little luck, this would be a one-off occurrence. Meanwhile, the engineers had realised that the cables connected on the earlier EVA had been misconfigured due to incorrect documentation; the simple solution was to reinstall the DDCU inside Destiny that was supplying the power.

During a rehearsal on 5 March of how the SSRMS would lift the central segment of the ITS from STS-110's bay and mount it on Destiny, the arm's primary system suffered a malfunction that suggested a 'short' in the brake circuit of the roll joint of its 'wrist', inhibiting the release of the brake. On the 13th, the Lab Cradle Assembly on Destiny that would be used during the installation was inspected using the arm's TV camera. The following day, a similar inspection was made of the radiators on the P6, to seek any signs of discoloration or deformation. The SSRMS was parked in a position to monitor the departure of the Progress ship on the 19th, in case it shed a rubber seal, as, seemingly, had its predecessor. Its successor docked on the 24th. In addition to the usual supplies, this Progress delivered apparatus for the visit by a European Space Agency astronaut on the next taxi mission. Onufrienko's crew were told on the 24th that they would have to extend their tour of duty by a month or so because STS-111 was to be slipped beyond its nominal 6 May launch to allow time for its crew to prepare for an additional EVA task – the replacement of the faulty wrist-roll joint of the SSRMS. On the 29th, new software was uploaded to restrict the use of this joint. While this restriction would make certain motions awkward, the other six joints provided sufficient agility to undertake the task. On 2 April, in preparation for the Shuttle's visit, the metal oxide canisters of the EMU suits were cleansed of carbon dioxide in a 14-hour regeneration cycle, this time without stinking the place out.

INTEGRATED TRUSS STRUCTURE

When complete, the Integrated Truss Structure (ITS) will comprise nine segments of truss, individually referred to in numerical order running to port and to starboard.[41] The central three form the structural backbone that is designed to support the power generation, thermal control system radiators and external payload accommodations. The middle segment, which is known as Starboard Zero (S0), and was to be attached to Destiny, is 13.4 metres in length and has a hexagonal cross-section that spans 4.5 metres at its widest point – which is the greatest width the Shuttle's payload bay permits. The P1 and S1 segments, however, have half-hexagonal cross-sections, are 1.8 metres thick, and have a Radiator Assembly on their flat 'trailing' faces. Each such assembly has a trio of 3.3-metre-wide, 27.8-metre-long radiators on a beam that can be swivelled to place the radiators in shade to maximise the rejection of heat by the 300 kilograms of anhydrous ammonia that circulates in the EATCS. There are two external loops – one on each side of the ITS – and in addition to drawing heat from the internal water loops of the pressurised modules they cool the various subsystems housed in the trusses beneath the thermal blankets that protect the aluminium frames.[42] A twin-track rail runs along the leading face of the hexagonal framework for the MSS. Further outboard are the square cross-section trusses of the Solar Array Assemblies – in the final configuration, there will be two at each end, altogether providing 104 kilowatts. When complete, the ITS will span 100 metres from end to end. The final segment, the S6, which is a mirror image of the P6 that was temporarily mounted on the Z1, was delivered to the Space Station Processing Facility in December 2002.

STS-110 lifted off on 8 April 2002. Two days later, as Michael Bloomfield eased Atlantis in to dock at PMA 2 in a 'nose up' attitude, Bursch ceremoniously rang the

[41] The sequence was devised for Space Station Freedom. When the ITS was scaled down for the ISS the nomenclature was retained, but with gaps in the sequence corresponding to the deleted segments.

[42] One 1.5-metre by 3-metre rectangular unit inside the ITS contains a GPS antenna that is part of the Space Integrated GPS/Inertial Navigation System. It is the first use of GPS to enable a spacecraft to determine its position, speed and altitude independently of radar tracking from the ground.

ship's bell to mark the arrival of the first visitors on what was proving to be a lonely tour of duty. The 12.25-tonne S0 truss was supported in Atlantis's bay by four bay-wall trunnion pins and two keel pins. The hatches were opened immediately to allow the two crews to review the procedure for installing the S0 truss on Destiny. Early the next day, Ellen Ochoa and Bursch hoisted the S0 from the bay using the SSRMS, swung it clear of the bay and around and 'above' Destiny, then lowered it to engage the 'claw' of the Lab Cradle Assembly. During this four-hour operation, Stephen Frick manoeuvred the RMS to enable its TV camera to provide perspective views to the RWS in Destiny. Meanwhile, Rex Walheim and Steven Smith suited up in Quest and, a few minutes after the S0 was emplaced, they made their exit. Once at the S0, they deployed its two forward-facing struts and bolted these to 'hard points' on Destiny using pistol-grip tools. They then unstowed trays of avionics equipment on the S0 and connected power and data cables and ammonia lines to Destiny. Upon moving over to the MT, which was on the leading edge of the S0, they connected the primary Trailing Umbilical System that would reel out electrical and computer cables as the transporter moved along the rail track. Because they were running late, it was decided to defer installing two circuit breakers in the truss. During much of this time, Walheim had ridden the SSRMS – the first time that this arm had been used in this rôle. The next day, the two crews offloaded cargo from the mid-deck to the ISS and installed it in Destiny – this included the 'Arctic' freezer that was to be used to store processed materials samples awaiting retrieval by a Shuttle. Jerry Ross and Lee Morin went out on the 13th. Morin stood on the SSRMS and Ochoa and Walz took turns in moving him around. The first task was to deploy and bolt the two aft-facing struts of the S0 to complete its installation. Next, a variety of projecting metal rods and clamps that had supported the framework for launch were removed and conveniently stowed within it. After connecting the backup Trailing Umbilical System of the MT, Ross found that he could not remove a restraining bolt of the mechanism that was to sever the cables in the event that they became snagged. This had to be left in place, awaiting a remedial procedure. As soon as the spacewalkers were back inside, the engineers on the ground verified the connections of the primary and backup umbilical systems.

Smith and Walheim made the mission's third excursion from Quest on the 14th. Working independently, Smith connected power, data and video cables to enable the SSRMS to be operated from fixtures on the ITS. Walheim released the 'claw' from the S0, removed the Lab Cradle Assembly from Destiny, then picked up the deferred task from the first outing and installed two circuit breakers in the truss. Finally, they released the restraints on the MT to enable it to travel along its track. The next day, Walz, at the RWS in Destiny, issued a command to the MT to drive off its initial position to worksite-4, some 5.2 metres away. Upon arriving half an hour later, a software fault prevented it from automatically locking itself it place. The software was designed to move the MT with great precision from one worksite to another, then latch it onto the truss and plug into the power source. It is required to latch on with about 3 tonnes of force in order to serve as a stable platform. In weightlessness, the MT was 'lifting' off the track sufficiently to prevent sensors from contacting the magnetic location markers on the rail, fooling the software. Until revised software could be written and uploaded to restore the fully automated system, the latching commands had to be issued from the ground. Later in the day, the MT advanced to another worksite and was then returned to its launch station, having travelled a total of 22 metres. In the meantime, the Shuttle had topped up the oxygen and nitrogen of the High-Pressure Gas Assembly on Quest. As a maintenance chore, and using a tool delivered by Atlantis, an attempt was made to access the faulty valve that was turning the depressurisation of the Crew Lock into such a protracted operation, but when this failed to release the panel the repair effort was abandoned. On the final EVA of the mission, on the 16th, Ross and Morin finished outfitting the S0. Riding the SSRMS, Ross retrieved the 4.3-metre-long Airlock Spur – a narrow beam with handrails on each side to enable a spacewalker to move along it rapidly in hand-over-hand manner – and strung it from a handrail on the Equipment Lock over to the S0 to provide direct access to the ITS. They then installed several halogen spotlamps on Unity and Destiny. Each spotlamp produces an elliptical 'footprint' of illumination aimed to assist spacewalkers working in orbital darkness. As 'get ahead' tasks, circuit breakers and electrical converters were installed, a work platform was installed, a particle detector was deployed, the thermal insulation on one of the four navigational antennae on the truss was adjusted, and shock absorbers were fitted to the MT. Finally, the jammed bolt on the MT was reinspected. Its mission accomplished, Atlantis left on 17 April. The ISS was placed in partial hibernation on 20 April to enable Soyuz-TM 33 to be moved from Zarya's nadir to Pirs. Yuri Gidzenko, Roberto Vittori (an Italian astronaut with a European Space Agency research programme) and Mark Shuttleworth (the second tourist) docked in Soyuz-TM 34 on the 27th and departed in Soyuz-TM 33 on 4 May.

The launch of STS-111 had been postponed by three weeks to give its crew time to develop a procedure to replace the wrist-roll joint on the SSRMS. When Endeavour lifted off on 5 June its bay held the MPLM Leonardo, the MBS that was to be installed on the MT, and the new wrist-roll joint in a GAS can on the bay wall.[43] Although the soft-docking on the 7th was nominal, Ken Cockrell had to wait almost an hour for post-contact oscillations to subside before the docking ring could be retracted. As soon as the hatches were opened, Valeri Korzun, Peggy Whitson and Sergei Treschev exchanged Kazbek couch liners and Sokol suits with Onufrienko's crew, then set to work in transferring priority equipment, supplies and experiments, including two new suits and tools to be used during the mission's three spacewalks. As Leonardo was being berthed on Unity's nadir the next morning, those on the ISS were alarmed to hear what Walz later described as a "growling vibration" as the spin-bearing of CMG 1 seized and rapidly spun down. Because there had been no prior indication of distress, it was probably a lubrication problem. The ISS can manoeuvre efficiently with three CMGs and maintain its attitude with just two. The fourth had been included for redundancy. In flight director Paul Hill's opinion, its loss was "a serious complication for long-term operations". Although doing so would be "a big deal", there was little option than to have spacewalkers access the Z1 and replace the

[43] Leonardo's 3,660 kilograms of cargo included EXPRESS rack 3, the European Space Agency's Microgravity Science Glovebox rack with two furnaces for metallurgical research, eight resupply stowage racks and five resupply stowage platforms.

failed unit. The only available spare was assigned to STS-114, which was due to fly early in the new year.

Franklin Chang-Diaz and Phillipe Perrin emerged from Quest on the 9th for the mission's first EVA. Cockrell manoeuvred the RMS to provide perspective views to assist Whitson and Korzun at the RWS in Destiny, while Paul Lockhart coordinated from the flight deck. Once a Power and Data Grapple Fixture had been installed half way up the P6 to enable the SSRMS to operate from that truss in the future, Chang-Diaz mounted a foot-restraint on the SSRMS and was swung down to retrieve a bundle of micrometeoroid shields from the bay. As Chang-Diaz was raised, Perrin hitched a ride, and they attached the bundle to PMA 1 for collection by the ISS crew and installation on Zvezda as a step towards upgrading that vehicle's shielding to the standard set by the GAO. Chang-Diaz visually inspected CMG 1 in the Z1, but saw no sign of an external reason for its problem. Once he had dismounted the SSRMS, Whitson grasped the MBS in the bay to let its heaters draw power through the arm. After the spacewalkers had removed the thermal blankets, Cockrell commanded the latches to release, and Whitson lifted the 1.5-tonne, 5.7 $\times$ 4.5 $\times$ 2.9-metre unit, swung it around, and positioned it 1 metre in front of the MT, where it was left to achieve thermal stability. The next day, Whitson, assisted by the view from a TV camera on the underside of its own Common Attach System frame, mated the MBS to the MT. Chang-Diaz and Perrin then exited Quest. After connecting the primary and backup power, data and video cables between the MT and MBS, they utilised a power wrench to engage four bolts to complete the attachment process. The Payload ORU Accommodation – essentially a Latching End-Effector to carry a payload – was then installed on the MBS. The TV camera on the MBS was relocated to the top of a mast – when operational, this camera will provide the RWS operator with an *in-situ* perspective view. Finally, photographs were taken of the MT–MBS to enable the Canadian designers to assess their handiwork. The SSRMS had retained hold of the MBS throughout to supply it with power via its grapple fixture, but when it was confirmed to be drawing power from the MT (which was in turn drawing power from the S0 truss) the SSRMS was withdrawn. The three crews spent most of the 12th unloading cargo from Leonardo. The next day, Chang-Diaz and Perrin set out to repair the SSRMS. With the arm in a convenient orientation for access, they detached the Latching End-Effector and tied it to a handrail on Destiny to ensure that it did not float away. Then six bolts were released to detach the wrist-roll joint, and another to release its power, data and video umbilical. Riding the RMS, Perrin then stowed the malfunctioning joint in Endeavour's bay for return to Earth and returned with its replacement. Once the SSRMS had been re-assembled, Bursch verified the restoration of its functionality. A later examination of the failed joint confirmed the diagnosis that a tiny fragment of wire had formed a short circuit. The repair had been left until the mission's primary objectives had been accomplished, just in case the operation left the arm inoperable. Later in the day, with its base still on Destiny, the SSRMS re-engaged the Power and Data Grapple Fixture on the MBS, but when control was shifted to the MBS an anomaly was noted, and it was decided to leave it in this configuration. As soon as the problem could be resolved, the SSRMS would 'walk' off Destiny onto the MT–MBS, where it was to operate

An STS-111 spacewalker's view of Endeavour docked at PMA 2, with the S0 truss in place on Destiny.

during the extension of the ITS. After Leonardo had been loaded with 2 tonnes of unwanted items and the retiring crew's research results, it was retrieved by the RMS and Endeavour departed on 15 June.

Korzun, Whitson and Treschev began their tour of duty by servicing Zvezda's toilet and undertaking routine maintenance on the life-support system. After being filled with trash, the Progress undocked on the 25th. When telemetry from CMG 2 went off-line on the 27th there was some concern, but the gyroscope was unaffected and the data stream was soon restored. Nevertheless, the fact that a second CMG was showing signs of distress was considered ominous. The Progress arrived on the 29th with some 2,580 kilograms of cargo to ensure that the ISS-5 crew had sufficient life-support.

When Atlantis landed after STS-110, the intention was to turn it around for launch on 22 August as STS-112. However, in routine servicing on 25 June cracks of up to 8 millimetres in length were found in one of the 'liners' designed to prevent turbulence in the flow of liquid hydrogen into the SSMEs. When similar cracks were found in Discovery, the fleet was grounded pending further investigation. In fact, it was a generic problem. However, the cracks did not appear to be related to the age of the apparatus or to cumulative use, and it was noteworthy that while the liners on Columbia – the oldest vehicle in the fleet – were stainless steel, the others were made of iconel alloy. If, as seemed likely, the cracks were due to poor welding, then it was possible that they had always been present and were not growing. Nevertheless, the risk of a piece of metal breaking off and being ingested by an engine, with potentially catastrophic consequences, was unacceptable. It was concluded that the cracks were most likely caused by high cycle-fatigue resulting from a combination of vibration, thermal and acoustic factors. In early August it was decided to reweld and polish the faulty flow liners and, if this proved satisfactory, to resume operations in early October. The immediate impact of the grounding was that STS-107, an independent science flight by Columbia that had been set for July, was slipped into the new year to make way for Atlantis as STS-112 in October and Endeavour as STS-113 in November, both of which were to deliver segments of the ITS to the ISS. The three cracks in the feed pipe for SSME-1 in Atlantis were repaired on 10 August, and on the 18th, with all three engines installed, the launch was rescheduled for 7 October.

Meanwhile, on 9 July a software patch was uplinked to overcome the problem encountered during STS-111's visit in trying to transfer RWS control to Power and Data Grapple Fixture-1 on the MBS, and the following day Whitson successfully 'walked' the SSRMS off Destiny, where it had been based ever since being installed by STS-100. On the 12th, the MT–MBS was driven to worksite-4, near one end of the S0 and the SSRMS performed a full rehearsal of retrieving a truss segment from a Shuttle's bay and mounting it on the end of the S0. On 1 August, the SSRMS was contorted to enable its TV camera to peer into the orifice of the POA on the MBS in order to visually verify the functionality of its End-Effector. It was then repositioned to observe the EVA from Pirs on the 16th. As the forward docking compartment of Zvezda was to be used as an alcove, Treschev went to Destiny to monitor procedures using the SSRMS. After Korzun and Whitson had donned their Orlan suits, depressurised the airlock and disconnected their umbilicals, the caution and warning alarms sounded in both suits to indicate that no oxygen was flowing from the backpack tanks, so they reconnected their umbilicals. When the airlock had been repressurised, Korzun realised that in preparing the suits he had neglected to open the oxygen valves! When they finally exited Pirs, they were almost two hours behind schedule. Korzun extended one of the Strela cranes along the length of Zarya, and Whitson then moved along it in hand-over-hand manner and tied the far end to a handrail to anchor it in place. Korzun then joined her. Having retrieved the bundle of micrometeoroid shields that had been left on PMA 1 by STS-111, and tied it to the crane, Korzun returned to the operator's post and swung the crane, with Whitson on its end, around to Zvezda. If they had been on schedule, this would have been a breathtaking ride for Whitson, but the transfer was conducted in orbital darkness and the ISS was only dimly visible in the beam of the spotlamp on the SSRMS. The six débris shields, each of which was sized and shaped to suit a specific point, were affixed to handrails on the module's narrow conical section.[44] The start of the next excursion on the 26th was delayed by 30 minutes because the pressure equalisation valve in Zarya's hatch bled air into Zvezda's forward docking compartment, which was once again serving as an alcove, frustrating the depressurisation. Whitson reset the valve to terminate the leak. On finally emerging, Korzun and Treschev mounted a pre-assembled frame on Zarya on which to store apparatus for future EVAs, and installed additional fairleads to hold tethers on future spacewalks. They made their way onto Zvezda and mounted a TV camera on a boom to document the retrieval of one of the Japanese exposure cassettes. They then replaced the Kromka package (in place to determine the effectiveness of the deflectors which had been fitted to the thrusters) and installed the final two ham radio antennae. It was all quite routine. On 4 September the SSRMS flawlessly executed another rehearsal of installing the next segment of the ITS, and much of the following week was spent preparing Quest and the EMU suits for use by the next Shuttle team. On the 16th, in a bid to counter the growing popular belief that the ISS crew were so busy 'maintaining' the ISS that they had no time to perform scientific experiments, NASA named Whitson as Destiny's 'science officer'. One of the Beta Gimbal Assembly motors stalled on the 18th, leaving its Solar Array Wing poorly oriented to generate power, but it responded to the command to restart. The Progress departed on the 24th. After a longer than usual rendezvous, during which Kurs trials were conducted, the next Progress arrived on the 29th with a cargo that included 80 kilograms of research apparatus for a forthcoming visit by a European Space Agency astronaut.

STS-112 lifted off on 7 October with the S1 segment of the ITS. Atlantis docked at PMA 2 on the 9th. The following day, Whitson grasped the S1 using the SSRMS, the umbilical for the heaters was commanded to withdraw, the launch restraints were released, and the 15-tonne, 13.5-metre-long half-hexagonal cross-section truss was lifted from the bay. There was a 10-hour window to install it and restore power. At each end of the S0 was a Segment-to-Segment Attachment System incorporating a remotely operated 'claw' and a trio of electrically driven bolt assemblies. With Jeff Ashby operating the RMS to provide perspective views to the RWS operator, the S1

[44] The plan is to affix 23 such shields.

was positioned to engage the 'claw' on the starboard end of the S0, and the bolts locked the two segments together. David Wolf and Piers Sellers then exited Quest and their activities were coordinated by Pamela Melroy and Fyodor Yurchikhin on Atlantis's flight deck. Sandra Magnus, operating the SSRMS, positioned Wolf to enable him to connect the power and data cables and fluid lines in umbilical trays on the upper surfaces of the trusses. Meanwhile, Sellers released the locks on the radiators on the Radiator Beam. Working together, Wolf and Sellers then deployed a second S-Band Antenna Support Assembly, which was stowed at the centre of the S1, between its keel pins. Once Magnus had manoeuvred him into position, Wolf used a pistol-grip tool to release four launch bolts and two mast bolts. After Sellers had removed the assembly from its launch position, Wolf held onto it while Magnus slewed him to its installation point, near the inboard end of the S1, where he held it in place to enable Sellers to release two clamps to mount it on the support bracket and lock it in position by tightening a stanchion bolt. While Wolf connected power and data cables, Sellers removed the antenna's shroud. Finally, Wolf rotated the gimbal locks away from the high-gain antenna. On the S1's extension of the rail track was a Crew and Equipment Translation Aid (CETA). After a number of designs had been tested on Shuttle missions to find an efficient and practical device that would enable astronauts to transport themselves and small payloads along the ITS, a 283-kilogram 2.5 × 2.4 × 0.9-metre 'cart' had been built that could be drawn along the track in hand-over-hand fashion by an astronaut standing on a foot-restraint. Two carts were to be provided, one for each side of the ITS. When not in use they would be connected to the MT, which would serve as the 'shunt engine' for the three-car train of the first 'railroad in space'. The next task was to release the CETA's launch restraints. This involved unfastening 24 bolts in a specific sequence, and the procedure had taken over 2 hours in hydrotank simulations. They began by releasing the locks on the parking brake, to enable this to fix the cart in position while they worked on it. When the job was finished, the brake was released and the cart was pushed along the rails to mate with the MT, on the S0 segment. Magnus then swung Wolf around to the keel of the S1, where he affixed the outboard nadir TV camera – one of a set that would provide the RWS operator with a sense of perspective. Meanwhile, Atlantis replenished the tanks of Quest's High-Pressure Gas Assembly.

Wolf and Sellers made a second spacewalk on the 12th, and this time Sellers rode the SSRMS. After removing insulation covers – 'booties' – from two Quick Disconnect (QD) fittings on ammonia lines at the base of the Z1 truss, Sellers installed small clamp-like devices known as Spool Positioning Devices (SPD) around the bodies of the QDs to hold them in a position that would prevent the internal seals and moving parts from leaking. In each case, he had to turn the collar of the QD to its 'unlocked' position, attach the circular section of the SPD to the QD and add a clamp-like device to tension it. SPDs were also fitted on QDs for ammonia lines at the interface between the Z1 and P6 trusses. Meanwhile, Wolf had made his way up to the S0 to prepare the CETA cart for service. After being swung over to the inboard end of the S1, Sellers installed umbilicals for the Ammonia Tank Assembly and Nitrogen Tank Assembly. Another exterior TV camera was then installed, this

On STS-112 David Wolf rides the SSRMS holding a TV camera that he is to install on the nadir of the S1 segment of the Integrated Truss Structure. (Notice that one of the spotlamps is already in place.)

time on Destiny. At this point, Sellers retrieved a bag of SPDs, dismounted the SSRMS, made his way to the inboard end of the S1 and installed them on the ammonia lines of Radiator Beam Valve Module-1. Meanwhile, Wolf had collected another bag of SPDs, mounted the SSRMS, and been positioned to install them on the ammonia lines of RBVM-6 at the outboard end of the S1.[45] In all, 22 SPDs were installed; two others that were to have been installed were retained when it was realised that the ammonia connectors were of an incompatible configuration – showing another frustrating example of incorrect documentation. Finally, the launch restraints on the Radiator Beam were released to enable it to rotate. The next day, the Thermal Radiator Rotary Joint of the middle radiator was powered up and rotated 90 degrees, but when the radiator was commanded to deploy, the protective logic that monitored the electric current drawn by the deployment mechanism aborted the action – the tolerances that were programmed in on the basis of ground tests were too strict. After a software patch was uplinked, the radiator was successfully unfolded in concertina fashion to its full 27.8-metre length over a 10-minute interval on the 14th. The other two radiators on the Radiator Beam were to remain stowed until later in the assembly sequence, when the ammonia loops of the External Active Thermal Control System were activated.[46] On the 14th, Wolf and Sellers went out again. A bolt securing the backup cable cutter had jammed during the installation of the MT by the STS-110 mission, and it had been decided to replace the Interface Umbilical Assembly that formed part of this Trailing Umbilical System. After Wolf had disconnected the TUS cable, Sellers – taking care not to crimp the cable – kept it under tension. Before the IUA could be replaced, the cable cutter had to be removed, and this was achieved without incident. The next item on their list was to install jumpers to enable ammonia coolant to flow between the S0 and S1 trusses, installing SPDs on the QDs. Another task was to release the two 'drag links' that had served as launch restraints for the S1 truss and the keel pins that had supported the truss in the bay. These were stowed inside the S1 framework. After dismounting the SSRMS, Sellers joined Wolf in fitting additional SPDs on ammonia lines. The final task was to ensure that the remotely operated 'claw' and bolts of the Segment-to-Segment Attachment System at the end of the S1 were ready to accept the next starboard segment of the ITS.

Meanwhile, during these days of external activity Magnus, acting as loadmaster, had supervised the transfer of some 800 kilograms of cargo from the mid-deck to the ISS and the retrieval of items for return to Earth. Atlantis undocked on the 16th and flew a loop around the ISS to document its latest addition, then departed. On the 24th, Whitson rehearsed how the SSRMS would install the P1 segment of the ITS –

[45] The Radiator Beam Valve Modules control the flow of ammonia through the radiators on the ITS, and provide pressure and temperature data and valve-actuation status/data.

[46] On the plan in force before the fleet was grounded by the loss of Columbia, the port-side EATCS was to be activated by STS-116 (assembly flight 12A.1), by which time the P3/4 and P5 segments of the ITS would have been installed. The coolant loop would be filled from the Early Ammonia Servicer. With the Solar Array Wings on the P5 segment functioning, the power distribution would be reconfigured and the port-side wing of the P6 retracted. Later, the process would be repeated for the starboard EATCS. With both wings retracted, the P6 would be relocated to complete the port-side of the ITS.

in this case the RMS would retrieve the truss from the bay and 'hand' it over to the SSRMS for attachment.

The fourth taxi mission introduced a new version of the Soyuz spacecraft – the 'A' in the 'TMA' designation stood for 'anthropometric', and indicated that it could accommodate a wider range of body sizes, ranging in height from 1.5 to 1.9 metres and in mass from 50 to 95 kilograms.[47] This eliminated the restriction on NASA in assigning astronauts to ISS crews. Soyuz-TMA 1's docking at Pirs on 1 November was monitored by the new external TV on the underside of the ITS. It brought Sergei Zalyotin, Yuri Lonchakov and Frank DeWinne – a Belgian astronaut flying for the European Space Agency. Upon departing in Soyuz-TM 34 a week later, they made Russia's first night-time landing for a decade. Endeavour was launched on the 23rd as STS-113, and docked on the 25th. As soon as the hatches were opened, Ken Bowersox, Nikolai Budarin and Don Pettit exchanged Kazbek couch liners and Sokol suits with the retiring crew. Pettit was a late addition to the new ISS crew. NASA medics had expressed concern at the total exposure to radiation that Don Thomas would accrue by following four Shuttle missions with a tour of duty on the ISS. By the time of his replacement, the crew's personalised clothing had already been delivered to the ISS, and Pettit joked that he had been selected simply in order that the shirts bearing the name tag 'Don' would not cause confusion!

On the 26th, Jim Wetherbee used the RMS to lift the P1 segment of the ITS out of the bay. At the RWS in Destiny, Bowersox and Whitson swung the SSRMS down and grasped a second grapple fixture on the P1. Once the RMS had let go, the SSRMS positioned it to engage the 'claw' of the Segment-to-Segment Attachment System at the end of the S0 and then activated the electrically operated bolts to complete the connection. A short time later, John Herrington and Michael Lopez-Alegria exited Quest. After traversing the Airlock Spur onto the S0, Lopez-Alegria connected cables from the S0 to the P1 and fitted SPDs on the linking ammonia lines. Meanwhile, Herrington released the launch restraints on the CETA cart on the P1. Then they jointly removed the drag links and stowed them inside the truss. Their final chore was to place a Wireless video system External Transceiver Assembly (WETA) on Unity – the first of several that were to be installed around the ISS to relay the video feed from the helmet cameras on the suits in the absence of a docked Shuttle. The spacewalkers and the two robotic arms were coordinated by Pettit in Destiny and by Paul Lockhart on Endeavour's flight deck. On venturing out again on the 28th, Herrington and Lopez-Alegria installed 'jumpers' for a pair of ammonia lines from the S0 to the P1, removed the keel pins and stowed them inside the truss, fitted a second WETA antenna (this one on the P1) and released the launch restraints on the radiator assembly on P1. The next task was to relocate the CETA from the P1 to the S1 in order to permit the MT to travel right to the end of the P1 truss as it extended the ITS in that direction. Once Herrington had mounted the SSRMS, which was operated by Whitson, he disconnected the 283-kilogram cart from the track and held it while she swung him over to the S1, where it was remounted and

[47] The 'TM' designation indicated a 'modified' form of the 'transport' variant of this most versatile spacecraft.

'hooked' to its counterpart, which was in turn connected to the MT, forming a small 'train' on the track. Finally, they inspected the WETA that they had placed on Unity, and reconnected one of its cables. On the 30th, the MT was commanded to move from worksite-4 on the S0 to worksite-7 on the P1, at which time the SSRMS was to 'walk' off Destiny onto the MBS to enable Whitson and Pettit to manoeuvre Herrington on the final EVA, after which it would return to Destiny, but the MT stalled about 3 metres short of its intended destination. On starting the EVA, Herrington inspected the MT to find out why it had stopped and found that its umbilical had snagged a stowed UHF antenna, so he deployed the antenna, releasing the MT, which finally reached worksite-7 and securely latched itself down. In the meantime, flight controllers had reprioritised the EVA to ensure that the most urgent tasks were tackled first, and when Herrington opined that he could work without the SSRMS, this was left on Destiny. Coordinating, Lockhart provided visual cues as the spacewalkers moved around the truss. After installing an additional 33 SPDs at various sites (including the Radiator Beam Valve Modules on the P1, the Umbilical Tray on the Z1, and the interface between the Z1 and the P6) they connected lines to the Ammonia Tank Assembly and Nitrogen Tank Assembly on the P1. Although a start was made on releasing the launch restraints on the radiator assembly, they ran out of time and the job was left to an excursion by the ISS crew. Meanwhile, Korzun and Treschev continued the handover activities with Bowersox and Budarin.

Shortly after Endeavour departed on 2 December, having restored the ISS to a symmetrical configuration, the spin rotor on CMG 4 began to draw an unusually high current, possibly due to a change in its lubrication state. CMG 1 (which had seized) was due to be replaced by STS-114 in March 2003 – which was none too soon now that units-2 and -4 had begun to display intermittent signs of distress. Bowersox's crew were expecting a lonely tour of duty, because no Soyuz or Shuttle flights were scheduled, and there would be only one Progress resupply ship. On the 5th, the EMU suits were re-sized in preparation for an EVA by Bowersox and Budarin on the 12th, but on the 10th this was postponed to early in the new year, and on 6 January it was announced that Pettit was to replace Budarin. On 5 December the NASA medics monitoring Budarin's cardiovascular telemetry as he exercised had disqualified him from external activity, and although the Russians pointed out that Budarin had made eight spacewalks from Mir, the planned EVA was to be made using EMUs and it was NASA's decision. As a consequence of this switch, the spacewalkers would have to work without the assistance of the SSRMS as Budarin had not been trained to operate it. Nevertheless, the SSRMS could be pre-positioned to enable Houston to monitor the excursion through its TV camera. On the 15th, Houston powered up the MT, which had been at the far end of the P1 truss (worksite-7) since STS-113's departure, and commanded it to return to the centre of the S0 truss (worksite-4) to protect its trailing umbilical from possible micrometeoroid damage – a move that took 20 minutes. The EVA from Quest later in the day was delayed by about 20 minutes because a strap on the interior thermal hatch cover interfered with the rotation of the crank handle of the hatch. On finally emerging, Bowersox and Pettit made their way onto the P1 to complete the job of releasing the launch restraints on its radiator assembly. They then watched as

John Herrington, riding the SSRMS, relocates the CETA cart from the P1 to the S1 segment of the Integrated Truss Structure.

On STS-113 Michael Lopez-Alegria and John Herrington work on the P1 segment of the Integrated Truss Structure.

Houston commanded the middle radiator to unfold, which it did over a 10-minute interval. As their attempt to deploy a stanchion on one of the CETA carts to support a lamp was frustrated by a balky pin, this task was deferred to a future outing. Moving to Unity, Pettit used a length of Kapton tape to retrieve small specks of fine grit from a sealing ring on the nadir CBM on which the MPLMs were temporarily berthed. As the SSRMS was unavailable, they improvised: Bowersox secured a PFR and stood on it, then grasped Pettit by the legs and held him in front of the otherwise inaccessible CBM to enable him to work. Finally, they made their way around Unity to the Z1 and onto the P6, to take a reading of the quantity gauge on the Early Ammonia Servicer – at 98.4 per cent full, the tank was evidently not leaking. On their return to Quest, they used scissors to cut away the awkward strap. It had been just another spacewalk. Despite early doubts about the number of hours of external activity assigned to station assembly and maintenance, in excess of 300 hours had been carried out so far.[48]

By the end of 2002, the only part of the 'US core' for the ISS that had yet to be delivered to the Space Station Processing Facility was (the as yet unnamed) Node 2, and that was only a few months away from completion. The remainder was either in space or awaiting launch, and the assembly was going remarkably well. STS-114 was due to replace the station's crew in March 2003, after Columbia had flown its much-delayed science mission.

CRISIS

STS-107 lifted off on 16 January 2003. On the 27th, despite being in very different orbits, the ISS crew were able to snatch a brief conversation with the 'duty shift' on Columbia. On the 27th the two crews marked the anniversaries of the losses of the Apollo 1 crew on 27 January 1967 and the Challenger crew on 28 January 1986. On the 31st, STS-107 wrapped up its science programme, preparatory to returning to Earth the next day. Columbia broke up during re-entry. As Bowersox's crew were preparing to monitor the transition to 'free drift' immediately prior to the undocking of the Progress ship, and the reinstatement of positive attitude control following its departure, they were informed of Columbia's loss. The next Progress was already on the pad, and it was decided to launch it rather than delay it pending a consideration of the implications of the tragedy. It lifted off on 2 February and docked on the 4th with 2,750 kilograms of assorted cargo.[49] Later in the day, video of the Memorial Service held at the Johnson Space Center for the Columbia crew was uplinked to the ISS, whereupon Bowersox rang the ship's bell seven times in tribute – once for each

[48] Note that although a pair of spacewalkers might spend six hours outside, this counts as six hours of external activity, not 12 hours.

[49] In addition to 870 kilograms of propellant that was to be pumped into Zvezda, 250 kilograms of the Progress's own propellant was reserved for docked operations. It would also pump on board 70 kilograms of potable water and 50 kilograms of oxygen. The 1,328 kilograms of 'dry' cargo would be unloaded manually.

lost colleague. On the 5th, they began a comprehensive audit of their consumables to determine which type of item was in shortest supply. Since the Progress ships had been boosting the ISS's orbit using their own engines, Zvezda's tanks had propellant for a year's nominal operations. In addition to the Elektron that liberated oxygen from water, there was a stock of Solid-Fuel Oxygen Generators in the vestibule beneath the Z1 truss and several months supply of oxygen in the tanks of the High-Pressure Gas Assembly on Quest. In addition to the Vozdukh carbon dioxide scrubber, there were lithium hydroxide canisters. The limiting factor, however, was water. As PMA 3 was no longer serving as a transfer tunnel, it had been turned into a storeroom for the bags of water left by Shuttles – this being a by-product of their fuel cells. If the condenser of Zvezda's air-conditioning system remained operational, the water would last until the end of June, when the next Progress was due to arrive.

In mid-February the launch of STS-114, which had been scheduled for 1 March, was cancelled. The Russian experience with their previous space stations was that a station without a crew could all too easily be overwhelmed by a problem that would readily have been rectifiable if a crew had been on board. With the Shuttle probably grounded for at least a year, it was deemed essential that the ISS remain inhabited. Bowersox's crew offered to remain on board for as long as was necessary to resume Shuttle flights, but their Soyuz would 'expire' at the end of April. Like a Progress, a Soyuz ship was capable of automatic operation, and one option was to dispatch a succession of spacecraft until the Shuttle resumed service. In light of the limited logistics capability, it was decided to fly crews of one NASA astronaut and one Russian cosmonaut, alternating the command rôle. Accordingly, it was decided that Bowersox's crew would return in Soyuz-TMA 1 when it reached the end of its six-month life. The visitors who were to have replaced it with Soyuz-TMA 2 were stood down, and the spacecraft commandeered.[50] Yuri Malenchenko and Ed Lu would fly as the ISS-7 crew, leaving their former crewmate Alexander Kaleri on the ground. With the tour of duty extended to the maximum six months permitted by the service life of the Soyuz, there would be only two flights per year. Researchers for the European Space Agency would be allowed to accompany handovers, but tourists would be not be permitted. Logistical support would now be totally reliant on Progress ships, but the cash-starved Russian Space Agency had severely cut production.[51] While it would involve a lengthy lead time to do so, production *could* be ramped up, but – the Russians enquired – who was going to pay? A ruling by the Congress that prohibited NASA from sending money to Russia was still in force. In a radical proposal, Russia invited Europe to contribute to the production of Progress spacecraft in return for flight opportunities that were nominally assigned to Russians.

Flying two-person 'caretaker' crews raised an interesting issue: NASA had previously said that simply maintaining the ISS consumed 2.5 full-time assignments

[50] Gennadi Padalka had been set to deliver Pedro Duque, a Spanish astronaut who was to undertake a European Space Agency programme, and, if one was ready, another space tourist.

[51] The Russian Space Agency's commitment for 2003 was to produce two Soyuz and three Progress ships, with launch vehicles, and the production line was set to produce a similar number in 2004.

(a figure that had fuelled criticism of the low level of scientific research in Destiny). In retrospect, of course, it was apparent that this figure had been overstated. NASA had scheduled five Shuttle missions for 2003. It would not be possible to continue to outfit Destiny with racks. However, although vital stores would have priority in the Progress ships, it would be possible to deliver small experiments. As it would not be possible to offload bulky unwanted items, these would simply have to be stored for eventual return to Earth.

One issue was whether EVAs could be conducted when the ISS was operated by a crew of two – a simple fault might escalate into an emergency if there was no one inside to take prompt action. It was therefore decided that Bowersox's crew should complete all outstanding EVA tasks prior to departing. For the last several weeks, a power controller module on the MT had been misbehaving. Although there would be no requirement to use the MT until Shuttle flights resumed the assembly process, if this was not replaced while there were three people on board it would eventually have to be addressed by two. On 8 April Bowersox and Pettit exited Quest. While Pettit replaced the troublesome power controller, Bowersox disabled the electrically driven bolts of the Segment-to-Segment Attachment Systems on the S0 to guard against an electrical malfunction casting loose the S1 and the P1 trusses. He then inspected a faulty heater cable on the Nitrogen Tank Assembly on the P1, but could see nothing awry. Meeting at the Z1 truss, they rerouted cables to ensure that a single power failure could not disable both CMG 2 and CMG 3. Two SPDs were installed on the heat exchanger on Destiny. After moving onto the S1, they secured a thermal cover on a Radiator Beam Valve Module and turned their attention to the jammed stanchion on the CETA cart: once Pettit had released it from its stowed position using a hammer, they erected the stanchion and installed its lamp. Having enjoyed the sheer pleasure of hammering something, they returned to Quest. On the 15th one of the CMGs suffered a dropout in telemetry but, as previously, it was restored by recycling the power. Nevertheless, these intermittent faults were a cause for concern.

On 23 April, Budarin powered up Soyuz-TMA 1 and tested its systems, and on the 25th he reviewed the undocking and descent procedures. A few hours later, more or less as had been planned for the taxi mission, Soyuz-TMA 2 lifted off. The third seat was occupied by a package of vital stores. Following the automated rendezvous on the 28th, Malenchenko took over and stood off at about 200 metres from the ISS to enable Pettit to shoot a series of photographs by a digital camera with a telephoto lens. He then he docked manually on Zarya's nadir. The images were downloaded to help engineers to assess the degree of detail that similar shots of a Shuttle would show, as it was likely to become standard practice for an approaching Shuttle to display its heat shield for inspection for evidence of damage.

With Budarin at the controls, Soyuz-TMA 1 undocked from Pirs on 3 May. The 4-minute 18-second de-orbit burn was initiated at 21:12. Incorrect settings confused the guidance computer – instead of pursuing a comfortable shallow descent, it dived steeply into the atmosphere on a trajectory that subjected the heat shield to higher temperatures and imposed a deceleration load on its occupants of 8 *g* – more than twice the expected 3 times the force of the Earth's gravity. Bowersox wryly observed later that his "eyes got very wide" as he pondered that they might be about to burn up, as had Columbia's crew. If all had been going to plan, they would have landed at 22:07 but the steeper trajectory resulted in a landing 3 minutes early and 450 kilometres short of the landing zone in northern Kazakhstan. On the TMA variant, the retro-rockets that fired immediately prior to making contact with the ground had been improved to provide a softer landing, but they misfired and the capsule hit the ground with such force that Pettit injured his shoulder. Because the radio antennae had been damaged, neither the crew nor the recovery force in the target zone realised that they were not where they were supposed to be. It took rescue forces about two hours to locate their position, and another two hours to reach them, by which time they had scrambled out. Bowersox and Budarin rushed to welcome their rescuers, but Pettit remained sitting on the steppe until he was carried on a stretcher to the helicopter. On 6 May the charred capsule was returned to the manufacturer for examination. On the 14th it was reported that the flight control system had recognised an unusual dynamic condition and, correctly, pursued a ballistic trajectory. Although the crew were not to blame, they *had* made mistakes – for example, they had inadvertently switched on the Kurs system. In fact, the descent had been well within the capsule's operational parameters – as had been shown on several occasions.[52] The commission called for changes to: (1) minimise the scope for the crew entering incorrect commands into the computer; (2) improve on-board documentation of procedures for future crews; and (3) issue each crew with a satellite telephone that incorporated a GPS system to assist in locating a capsule after an off-target landing.[53]

In a change of procedure reflecting the strained logistical situation of the ISS, the next Progress resupply ship docked at Pirs on 11 June, and for the first time the ISS had two such craft in place. The one on Zvezda was retained to continue to provide orbital boost – thereby saving the propellant in Zvezda's tanks – and to continue to serve as a trash can. The cargo included spare parts for environmental systems in both the US and Russian segments, food and, most importantly, 340 kilograms of water – four times the usual quantity. In July, Malenchenko and Lu participated in tests designed to assess whether the SSRMS could be remotely operated from the ground. On 20 October, Michael Foale and Alexander Kaleri arrived in Soyuz-TMA 3, with Pedro Duque, a Spanish astronaut who was to work on the European Space Agency's programme during the handover and retire with Lu and Malenchenko. On 8 November, one of the CMGs showed signs of distress, drawing current irregularly and issuing vibrations. A CMG was too big to pass through the hatch of a Progress ferry, so there was no option except to wait for the Shuttle to resume flying. A few days later, Foale and Kaleri serviced the Orlan suits to prepare for a spacewalk that might be needed in February 2004 to configure Zvezda's rear docking port to accept the prototype of the Automated Transport Vehicle that the European Space Agency hoped to dispatch later in the year. As part of the planning for this spacewalk, a procedure was developed for re-entering the ISS via the orbital module of the Soyuz

[52] In addition to several unpiloted craft that made ballistic returns, Soyuz 1 made a ballistic return but its parachute failed, killing Vladimir Komarov. When Soyuz 5's propulsion module failed to separate properly after the de-orbit burn, it made a ballistic re-entry. And Soyuz 33 made a ballistic descent without incident after having to use its backup engine for the de-orbit burn.

[53] A GPS/satellite telephone was ferried up to Soyuz-TMA 2.

A picturesque view of the ISS at the time of the Shuttle's grounding by the loss of Columbia featuring the central three segments of the Integrated Truss Structure.

in the event that it proved impossible to repressurise Pirs. In April, Foale and Kaleri handed over to Michael Fincke and Gennadi Padalka, with Dutch astronaut André Kuipers continuing the European Space Agency's research during the handover. With luck, they or their successors will host STS-114, which will deliver overdue logistics and replace the dead CMG in the Z1 truss. If everything goes well, the next few missions should complete the 'US Core' of the ISS.

Table 11.5 Spacewalks for assembling and maintaining the ISS

Astronauts	Airlock	Start	Date	Finish	Hours
Ross and Newman	ODS	1710 EST	7 Dec 1998	0031 EST	7.30
Ross and Newman	ODS	1530 EST	9 Dec 1998	2235 EST	7.01
Ross and Newman	ODS	1533 EST	12 Dec 1998	2132 EST	6.99
Jernigan and Barry	ODS	2256 EDT	29 May 1999	0651 EDT	7.92
Voss and Williams	ODS	2148 EDT	21 May 2000	0432 EDT	6.74
Malenchenko and Lu	ODS	0047 EDT	11 Sep 2000	0701 EDT	6.24
Chiao and McArthur	ODS	1027 EDT	15 Oct 2000	1655 EDT	6.47
Wisoff and Lopez-Alegria	ODS	1015 EDT	16 Oct 2000	1722 EDT	7.12
Chiao and McArthur	ODS	1030 EDT	17 Oct 2000	1718 EDT	6.80
Wisoff and Lopez-Alegria	ODS	1100 EDT	18 Oct 2000	1756 EDT	6.93
Tanner and Noriega	ODS	1335 EST	3 Dec 2000	2108 EST	7.55
Tanner and Noriega	ODS	1221 EST	5 Dec 2000	1858 EST	6.62
Tanner and Noriega	ODS	1113 EST	7 Dec 2000	1623 EST	5.17
Curbeam and Jones	ODS	1050 EST	10 Feb 2001	1824 EST	7.57
Curbeam and Jones	ODS	1040 EST	12 Feb 2001	1749 EST	6.16
Curbeam and Jones	ODS	0948 EST	14 Feb 2001	1513 EST	5.42
Helms and Voss	ODS	0012 EST	11 Mar 2001	0908 EST	8.93
Richards and Thomas	ODS	0023 EST	13 Mar 2001	0644 EST	6.34
Hadfield and Parazynski	ODS	0745 EDT	22 Apr 2001	1455 EDT	7.17
Hadfield and Parazynski	ODS	0834 EDT	24 Apr 2001	1615 EDT	7.67
Usachev and Voss	Zvezda	0921 EDT	8 Jun 2001	0940 EDT	0.32
Gernhardt and Reilly	ODS	2310 EDT	14 Jul 2001	0509 EDT	5.99
Gernhardt and Reilly	ODS	2304 EDT	17 Jul 2001	0533 EDT	6.49
Gernhardt and Reilly	Quest	0035 EDT	21 Jul 2001	0437 EDT	4.03
Barry and Forrester	ODS	0958 EDT	16 Aug 2001	1614 EDT	6.27
Barry and Forrester	ODS	0942 EDT	18 Aug 2001	1511 EDT	5.49
Dezhurov and Tyurin	Pirs	1023 EDT	8 Oct 2001	1521 EDT	4.97
Dezhurov and Tyurin	Pirs	0517 EDT	15 Oct 2001	1109 EDT	5.87
Dezhurov and Culbertson	Pirs	1641 EST	12 Nov 2001	2145 EST	5.07
Dezhurov and Tyurin	Pirs	0820 EST	3 Dec 2001	1106 EST	2.78
Godwin and Tani	ODS	1252 EST	10 Dec 2001	1704 EST	4.20
Onufrienko and Walz	Pirs	1559 EST	14 Jan 2002	2202 EST	6.05
Onufrienko and Bursch	Pirs	1019 EST	25 Jan 2002	1618 EST	5.99
Walz and Bursch	Quest	0638 EST	20 Feb 2002	1225 EST	5.82
Smith and Walheim	Quest	1036 EDT	11 Apr 2002	1824 EDT	7.80
Ross and Morin	Quest	1009 EDT	13 Apr 2002	1739 EDT	7.50
Smith and Walheim	Quest	0948 EDT	14 Apr 2002	1615 EDT	6.45

Astronauts	Airlock	Start	Date	Finish	Hours
Ross and Morin	Quest	1029 EDT	16 Apr 2002	1706 EDT	6.62
Chang-Diaz and Perrin	Quest	1127 EDT	9 Jun 2002	1841 EDT	7.24
Chang-Diaz and Perrin	Quest	1120 EDT	11 June 2002	1620 EDT	5.00
Chang-Diaz and Perrin	Quest	1116 EDT	13 Jun 2002	1833 EDT	7.28
Korzun and Whitson	Pirs	0523 EDT	16 Aug 2002	0948 EDT	4.42
Korzun and Treschev	Pirs	0127 EDT	26 Aug 2002	0648 EDT	5.35
Wolf and Sellers	Quest	1121 EDT	10 Oct 2002	1822 EDT	7.01
Wolf and Sellers	Quest	1031 EDT	12 Oct 2002	1635 EDT	6.07
Wolf and Sellers	Quest	1011 EDT	14 Oct 2002	1647 EDT	6.70
Lopez-Alegria and Herrington	Quest	1429 EST	26 Nov 2002	2135 EST	6.75
Lopez-Alegria and Herrington	Quest	1336 EST	28 Nov 2002	1946 EST	6.17
Lopez-Alegria and Herrington	Quest	1425 EST	30 Nov 2002	2125 EST	7.00
Bowersox and Pettit	Quest	0750 EST	15 Jan 2003	1441 EST	6.85
Bowerson and Pettit	Quest	0840 EDT	8 Apr 2003	1506 EDT	6.43

Note: The total time for the spacewalks in this table is just short of 320 hours.

AN ASSESSMENT

The original rationale for the Shuttle was to assemble a space station, and in this rôle it has excelled. Although the schedule slipped for a variety of reasons, and from time to time 'extra' missions were flown to address specific contingencies, this 'case study' shows that if the issues of schedule and cost (which are not independent factors) are ignored, the Shuttle fleet *was* able to pursue a highly integrated sequence of missions, and the assembly of such a complicated structure – the most ambitious of orbital construction projects – was feasible.

Thousands of people have been involved in planning, designing and fabricating the hardware in preparation for launch. When the fleet was grounded by the loss of Columbia, the remainder of the hardware was in the Space Station Processing Facility at the Kennedy Space Center. Once the Shuttle resumes flying, completing the station should pose no 'show-stopping' surprises.

The 'stars' of the assembly process are the spacewalkers who, during 320 hours of EVA to date, have installed antennae and sensors (saving the cost of developing self-deploying systems), connected cables and plumbing between the ITS segments, rectified the tensioning system on the P6 blankets, installed thermal covers on the Beta Gimbal Assemblies, rectified a jammed bolt on the MT that would otherwise have degraded its redundancy, installed micrometeoroid shielding that could not have been fitted prior to launch (as that would have made the modules too heavy for the Proton rocket to lift) and extracted a rubber ring that was fouling a docking mechanism. And as soon as the Shuttle returns to the ISS they will 'save the station' by replacing the failed CMG in the Z1 truss. In a very real sense, the ISS has been 'hand built' by *homo spacewalker*.

A spacewalker with a toolkit – what spaceflight is really all about!

12

The loss of Columbia

A NEW MODULE

After servicing the Hubble Space Telescope in March 2002 Columbia was to have been turned around to fly STS-107 in July, but when cracks in the 'flow liners' of the SSMEs of all four orbiters were discovered in June the fleet was grounded, and in August, when it became evident that operations would be able to resume in October, STS-107 was postponed into the new year to give priority to two ISS missions.

The payload for Columbia's 16-day STS-107 mission was Spacehab's Research Double Module – a new facility, similar to the logistics carrier but with both halves powered. With a total payload of 4 tonnes, it had twice the capacity of the 'single' module last flown on STS-95 – the mission which had made headlines for its crew including John Glenn. It was fitted with facilities for both standard racks and for mid-deck lockers. In this case NASA had provided 82 per cent of the research payload and Spacehab sold the rest commercially. It was an international payload of 80 microgravity experiments on a wide range of topics, including combustion, fluid flow, zeolite crystal growth, cell cultures (one investigating prostate cancer), protein crystal growth (also cancer-related), and a test of a water-recycling system prior to its installation on the ISS. Columbia was to have deployed Triana, a spacecraft that would have manoeuvred into a stationary point directly up-Sun of the Earth to provide continuous views of the sunlit hemisphere that would be made available on the internet in real-time, but owing to administrative issues it was deleted from the manifest. Instead, a Hitchhiker was fitted aft of the module. When the science team was announced in 2000, it raised public interest by including an Israeli fighter pilot, Ilan Ramon, as a payload specialist. The flight crew had yet to be chosen. There was speculation that Susan Still might command, with Pam Melroy as pilot, but Rick Husband and William McCool were eventually assigned.

The STS-107 crew emerge for the drive to the pad.

FINAL FLIGHT

Although the ascent for Columbia's 28th mission on 16 January 2003 seemed to be uneventful, a routine examination the following day of the film from the long-range tracking cameras revealed that a large piece of foam had detached from the ET and fragmented upon striking the belly of the orbiter. The foam came from the base of the left leg of the bipod that supports the orbiter's nose. The leading edge of the mount is protected from aerodynamic heating by a 'ramp' of foam, which is applied by hand. On occasion, where new foam had been applied over earlier foam without forming a firm bond this process left a 'flaw', and if water seeped into this void it would freeze when the ET was loaded with cryogenic propellants, and the expanding ice would open up the flaw and make it possible for the vibrations and aerodynamic stress of launch to detach a piece of foam. While foam strikes had occurred before without serious consequences, an investigation was initiated to determine the likely extent of the damage to Columbia's thermal protection tiles.

Meanwhile, Columbia's crew had split into two shifts to pursue round-the-clock operations. The function of the Mediterranean Israeli Dust Experiment – which was conducted by Ramon and was carried on the Hitchhiker bridge – was to measure the distribution of aerosols over the Middle East, Mediterranean and North Africa. It also captured the first-ever images from orbit of a 'sprite', a powerful lightning-like discharge from a thunderstorm up into the ionosphere. On 27 January Columbia had a brief conversation with their colleagues on the ISS, which was in a more steeply inclined orbit. Having successfully concluded their experiments, it was a happy crew that suited up on Saturday 1 February to return to Earth. After the de-orbit burn at 08:15 EST they expected to touchdown at the Kennedy Space Center an hour later.

As Columbia crossed the California coast, heading east, its computer initiated the first of a series of banking S-turns to bleed off energy. Over the western states, flight controllers monitoring its telemetry began to see anomalous readings from an ever-increasing number of sensors prior to their readings dropping off-scale, suggesting an instrumentation fault. This was a puzzle, as there was no obvious commonality that could disable such a wide variety of sensors. After the temperature in the left landing gear well rose, and Husband saw a computer message warning that the tyres on both wheels in that well had lost pressure, his report to Houston of this warning was cut off in mid-sentence. At the same time, the flow of telemetry to the flight controllers' consoles abruptly ceased. As the controllers pored over their last

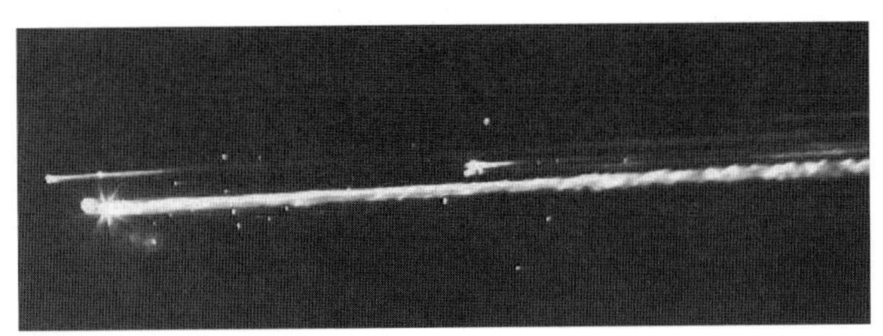

An amateur photographer snapped Columbia disintegrating high above Texas.

snapshot of data, seeking a reason for the anomalous readings, a Dallas TV station reported that there were trails of débris streaking overhead, accompanied by a sustained crackling sonic boom: Columbia had disintegrated. As reports of smoking débris saturated 911 lines, emergency personnel realised that the objects falling from the sky were fragments of the Shuttle.

THE INVESTIGATION

In executing the Shuttle Contingency Action Plan that had been established after the loss of Challenger in 1986, NASA formally notified the astronauts' families, the President and members of Congress. President Bush then telephoned Ariel Sharon, the Israeli Prime Minister, to inform him of the loss of Columbia crew member Ilan Ramon. Within minutes of the accident, the NASA Mishap Investigation Team was activated to coordinate débris recovery efforts with local, state and federal agencies. The International Space Station and Space Shuttle Mishap Interagency Investigation Board was activated under the chairmanship of retired Admiral Harold Gehman. In declaring East Texas to be a federal disaster area President Bush enabled teams from the Federal Emergency Management Agency and the Environmental Protection Agency to begin searching for and recovering as much of the débris as possible. Later that day, the President addressed the nation, saying, "The Columbia is lost. There are no survivors."

In a press briefing several hours after the incident, Ron Dittemore, the Shuttle programme manager, said that the only indications were sensor dropouts, for which there was no obvious commonality. Apart from the anomalous warning from the left wheel bay that the tyres had deflated – which was surely a spurious reading – the crew had raised no other issues. There had been no indications of excessive heat. When asked about the foam that had detached from the ET at T+82 seconds and had apparently broken up on striking the left wing, he said that an engineering analysis of the film had concluded that there was no cause for concern. Nevertheless, it could hardly be a coincidence that the anomalous sensor readings had come from the wing that had suffered the foam strike. In fact, pieces of foam had been seen to fall from the bipod mount on *four* previous missions without causing significant damage. The point of reference for the analysis by Boeing was STS-50 in 1992. A piece of foam had 'grazed' Columbia's belly at an angle of 3 degrees, etching a furrow in the tiles some 23 centimetres in length, 10 centimetres in width and 4 centimetres in depth. No adverse consequences had resulted. The analysis in the second week determined that the energy of STS-107's impact would have been 5 to 16 times greater owing to the angle of the strike being steeper, and would have caused greater damage. This was the largest piece of foam ever to strike an orbiter, but the analysis still concluded that this should not pose a serious threat. However, some engineers at NASA's Langley Research Center were critical of the analysis technique. Robert Daugherty had suggested in e-mails to colleagues in Houston that flight controllers should be alert to the possibility that they might see off-nominal readings from certain sensors during re-entry. If they suspected that a

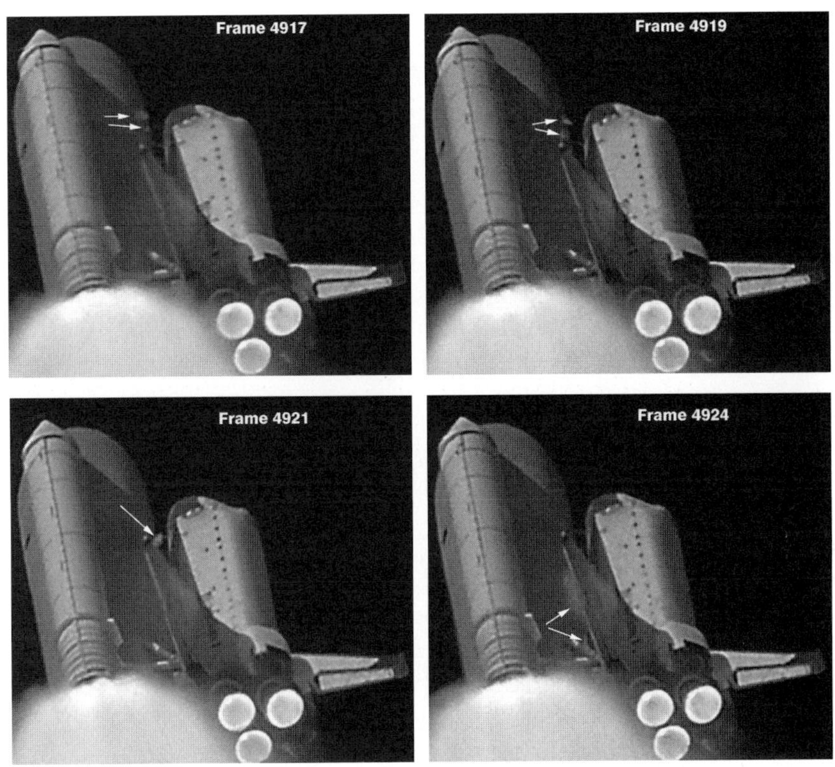

A sequence showing how pieces of foam (arrowed) detached from the bipod mount on the External Tank struck Columbia's left wing during STS-107's ascent.

tyre had deflated they should not hesitate to recommend that the crew fly offshore and bale out, leaving Columbia to ditch in the Atlantic. These e-mails were made public in mid-February. NASA administrator Sean O'Keefe said he was "puzzled" that these engineers had not called the confidential safety 'hot line' that had been introduced after the loss of Challenger. The Department of Defense had offered to use one of its assets to inspect Columbia in orbit, but because the foam impact was not thought to have caused serious damage this offer was turned down. The orbiter had been photographed by the AMOS site in Hawaii, but unfortunately had not been showing its belly. As was common on microgravity science missions, the vehicle was flying with its bay facing the ground.[1] Prior to an in-flight press conference towards the end of the mission, flight director Steve Stich had e-mailed Husband to inform him that, should a reporter ask about it, NASA had released video showing the ET shedding foam at T + 82 seconds. Stich assured Husband that there was no cause for alarm: "We have seen this on several other flights and there is absolutely no concern for re-entry."

Immediately after the accident, Air Force Space Command initiated an in-depth

[1] If only Columbia had been the other way up! Just as if only the flight director had been watching the 'other' video feed of Challenger's final launch!

A picture of STS-107's lift off showing the bipod mount on the External Tank and the Reinforced Carbon–Carbon panel on the leading edge of the left wing which was damaged by a foam strike.

review of its Space Surveillance Network radar data to determine whether there were any anomalies during the mission. It found that a radar in Albuquerque had observed an object drift away from Columbia on its second day in orbit, shortly after it performed a change of orientation; this object re-entered the atmosphere two days later.

In the days after the disaster, some 3,000 people contacted NASA to offer first-hand reports, still photographs and videos documenting all but a minute or two of Columbia's path over the United States. Paul Hill, a missions operation director for the Shuttle, led the team that analysed some 12,000 videos and images.

The search for débris extended west from Texas across New Mexico, Arizona, Utah, Nevada and California. The course of many pieces of débris had been tracked by air traffic control radars. Most of the débris was southeast of Dallas. A helmet from a launch and entry suit was found resting upright on the ground, charred but intact, and body parts were recovered from widely scattered sites. The recovery teams used the Global Positioning System to mark the location of each piece, and a map was compiled. This revealed how the ship had broken up. A striking pattern in the débris trail was evident: most of the early fragments of RCC were from the left wing outboard of RCC #8, then the trail switched to inboard parts of that wing and the right wing. This confirmed the suspicion that the problem had started in the left wing, and the orbiter had disintegrated when the wing suffered structural failure.

Even as late as mid-March, the Board did not favour a specific theory. The foam strike was under consideration together with the possibility of an impact by either a

A view of STS-107 in orbit on 28 January 2003, taken by the US Air Force's Maui Optical Site (AMOS) in Hawaii.

micrometeoroid or a piece of orbital débris. There simply was not enough evidence in the telemetry to determine the cause conclusively. This situation changed with the recovery on 19 March near Hemphill in Texas of the 'black box' that had recorded data from a large number of sensors whose output was not transmitted as telemetry. This Modular Auxiliary Data System had been installed in Columbia for the 'Orbiter Experiment', which was designed to monitor the vehicle's flight characteristics. Its data showed that Columbia's computer had been working to overcome a developing yaw long before the crew or the flight controllers were aware that something was wrong. The fact that the temperature in the left wheel well had risen tied in with the telemetry indicating that the tyres had deflated. Attention therefore focused on what would have happened if the foam had struck the door of the well, breaching its seal and letting *in* plasma. However, once the launch film was computer-enhanced in early May, the trajectory of the foam was refined and it was realised that it had hit the *leading edge* of the wing, outboard of the wheel well. The crucial realisation was that the foam was *more* dangerous than a chunk of ice because the lightweight foam would *decelerate* rapidly in the airstream, which would in turn increase the energy of the impact when the Shuttle slammed into it. A block of ice would have continued to climb with the Shuttle and so, although denser than foam, ice would not have caused such a violent impact. Consideration was given to whether the foam loosened one of the 'carrier panels' that connected the RCC panels on the leading edge to the tiles on the orbiter's belly. Attention then turned to whether the foam dislodged one of the narrow T-seals located between the RCC panels sufficiently to let plasma penetrate. The first anomalous sensor reading on re-entry was from a strain gauge on the spar behind an RCC panel. On 29 May, using the best estimate of the foam's trajectory, the Southwest Research Institute in San Antonio, Texas, began a series of tests to recreate the foam strike. A 10-metre-long gas-powered gun was used to fire a 0.76-kilogram piece of foam – similar to the piece that hit Columbia – into a fibreglass mock-up of the leading edge of a wing at the calculated speed of 850 kilometres per hour. On one test the foam hit a T-seal between two panels. "Things spread apart," reported Scott Hubbard, director of NASA's Ames Research Center and a member of the Board. "We deformed the T-seal." In fact, the impact opened a 6-millimetre-wide gap between RCC panels #6 and #7 that extended for over half a metre. After a check of the records, it was found that Atlantis had suffered a breach between two RCC panels during its descent as STS-101 in 2000. It was Atlantis's first flight since its 1997 refit, and the problem was traced to an incorrectly installed seal that had become dislodged during re-entry, opening a 6-centimetre-wide gap, and the heat was sufficient for titanium and iconel components behind the leading-edge panels to char and scorch and for other parts to become coated with a glassy substance. Despite this, Atlantis had landed without incident, so a damaged T-seal was evidently not fatal. Since, in some of the foam tests, the peak loads had been sufficient to break an RCC panel, the mock-up was fitted with real RCC panels, and the test on 7 July provided the proverbial 'smoking gun'. To the horror of the team, the foam made a hole in the RCC panel some 40-centimetres wide, and even damaged some of the measuring equipment. High-speed video caught an initial rip

 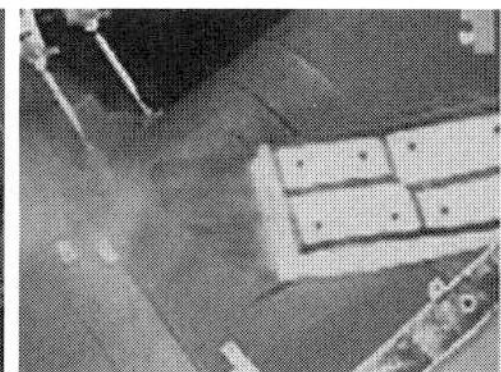  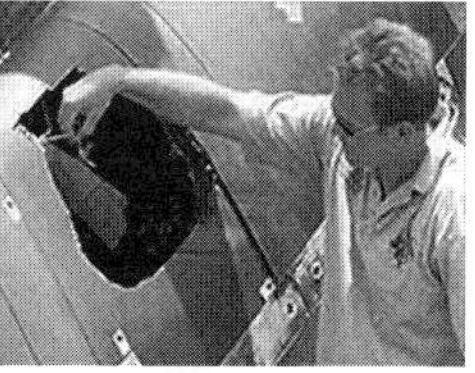

A sequence showing the 'smoking gun' test in which a piece of foam smashed a hole in a Reinforced Carbon–Carbon panel such as used on the leading edge of Columbia's wing.

that, as Hubbard told reporters, "tore all the way across the panel and produced the very ragged hole".[2] He concluded that a strike on the leading edge was "the direct cause of the accident".

Columbia had re-entered the atmosphere with a gaping hole in the leading edge of its left wing, allowing plasma to enter the wing's interior. As the entry proceeded, Columbia's computer sensed the aerodynamic drag on the left side of the vehicle and commands were sent to adjust the trim utilising the elevons at the rear of each wing. This drag steadily increased as Columbia streaked across the southwestern US. The wiring for the failed sensors at the rear of the wing was routed just outboard of, and then in front of, the left main landing gear wheel well. The plasma severed the wiring, and this was the commonality that explained the sensor dropouts. As the major internal support structures in the mid-wing were constructed from an aluminium alloy that melts at 1,200 °F, it is evident that the truss tubes in the mid-wing melted. This led to a catastrophic structural failure, causing the section of the wing outboard of RCC panel #8 to separate, at which point the aerodynamic forces tore the vehicle apart. With this insight, it is possible to relate Columbia's final descent in detail.

FATAL RE-ENTRY

As Columbia descended through 500,000 feet, somewhat to the west of Hawaii, the flight deck crew started to shoot a video. Kalpana Chawla and Laurel Clark were seated behind Rick Husband and William McCool respectively. They were wearing their orange launch and entry suits, but with their gloves off and helmet visors up. Michael Anderson, David Brown and Ilan Ramon were on the lower deck (and not seen in the video). They were all in good spirits, laughing and joking as the computer flew the craft.

The 'entry interface' was at 08:44:09. Arbitrarily defined as 400,000 feet, this is when a spacecraft enters the atmosphere. Columbia was heading east over the Pacific at a speed of Mach 24.56. It was oriented forward with its nose up so that its belly would bear the heat produced as the rarefied air was compressed and ionised. As the plasma emitted light, Clark aimed the video camera up through her overhead

[2] In addition, this test established that while much of the panel broke off, a large piece could have been caught within the panel's hollow interior during the ascent, and later nudged out by a thruster firing. This accounted for the 'mystery object' observed moving away from Columbia on its second day in orbit.

window to document the bright flashes in the ship's wake. McCool assured Ramon – who, being on the mid-deck, had no view – that the glow around the nose was "pretty good". After pointing out that it looked "like a blast furnace", Husband emphasised: "You definitely don't want to be outside now." The video ends abruptly some three and a half minutes after the entry interface. It would seem that filming continued, but the rest of the tape was destroyed in the disaster which – unbeknown to either the crew or to Mission Control – was *already* unfolding.

In the first six minutes, the temperature of the leading edges of the wings, which were protected by the RCC panels, rose progressively to an estimated 2,500 °F. The on-board data recorder noted the first off-nominal sensor reading at 08:48:39, when a strain gauge on the leading edge spar immediately behind panel #9 on the left wing (which was adjacent to and outboard of #8, the one damaged by the foam) began to show signs of stress. Twenty seconds later, a sensor in front of the spar behind #9 began to report an off-nominal increase. At 08:49:32 Columbia's computer initiated the first of a series of four sweeping turns, the first being a steep bank of 70 degrees to the right, designed to reduce energy. At 08:50:53, travelling at Mach 24.1 and at approximately 243,000 feet, Columbia entered a period of maximum thermal stress that would last 10 minutes. At 08:51:14 a sensor *inside* the wing, in line with panel #9, began to show an unusual temperature increase. The breach in the leading edge had enabled a jet of plasma to penetrate the interior of the wing, and it acted like a blowtorch. Five seconds later, as Columbia neared the California coast, an infrared sensor on a Defense Support Program missile early warning satellite saw anomalous activity in Columbia's plasma trail. In an effort to hold the vehicle on course against the aerodynamic drag from the damaged left wing, Columbia's computer was firing

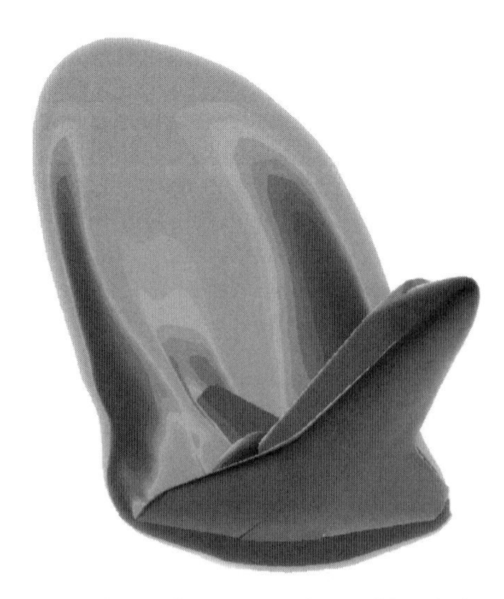

A graphic showing how the plasma flows around an orbiter (oriented with its nose high) as it enters the atmosphere. The thermal stress is borne by the belly, nose and the leading edge of the wings.

the reaction control system thrusters. At 08:52:19 the temperature sensor in front of the left wing spar behind panel #9 – the one that had been first to show a rise – went off-scale low, apparently because the plasma jet had damaged the wiring carrying its signal to the recorder. Some 20 seconds later, sensors in the left wheel well, inboard of the breach in the wing, began to show rising temperatures. A few seconds later, the sensor behind panel #9 – the sensor that had been the first to show a rise in the temperature inside the wing – went off-scale low, followed several seconds later by a sensor measuring the skin temperature on the underside of the left inboard elevon. At 08:53:09 a temperature sensor on the forward face of the left OMS pod began to rise sharply (it ultimately reached 1,200 °F; the normal maximum would be about 60 °F). A few seconds later, four temperature sensors on the hydraulic lines to both the inboard and outboard elevons of the left wing went off-scale low. Because these sensors were telemetered, the first people in Mission Control to realise that Columbia was in trouble were Ken Smith and David Lechner, backroom engineers supporting Jeffrey Kling at the MMACS console in the control room.[3]

"What in the world?" exclaimed one of them as the sensors began to go off line.

"This isn't funny," Kling replied. "On the left side."

"On the left side," the engineer confirmed.

Kling called flight director LeRoy Cain. "I've just lost *four* separate temperature transducers on the left side of the vehicle, hydraulic return temperatures – two of them on system 1 and one in each of systems 2 and 3." Then he added, "To the left outboard and left inboard elevon."

Cain, seeking a commonality, asked, "Okay, is there anything common to them? DSC [discrete signal conditioner] or MDM [multiplexer–demultiplexer] or anything? I mean, you're telling me you lost them all at exactly the same time?"

"No, not exactly," Kling replied. "They were within probably four or five seconds of each other."

"Where is that instrumentation located?"

"All four of them are located in the aft part of the left wing, right in front of the elevons – elevon actuators – and there is no commonality."

"No commonality," Cain slowly enunciated, wondering what was going on, and suddenly recalling the fact that the left wing had been struck by foam at launch.

Kling then spoke to his engineers: "Keep an eye on the left side. Everything else ... looks like the rest of the temperatures are good?"

"Yep," he was assured. In fact, more sensors in the vicinity of the left wheel well were now registering rising temperatures, but the data was not telemetered. If it had been, he would have immediately realised what was going on. The wiring from the temperature sensors on the elevon hydraulics lines was routed close to the outboard side, and around in front of the wheel well, within which the temperature was rising, and the wheel well was just inboard of the breach in the wing. The commonality was the plasma jet that was destroying cables. Worse, and yet to make its effect evident, was the fact that the jet was also melting the wing's structural frames. Although the drag from the damaged left wing was increasing,

[3] The Maintenance, Mechanical, Arm and Crew Systems, known on the intercom loop as 'em-aks'.

the computer was managing to counter it; neither the crew nor Mission Control was aware of its valiant efforts, however.

On his private loop, Kling began to ponder the foam strike. "Those [sensors] are really spread too far apart to be picking up ... like débris damage or anything to both of them, unless they cut a big swath ... have to be *two* hits."

Meanwhile, at the urging of James Hartsfield, the Public Affairs Officer in Mission Control, some residents of San Francisco and Sacramento had gone out to witness the "spectacular view" of Columbia passing overhead in the pre-dawn sky, many of them taking their cameras. As Columbia went 'feet dry' at 08:53:26, travelling at Mach 23 and 231,600 feet, it appeared to observers on the ground as a bright spot of light moving rapidly across the sky. Observers at the Lick Observatory on Mount Hamilton near San Jose caught its passage on video. At 08:53:44 the plasma around Columbia suddenly brightened and the resulting streak in the trail gave the observers the impression that the vehicle was shedding débris. After four similar events in the next 23 seconds, there was a particularly bright flash seconds after crossing the state line, and witnesses observed another 18 similar events in the next four minutes as the ship streaked over Nevada, Utah, Arizona, New Mexico and on across Texas.

At 08:55:55, as Columbia crossed the Utah–Arizona state line travelling at Mach 21.5 at 222,100 feet, it emerged into daylight.

In Mission Control, flight director Cain was still mystified. "MMACS, tell me again which systems they're for."

"That's all three hydraulic systems, two of them are to the left outboard elevon and two of them to the left inboard," Kling reiterated.

"Okay, I got you."

At 08:56:30, 220,000 feet over Arizona and travelling at Mach 20.9, Columbia's computer initiated the first roll reversal, to switch from a right to left bank.

"Everything look good to you?" Cain asked of Guidance, Navigation and Control officer, Mike Sarafin. "Control and rates and everything is nominal – right?"

"Control's been stable through the rolls that we've done so far," Sarafin replied. "We have good trims. I don't see anything out of the ordinary."

The firing of the thrusters by Columbia's computer to overcome the drag from the left wing was not evident to Mission Control.

At that very moment Columbia was passing just north of Albuquerque, New Mexico, where engineers at the Starfire Optical Range at Kirtland Air Force Base

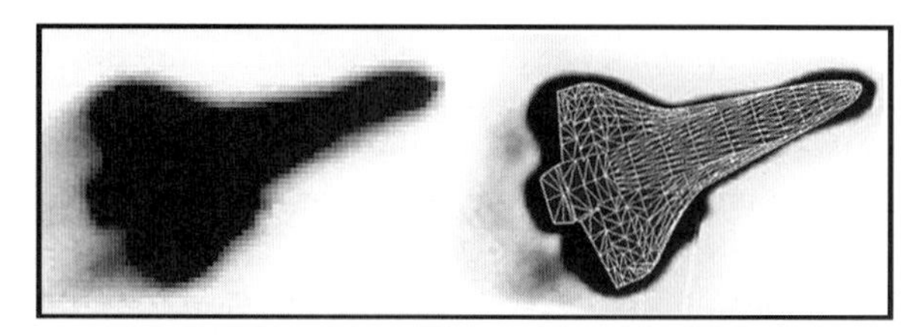

The Starfire Optical Range at Kirtland Air Force Base in New Mexico snapped this picture of Columbia engulfed in plasma, with an anomalous plume trailing behind its left wing.

had fitted a small telescope with an imaging system to track Columbia's re-entry. Their picture showed a plume trailing behind the left wing, whose leading edge seemed to have an irregular profile. In fact, the shape in the picture was not the outline of the vehicle, but the dense plasma surrounding it. Analysts would later conclude that the distortion evident in the image came from the modification and interaction of shock waves due to the damaged leading edge.

Meanwhile, Cain's focus was still on Kling's systems. "All other indications for your hydraulic system indications are good?"

"They're all good," Kling confirmed. In fact, the Auxiliary Power Units that drive the hydraulics system were functioning perfectly. "We've had good quantities all the way across."

"And when you say you lost these, are you saying that they went to zero?"

"All four of them are off-scale low," Kling specified.

"Off-scale low."

As Kling was speaking, one of his support team pointed out, "Staggered, too." Kling echoed this on the flight director's loop. "And they were all staggered. They were, like I said, within several seconds of each other."

"Okay."

It was now 08:58:00.

Kling resumed his discussion with his engineers. "All the rest of the telemetry as we can see is still looking good?"

"Yeah."

"Great."

At this point, Columbia's computer sharply increased the elevon trim to counter the increasing aerodynamic drag. Nevertheless, the tendency to yaw was becoming pronounced.

At 08:58:20, Columbia crossed into Texas. It was at 209,800 feet and travelling at Mach 19.5, and nearing the end of the period of maximum thermal stresses. At about this time it shed a TPS tile. This would subsequently be the most westerly item of débris recovered when it was found in a field in Littlefield, Texas, just northwest of Lubbock. A few seconds later the sensors reporting the pressures in the tyres of the left main gear went off-scale low. At this point, Columbia's Backup Flight System displayed four tyre-pressure fault messages on a cockpit display. This was the first indication to the crew that they had a problem. Unaware that Mission Control was seeing puzzling off-scale low indications, Husband and McCool would have been in two minds as to whether these messages represented an instrumentation fault. If the tyres really had deflated, then they would be facing baling out and ditching Columbia in the Atlantic. Nevertheless, there was no evidence of concern in his voice when Husband called down to report the main landing gear tyre pressure readings, but he was cut off, possibly because the tail fin blocked the line of sight to the TDRS relay from an antenna on top of the crew compartment. "And, uh, Hou(ston)..."

Meanwhile, Kling was reporting this problem to the flight director. "We just lost tyre pressure on the left outboard and left inboard – both tyres."

Cain told the CapCom, astronaut Charles Hobaugh, to let the crew know that Mission Control was aware of the tyre pressures, but the most recent transmission had cut out. As Hobaugh made the call, Cain asked Kling: "Is it instrumentation, MMACS? Gotta be..."

Kling was speaking to his engineers: "I am not believing this."

"No," one replied.

At 08:59:32 Columbia was approaching Dallas, Texas, at 200,700 feet travelling at Mach 18. At that time, Husband responded to Hobaugh's call, but he was cut off once again: "Roger, uh buh..." A momentary interruption had been expected as the Backup Flight System computer switched between antennae for a better line of sight to the TDRS relay, but the signal did not resume at the expected strength. Although telemetry was still being transmitted and recorded at the ground site, it was of such low quality that the processing system ceased to forward it to the flight controllers, who were denied any further insight into what was going on. However, subsequent analysis showed that after five seconds the telemetry was cut off, only to resume for two seconds some 25 seconds later, then fell silent. The on-board recorder registered the sounding of the Master Alarm in the cockpit some two and a half seconds after Husband began his call. Although the computer was vigorously firing the thrusters in a valiant effort to counter the increasing leftward yaw, it was losing the battle.

The final burst of telemetry indicated a rapid change in the lift-to-drag ratio, and the vehicle was in an "uncommanded orientation" and rapidly yawing to one side. In fact, the plasma jet had so eaten away at its structural members that a large section – if not all – of the left wing was missing. The left OMS pod showed signs of severe damage, and there was a propellant leak. The aft engine compartment, fuselage, right wing and crew compartment were intact. All three fuel cells were active, as were the APUs powering the hydraulics, and the life support system was still functioning but the cooling system had shut down. By this point, the crew would have known that Columbia had suffered a catastrophic failure, but there was nothing they could do. Nevertheless, in those last two seconds of ragged data there were indications that the autopilot had been commanded to disengage – either deliberately or as a result of one of the pilots moving the stick in an effort to counter the developing yaw. After the loss of downlinked telemetry at 09:00:04.826, the on-board tape recorder continued to operate for a further 13 seconds, drawing power from the fuel cells located beneath the floor of the payload bay. At 09:00:18, it stopped. It was located near the middle of the deck, in the pressurised cabin, therefore its continuing ability to receive inputs from sensors around the vehicle is a strong indication that the cabin was still intact. It probably ceased working when the electrical system failed when the fuselage disintegrated. The videos shot on the ground at this time support this conclusion. The reinforced crew compartment probably survived the breakup intact (much as Challenger's did when that vehicle was lost in 1986). The recorded data shows that the crew experienced no fatal accelerations prior to the moment that the fuselage failed. As the crew compartment plunged Earthward in its fiery cocoon of plasma, the astronauts almost certainly had time to understand their fate. They lived for almost a minute after Husband's final – interrupted – call. Engineers concluded that they died when the crew compartment, which was probably the only part of the vehicle to survive the initial breakup intact, finally broke open by being buffeted by increasingly extreme aerodynamic forces as it plunged steeply into the thickening atmosphere.

With the loss of signal, Mission Control was unaware of Columbia's fate. As the loss of signal was much longer than expected for an antenna-switch, Laura Hoppe, the Instrumentation and Communications Officer, updated the flight director: "Just taking a few hits here. We're right up on top of the tail. Not too bad."

As soon as Hoppe was off his loop, Cain returned to Kling. "And there's no commonality between all these tyre pressure instrumentations and the hydraulic return instrumentations?"

"No sir, there's not."

As the flight controllers studied the last data received from Columbia, they each saw anomalies in their own specific systems. No one could see the 'big picture', but by now Guidance Officer Sarafin was having doubts that it was an instrumentation problem, and told Cain that if there was a problem with the vehicle then they should be prepared. "If we have any reason to suspect any sort of controllability issue, I would keep the control cards handy on page 4-dash-13."

By 09:03, when Columbia ought to have been within UHF range, the CapCom was repeatedly calling 'in the blind', but there was no response. The inflections in the Public Affairs Officer's reports revealed his unease with the situation. When the radar at the Kennedy Space Center failed to pick up Columbia making the predicted approach, it switched to 'search' mode to look for it in case it was off course, but with no success.

At 09:12:40, by which time Columbia should have been lining up on the Shuttle Landing Facility, Phil Engelauf of the Mission Operations Directorate received a call on his cellular phone from someone who had just seen a Dallas TV station showing a video tape of multiple contrails, indicating that Columbia had disintegrated during re-entry. He informed astronaut Ellen Ochoa, who was sitting on his right, then leaned over his console to inform Cain, who, after taking a moment to digest this shocking news, declared a 'contingency'.

BOARD RECOMMENDATIONS

The conclusion of the Board's investigation was that Columbia had been doomed by the foam strike at T + 82 seconds. Gehman said that if it had been realised as soon as it reached orbit that Columbia had sustained such damaged as to rule out re-entry, then it *would* have been possible to launch Atlantis on a rescue mission. However, as Atlantis could not have been launched until about 12 February, Columbia's vital consumables would have to have been stretched to twice the planned duration. While no spacewalks were scheduled, Columbia carried suits for contingency EVAs. The 'simple' option would have been to retrieve the stranded crew a few at a time and abandon Columbia to its fate. Of course, any number of problems might have made this impossible but (as Gehman put it) "at least we'd have tried". A riskier option would have been to try to repair the damage, but even if this was done, it is likely that Husband and McCool would have insisted that their crewmates transfer to the safety of Atlantis before a re-entry was attempted. The essential fact was, however, that no one had realised that Columbia had been so critically damaged.

In addition to determining the technical reasons for the loss of Columbia, the Board also investigated the management of the Shuttle since 1999 when – as part of the privatisation effort intended to cut costs – the Boeing–Lockheed United Space Alliance was contracted to manage ground processing, paying particular attention to the inspection, testing and quality assurance procedures on turnarounds. The Board was critical of NASA for treating the shedding of foam by the ET as a *maintenance* rather than a safety issue. Asked why the leading-edge panels on the wings were not inspected between missions, O'Keefe pointed out that this would necessitate their removal, which would introduce the possibility of damage.

In its report, published in late August, the Board listed recommendations that it believed must be satisfied before the Shuttle could resume operations. In relation to the accident, these included:

- Eliminate débris-shedding by the ET; and while this will be difficult to avoid completely, it is unacceptable for *large* blocks to become detached. The first target will be the bipod mount. As Gehman put it, "No Shuttle is going to fly with a bipod ramp again."
- Increase the orbiter's ability to sustain minor débris damage by, for example, making the RCC more resistant to impacts.
- Conduct non-destructive tests to determine the structural integrity of RCC system components.
- For ISS missions, develop a capability to inspect and effect emergency repairs to the widest possible range of damage to both TPS tiles and RCC, taking advantage of the additional capabilities available when near to, or docked at, the station. For example, it should be possible for a rendezvousing Shuttle to manoeuvre to display itself for photography, with the digital imagery being downloaded to Earth for prompt examination. To effect repairs, in the early days, NASA had envisaged astronauts going out with a 'calking gun' to glue on patches where tiles had become detached or damaged, but this apparatus was never flown. Something similar may have to be developed.
- A comprehensive autonomous inspection and repair capability should be developed to satisfy the widest possible range of damage scenarios using appropriate assets and capabilities for an ISS mission that fails to achieve its correct orbit, fails to dock, or is damaged during or after undocking, or for a mission that is independent of the ISS (such as servicing the Hubble Space Telescope).
- The long-range tracking system at the Kennedy Space Center should be upgraded to provide a minimum of three useful views of a Shuttle from liftoff to at least SRB separation. Only two of the three long-range cameras for recording launches had been available for STS-107. If the operational status of these assets is included in the Launch Commit Criteria, this will effectively limit launches to daylight and clear skies.
- A method of downlinking high-resolution images of the orbiter's underside and the forward section of both wings, together with high-resolution views of the ET following separation, should be installed. When Atlantis launched in

Mission Control at the time that contact was lost with Columbia.

October 2002, it tested a 'webcam' mounted on the ET, positioned to peer down the stack in order to observe the separation of the SRBs. However, the efflux from the pyros that jettisoned the boosters smeared the lens, and the rest of the ascent was obscured. Nevertheless, if this system is improved and relocated to monitor the orbiter's belly throughout the ascent, the crew could be given 'clearance' upon reaching orbit to proceed with the planned mission or, if there was any damage, to take appropriate action.
- The Memorandum of Agreement with the National Imagery and Mapping Agency should be modified to make the inspection of a Shuttle soon after it achieves orbit a standard requirement.

Also, concerning the broader issues:

- Adopt a flight schedule that is consistent with available resources.
- Enhance the training of the Mission Management Team with contingencies that involve the potential loss of a Shuttle or crew, a variety of uncertainties and unknowns, and cooperation with appropriate NASA and contractor support organisations.
- Devise a plan for an independent Technical Engineering Authority, an independent safety programme, and a reorganised Space Shuttle Integration Office.

O'Keefe immediately accepted the report, "The findings and recommendations of the Columbia Accident Investigation Board will serve as NASA's blueprint. We have accepted the findings and will comply with the recommendations to the best of our ability." Conforming to the Board's requirements would be difficult, but it was hoped to resume operations within 18 months. The Return to Flight Task Group led jointly by former Apollo commander Tom Stafford and former Shuttle commander Dick Covey, had been formed in July and was already at work. But progress was slow, and in February 2004 NASA was obliged to reschedule STS-114 from late 2004 to early 2005.

13

Reflections

A SPACEPLANE

Prior to the dawning of the Space Age, the winged spaceplane, as a development of the aeroplane, possibly with rocket assistance at launch, was the 'obvious' route to spaceflight. However, a spaceplane was not technologically feasible due to (1) the shortfall of rocket power to attain orbital speed, (2) the lack of thermal protection to withstand hypersonic speeds during atmospheric re-entry, and (3) the inadequacy of contemporary electronics technology. The significance of this last obstacle is not so much that vacuum-tube electronics would not have been able to implement a fly-by-wire control system, but in the absence of computers on which to design such a vehicle it would have been necessary to employ the same development methodology as used for contemporary aircraft, in which a succession of prototypes were built and subjected to flight test – a process that would have been costly, time-consuming and risky for the test pilots. By the end of the 1960s, however, micro-electronics had matured, rocket thrust was readily available and a radiative thermal protection system was deemed feasible, so a reusable spaceplane was ordered to supersede the 'expendable' rocket.

NATURAL SELECTION

As originally conceived, the Shuttle was seen as the universal spacecraft that was to replace all previous satellite rockets and crew-carrying spacecraft but, for a variety of reasons, its rôle has evolved over the years, primarily by means of the imposition of constraints.

The Shuttle was specifically designed to provide access to low orbit, one of its primary rôles being to deploy satellites. Those destined for geostationary orbit were either to be fitted with kick motors for insertion into transfer orbit, or to be ferried into into their operating orbits by the purpose-designed IUS. A related rôle of repairing satellites *in situ* was demonstrated at an early stage with SolarMax and Leasat-3, as was the retrieval of satellites with the two HS-376s whose PAM motors failed. The fitting of a new kick motor to a stranded Intelsat demonstrated yet another variation on this theme. However, the plan for the Shuttle to recover satellites from high orbit (a case in point being Leasat-4) was undermined by the cancellation of the Orbital Manoeuvring Vehicle or 'space tug' that was to ferry satellites from the Shuttle to geostationary orbit and, if necessary, return them for servicing. Having designed the Shuttle as a 'space truck' to ferry satellites into low orbit, each deployment was an objective accomplished. Of course, if it were not for the Shuttle-only policy, many of these satellites would have gone up on expendable rockets, but the wide-bodied geostationary communications satellites, most of the classified satellites and the Hubble Space Telescope had been designed to exploit the orbiter's capacious payload bay. Although the Shuttle made feasible the free-flyer – a development that was welcomed by the scientific community – the interplanetary spacecraft deployed by the Shuttle paid a heavy price for their ride into low orbit.

During the early years of Shuttle operations, NASA demonstrated a remarkable ability to perform dynamic scheduling, not only by adding mundane tasks left over by one flight to the next, but also by making spacewalks of increasing complexity to rectify faults in satellites. This represented a significant operational accomplishment, because it was achieved within the context of a sustained drive to increase the flight rate. The low rate (at least in relation to the original plan) at which the Shuttle flew represented another constraint, because there were simply too many satellites – both government and commercial. The Shuttle could not carry the maximum five HS-376 or GPS satellites, even though they were to be deployed in orbit, because a full load would position the centre-of-mass too far forward in a landing after a launch abort. In fact, a Spacelab module represented a significant risk if the orbiter also had a full load of propellant in its forward-RCS cluster, which was why a test was ordered to demonstrate that excess propellant could be dumped during the descent. The folly of the Shuttle-only policy was highlighted by the scramble for the few remaining rockets in the aftermath of the loss of Challenger, in particular to launch two dozen GPS satellites for the Department of Defense. After the Challenger accident the Air Force lost interest. The decommissioning of the launch facility at Vandenberg denied NASA access to polar orbit, which ruled out missions involving meteorological and environmental satellites, including the planned repair of the ailing Landsat. The risk of a Centaur stage disgorging its cryogenic propellant in a rough landing following an aborted launch prompted the cancellation of this powerful stage, which impacted the planned interplanetary probes. The decision not to push the SSMEs to 109 per cent of their rated thrust significantly reduced the capacity to high-inclination orbits. The offloading of the commercial satellites to expendable launchers pre-empted the hope of the Shuttle generating revenue. Nevertheless, the switch to the mixed-fleet policy was a positive step that not only stripped the manifest of straightforward satellite deployments, but also went a long way towards relieving the pressure to attain the advertised flight rate. Although NASA initially hoped to be able to increase the flight rate, it soon acknowledged that it could sustain only six to eight flight per year, and this low flight rate further pared the manifest. Although the Shuttle would have to deploy satellites that could not reasonably be offloaded to expendables, it was to focus on tasks for which it was uniquely capable. As a result, the multifaceted Spacelab rose to the top of the manifest. Spacewalking became less common, but when it was scheduled it was no less ambitious, as demonstrated by the fitting of a kick motor to the Intelsat satellite and the repair of the Hubble Space Telescope. Despite some early scepticism, spacewalkers excelled at delicate work. In fact, the Shuttle proved to be an excellent platform for spacewalking, and although the RMS was invaluable in this respect, the most effective tool proved to be the human hand.

In retrospect, the most significant factor constraining the Shuttle's rôle had been imposed before its development was authorised. NASA had proposed the reusable transportation system in the context of a space station that it would assemble and service, but Congress denied funding for an orbital base. The Shuttle was to launch, achieve its objectives and return to Earth. This was fine for satellite deployments and for short-run observational and microgravity work using Spacelab, but because the Shuttle would have nowhere to go, all such activities would be limited to the time that it could fly in space, and even with the EDO upgrade it was limited to 16 days. This constraint on NASA's activities was partially lifted by the decision to install apparatus on Mir, which, as a modular orbital complex, proved to be a high-fidelity precursor to the ISS, which faced similar design and operational challenges. A fair assessment of the Shuttle must consider the value of the transportation system in the context of the station, as they are, in reality, elements in an integrated system.

A MATTER OF COST

The cost–benefit analysis that was used to 'prove' that the Shuttle would be a more effective long-term solution than the expendable rocket relied on a very high rate of flights that proved impracticable. Although it is common to cite this as a failure of the Shuttle to live up to its requirements, it is more realistic to acknowledge that the analysis was flawed in that it did not fully appreciate the factors that would dictate the operational pace. The only way to attain the envisaged weekly flight rate would have been to build a large fleet of orbiters, each of which would fly only two or three times per year, but the 'turnaround costs' would have undermined the economics of the analysis.

The cost of a Shuttle mission is difficult to define. One way to compute the cost for each mission is to start by dividing the total development cost by the expected flight rate in order to work out the amortisation overhead, add in a proportion of the annual operating costs, and then the specific development costs associated with the payload. Doing this, however, makes each flight appear considerably more costly than it actually is, because the cost of the Shuttle's development is in the past, and nothing will be saved by *not* flying. It was the amortisation of the development to the weekly flight rate that suggested that the Shuttle would, in the end, be more cost-effective than to continue with expendables, as long as the expendables were phased out. A small fleet of orbiters cannot run at that rate. This increases the amortisation overhead for each mission that is flown, but this is an artificial increase. Typically, it takes three to four months to turn an orbiter around and roll it back out to the pad.

The resultant operating overhead represents a genuine increase in the cost of a flight in comparison to the originally projected cost but, as experience over the years has shown, a higher-than-expected operating overhead is unavoidable. Nevertheless, this, too, is misleading, because 80 per cent of the annual operating costs are *fixed*. The cost of maintaining the facilities is largely independent of the flight rate. Hence, it is rather naïve simply to divide up the development cost and the annual operating cost to determine a 'basic cost' for each flight. Rather, the cost of mounting any specific mission must be computed on the basis of its directly incurred costs. In the extreme, this cost would be precisely the difference in the annual budget between making and *not* making the flight. But to argue this would also be naïve, because the objective of the system is to make flights possible. It is a matter of balance. There is a flight rate at which the Shuttle would be under-used, and the overhead prohibitive. Conversely, any given annual budget imposes a *maximum* number of flights. In assessing costs, it must also be noted that the cost of developing and operating a payload is not part of by the Shuttle's budget – the Hubble Space Telescope, for example, is accounted for independently. And, of course, when NASA was in the business of deploying commercial satellites (even though it was locked into an unrealistically low fee structure) it generated income to offset mission costs.

NASA therefore settled down to a readily sustainable rate of six to eight flights per year. With a fleet of four orbiters, it would be difficult to increase the flight rate beyond the once-a-month level. United Space Alliance – the contractor that handles the Shuttle's ground processing and provides pad services – argued to overturn the decision that prohibited the Shuttle from carrying commercial satellites. The company wished to exploit the 'spare' capacity in its turnaround process by flying such missions, and thereby operate the system at its peak rate. At such a low flight rate such 'private' missions would be feasible only because the Shuttle has matured as an operational system, and its turnaround has become subject to control, both in terms of time and cost. However, such 'private' flights were not authorised.

RISKY BUSINESS

In authorising the development of the Shuttle, President Nixon stated that its goal would be to 'routinise' access to low orbit, and it is clear that the Shuttle has indeed done so. Indeed, the Shuttle is the *most successful* spacecraft ever built to carry a human crew. It has carried more people, more often, in a shorter overall period, than the Soyuz, the workhorse of the Soviet–Russian programme, and has successfully performed a multiplicity of rôles. Nevertheless, in suffering two losses the Shuttle has killed 14 people.

Spaceflight is a risky activity, but how dangerous is it? In 1979, one NASA estimate said that 1 in 50 Shuttle flights would suffer catastrophic failure during the ascent and 1 in 100 would be lost in returning to Earth. All other failure modes were expected to be dealt with by planned abort options. In 1985, NASA's Public Affairs Office quoted the risk at 1 in 100,000. At this same time, the Air Force said 1 in 35, with the most likely cause of catastrophic failure being a fault with the solid rocket boosters. A few months later, Challenger was lost on the programme's 25th launch. In 1989, after the resumption of flights, NASA considered the risk of catastrophic failure leading to the loss of the vehicle at 1 in 168 – although it was hoped that the crew would manage to bale out. Unfortunately, when Columbia disintegrated during its re-entry in 2003, on the programme's 113th mission, there was no possibility of escape.

SUPERSEDING THE SHUTTLE

In the immediate aftermath of the loss of Challenger, America asked itself whether it was right and proper to risk the lives of astronauts to deploy satellites, and decided that it was not. In the resulting mixed-fleet policy, the Shuttle was complemented by expendable rockets. A decade later, the Shuttle was routinely visiting the Mir space station and it is now making regular visits to the ISS. Is the ferrying of supplies a fair use of the Shuttle? Is it more worthy to risk the lives of astronauts to deliver food, clothing, spare parts and water to a space station? With the exception of propellant (which is delivered by Progress tankers) the ISS is essentially reliant on the Shuttle for logistical support. With an upper stage capable of rendezvousing and docking, it would be feasible (and perhaps better) to launch this cargo on expendable rockets. The European Space Agency has developed the Automated Transfer Vehicle as an upper stage for its Ariane V rocket to supersede the Progress this rôle. Such logistics vehicles complement the Shuttle, and release it for more appropriate work for which it is uniquely capable, such as delivering large elements from which to build orbital facilities.

After the loss of Columbia, Admiral Gehman, the chairman of the Investigation Board, urged "a sense of urgency" in replacing the Shuttle for ferrying people to and from the station. In fact, for some time there had been calls to phase out the Shuttle and replace it with a *more reusable* spaceplane incorporating modern technologies that would be cheaper to operate. There is a distinction, however, between a *mature* and an *obsolete* technology. As in the case of the venerable Soyuz, the Shuttle was optimised for the rôles for which it was most suited. And with its ground processing privatised, the Shuttle was operating as efficiently as it ever would. Critics say that the basic design is obsolete, and should be replaced by a more advanced vehicle.[1] It is certainly the case that the configuration of the stack derives from the decisions made in the early 1970s, which were in turn dictated by budgetary constraints. In a very real sense (its costly development notwithstanding) the Shuttle was developed on a shoestring budget. It is also true that the airframes are ageing, but that is not to say that they are approaching the point of exhaustion. In this respect, it must be recalled that initially, and naïvely, the Shuttle was to fly on a weekly basis. As a result, each orbiter was designed to be capable of 100 flights. To a degree, the fact that they are flying less frequently serves to extend their service life. Also, the orbiters are not in

[1] On the other hand, it is often the case that as soon as a product becomes available, someone argues that it is obsolete and proposes an idea for something better!

the same condition as when they rolled off the production line, because there is an ongoing programme of upgrades. It is said that the computers are obsolete, but the pace at which computer technology develops is so rapid that this can be said of any product whose service life is expected to last decades. Over the years, the avionic systems, engines, pumps, auxiliary power units and thermal protection system have all been improved. Nevertheless, in parallel with its effort to keep the Shuttle flying beyond 2010, NASA has investigated a number of options for superseding it.

In this regard, it is worth reflecting on how the space programme was influenced by the design of the Shuttle. The abandonment of the Saturn V launch vehicle denied NASA the option of orbiting a 100-tonne payload. In theory, this heavy-lift would have been provided by 'Shuttle-C', but this use of Shuttle-based technology was always very low on the list of priorities, and when NASA set out to design a space station it did not consider launching a few large fully integrated components, each of which was comparable to Skylab, it instead developed a modular design like Mir. And as with Mir, whose modular components were constrained by the mass and volumetric limitations of the Proton launch vehicle, the components of the ISS were constrained by the Shuttle's payload bay. In fact, the constraints on Mir and the ISS are similar as their modules could not exceed about 4 metres in diameter and were limited to about 20 tonnes. However, Mir's modules required their own propulsion systems in order to rendezvous and dock, whereas the entire mass of a module delivered by the Shuttle can be specific to its rôle as part of the ISS. The size of the Shuttle's bay was defined by the Department of Defense. If the spaceplane that NASA had originally envisaged had been built, the granularity of the ISS would have been even smaller. The ISS is therefore very much the child of the Shuttle. What rôle should the space truck play once the assembly of the ISS is finished?

One point of view is that because NASA *has* the Shuttle, it must *use* it, but this logic is valid only while the Shuttle is NASA's *only* vehicle for human spaceflight. There are alternatives under development for logistical support and crew transfer. In addition to superseding the Progress, the Automated Transfer Vehicle will very likely be able to supersede the Shuttle in routinely resupplying the ISS. Although recent events have shown that in addition to serving as a 'lifeboat' the Soyuz spacecraft can revert to its original function as a crew transfer vehicle, it is really too small to supersede the Shuttle in this rôle. It has been suggested that NASA should develop a *small* crew transfer vehicle that can (a) be launched on a rocket booster, (b) glide back to land on a runway, and (c) be rapidly refurbished for further use. Ironically, there has already been considerable study on such a vehicle. Japan's long-term aspiration is to build a spaceplane named HOPE, and France would dearly like to build its Hermes spaceplane, but both developments have been put on hold due to cost. Overruns in the ISS's budget forced NASA to suspend the development of the X-38, the small 'lifting body' craft to supersede the Soyuz as the ISS's 'lifeboat'. In mid-2003, after the loss of Columbia, NASA reported that it would accelerate the development of an Orbital Space Plane to supersede the Shuttle as a crew transport.[2]

[2] It might be better if the various spacefaring nations *shared* the cost of developing a single design which *each* nation could use *as it* saw fit.

An artist's depiction of the proposed Orbital Space Plane.

Another artist's depiction of the proposed Orbital Space Plane.

However, there was considerable disagreement as to the best configuration for such a vehicle, and even if its development were accelerated, it would be unlikely to become available before 2010. But even if this spacecraft is introduced, and the Automated Transfer Vehicle sustains the ISS with logistics, there remains a rôle for which the Shuttle is uniquely capable. Mir operations were complicated by the fact that there was no means of offloading either large amounts of cargo or heavy or bulky items. Retiring the Shuttle before an alternative means of returning such items to Earth is provided will inevitably result in the ISS becoming as cluttered as Mir was towards the end of its life, with the consequent reduction in its operating efficiency.

The panacea in superseding the Shuttle is a vehicle that can lift off and fly into orbit without shedding any boosters or external tanks, glide back, land on a runway, be serviced and be ready to fly again within a matter of days. Such a Single-Stage To Orbit (SSTO) vehicle would be expected to slash the $25,000 that it costs to place each kilogram of payload into orbit using the Shuttle. However, the development of a vehicle of this type is going to be a technologically challenging, very costly and long-term venture. In 1996 Lockheed won an industry play-off to work with NASA to develop the X-33 to demonstrate – in subscale – an SSTO design that the company would later operate commercially as the 'VentureStar'. Although progress was made with the innovative 'aerospike' engine, the super-lightweight materials required for its integral propellant tanks proved elusive and, after consuming $1.3 billion, the project was terminated in March 2001. Two months later, NASA awarded the first of a multitude of contracts for its $5-billion Space Launch Initiative that was to

An artist's depiction of the X-33/VentureStar about to dock at the ISS.

develop the technologies to facilitate a reusable spaceplane launched on a fly-back booster. However, with the loss of Columbia on 1 February 2003, all plans were put on hold.

A year later, on 14 January 2004, President George W. Bush announced that the Shuttle would not be upgraded to serve beyond 2010; after completing the assembly of the ISS it would be retired. By then, NASA must have designed a crew transport vehicle as part of a larger programme designed to initiate a new phase in the human exploration of the Solar System.

The great irony of the Shuttle is that in its initial proposal NASA wanted such a vehicle to assemble a space station, but this has taken so long that as soon as the ISS is finished the Shuttle will be retired for reasons which derive from compromises imposed during its own design.

The Space Shuttle – a magnificent flying machine.

Shuttle Mission Log

The following table lists, for each Shuttle mission in chronological order: its sequence number; its international designation; the orbiter and its flight count; the launch and landing; the landing site and its use count; the duration of the flight; the orbital inclination; the crew, with their flight counts aboard the Shuttle (prior experience is not counted); and details of payloads and mission activities. Payload acronyms are explained in the next section.

#	Designation	Orbiter	Launch	Landing	Runway	Days	Incl
STS-1							
1	1981-34A	Columbia (1)	0700 EST 12 Apr	1021 PST 14 Apr	EAFB (1)	2.26	40
CDR:	John Young (1)						
PLT:	Bob Crippen (1)						
Payload:	OAST[1]						
STS-2							
2	1981-111A	Columbia (2)	1010 EST 12 Nov	1323 PST 14 Nov	EAFB (2)	2.26	38
CDR:	Joe Engle (1)						
PLT:	Dick Truly (1)						
Payload:	OAST, IECM, PDP, NOSL, HBT, OSTA 1						
STS-3							
3	1982-22A	Columbia (3)	1100 EST 22 Mar	0905 MST 30 Mar	WSTR (1)	8.00	38
CDR:	Jack Lousma (1)						
PLT:	Gordon Fullerton (1)						
Payload:	OAST, IECM, OSS 1, MLR, HBT, EEVT (CFES), GFVT (GAS), SSIP, PGU(LIGNIN)						
STS-4							
4	1982-65A	Columbia (4)	1100 EDT 27 Jun	0909 PDT 4 Jul	EAFB (3)	7.05	28
CDR:	Ken Mattingly (1)						
PLT:	Hank Hartsfield (1)						
Payload:	OAST, DOD82-1 (CIRRIS, UHS), MLR, CFES, IECM, SSIP, GAS, NOSL						
STS-5							
5	1982-110A	Columbia (5)	0719 EST 11 Nov	0633 PST 16 Nov	EAFB (4)	5.09	28

Table (cont)

#	Designation	Orbiter	Launch	Landing	Runway	Days	Incl
CDR:	Vance Brand (1)						
PLT:	Bob Overmyer (1)						
MS1:	Joe Allen (1,EV)						
MS2:	Bill Lenoir (1,EV)						
Payload:	OAST, GAS, SSIP						
Activity:	Deployed satellites SBS 3/PAM on 11 Nov and Anik C3/PAM on 12 Nov.						
STS-6							
6	1983-26A	Challenger (1)	1330 EST 4 Apr	1053 PST 9 Apr	EAFB (5)	5.00	28
CDR:	Paul Weitz (1)						
PLT:	Karol Bobko (1)						
MS1:	Donald Peterson (1,EV)						
MS2:	Story Musgrave (1,EV)						
Payload:	CFES, MLR, NOSL, GAS						
Activity:	Deployed TDRS-A/IUS on 4 Apr.						
Spacewalks:	Peterson and Musgrave						
STS-7							
7	1983-59A	Challenger (2)	0733 EDT 18 Jun	0657 PDT 24 Jun	EAFB (6)	6.10	28
CDR:	Bob Crippen (2)						
PLT:	Rick Hauck (1)						
MS1:	John Fabian (1,EV)						
MS2:	Sally Ride (1)						
MS3:	Norm Thagard (1,EV)						
Payload:	OSTA 2, MLR, CFES, GAS, OAST						
Activity:	Deployed Anik C2/PAM on 18 Jun, Palapa B1/PAM on 19 Jun. The SPAS free-flyer 1 was deployed and retrieved on 22 Jun.						
STS-8							
8	1983-89A	Challenger (3)	0232 EDT 30 Aug	0041 PDT 5 Sep	EAFB (7)	6.05	28
CDR:	Dick Truly (2,EV)						
PLT:	Dan Brandenstein (1)						
MS1:	Dale Gardner (1,EV)						
MS2:	Guion Bluford (1)						
MS3:	Bill Thornton (1)						
Payload:	PFTA, CFES, MLR, GAS, AEM, RME						
Activity:	Deployed Insat 1B/PAM on 31 Aug.						
STS-9							
9	1983-116A	Columbia (6)	1100 EST 28 Nov	1547 PST 8 Dec	EAFB (8)	10.33	57
CDR:	John Young (2,EV)						
PLT:	Brewster Shaw (1)						
MS1:	Owen Garriott (1,EV)						
MS2:	Bob Parker (1)						
PS1:	Byron Lichtenberg (1)						
PS2:	Ulf Merbold (1,ESA)						
Payload:	Spacelab 1						
STS-10/41B							
10	1984-11A	Challenger (4)	0800 EST 3 Feb	0716 EST 11 Feb	KSC (1)	7.97	28
CDR:	Vance Brand (2)						
PLT:	Hoot Gibson (1)						
MS1:	Bruce McCandless (1,EV)						
MS2:	Robert Stewart (1,EV)						
MS3:	Ron McNair (1)						

#	Designation	Orbiter	Launch	Landing	Runway	Days	Incl

Payload: MMU, IFE, RME, MLR, IRT, SSIP, GAS
Activity: Deployed Westar 6/PAM on 3 Feb and Palapa B2/PAM on 4 Feb. Deployment of SPAS 1A free-flyer abandoned due to RMS fault.
Spacewalks: McCandless and Stewart

STS-11/41C

11	1984-34A	Challenger (5)	0858 EST 6 Apr	0538 PST 13 Apr	EAFB (9)	6.98	28

CDR: Bob Crippen (3)
PLT: Dick Scobee (1)
MS1: Terry Hart (1)
MS2: James van Hoften (1,EV)
MS3: George Nelson (1,EV)
Payload: RME, IMAX, Cinema-360, SSIP
Activity: Deployed LDEF on 7 Apr. Retrieved SolarMax on 10 Apr, repaired it and released it on 12 Apr.
Spacewalks: Nelson and van Hoften

STS-12/41D

12	1984-93A	Discovery (1)	0841 EDT 30 Aug	0637 PDT 5 Sep	EAFB (10)	6.04	28

CDR: Hank Hartsfield (2)
PLT: Michael Coats (1)
MS1: Judy Resnik (1)
MS2: Steven Hawley (1,EV)
MS3: Mike Mullane (1,EV)
PS: Charles Walker (1)
Payload: OAST 1, LFC, CFES, RME, CLOUDS, IMAX, SSIP
Activity: Deployed SBS 4/PAM on 30 Aug, Leasat 2 on 31 Aug and Telstar 3C/PAM on 1 Sep.

STS-13/41G

13	1984-108A	Challenger (6)	0703 EDT 5 Oct	1226 EDT 13 Oct	KSC (2)	8.22	57

CDR: Bob Crippen (4)
PLT: Jon McBride (1)
MS1: Kathryn Sullivan (1,EV)
MS2: Sally Ride (2)
MS3: David Leestma (1,EV)
PS1: Paul Scully-Power (1)
PS2: Marc Garneau (1,Canada)
Payload: OSTA 3, CANEX 1, RME, TLD, APE, IMAX, GAS
Activity: Deployed ERBE on 5 Oct.
Spacewalks: Sullivan and Leestma

STS-14/51A

14	1984-113A	Discovery (2)	0715 EST 8 Nov	0700 EST 16 Nov	KSC (3)	7.99	28

CDR: Rick Hauck (2)
PLT: David Walker (1)
MS1: Joe Allen (2,EV1)
MS2: Anna Fisher (1)
MS3: Dale Gardner (2,EV2)
Payload: DMOS, RME
Activity: Deployed Anik D2/PAM on 9 Nov and Leasat 1 on 10 Nov. Palapa B2 was retrieved 12 Nov and Westar 6 was retrieved 14 Nov.
Spacewalks: Allen and Gardner

STS-15/51C

15	1985-10A	Discovery (3)	1450 EST 24 Jan	1623 EST 27 Jan	KSC (4)	3.06	28

CDR: Ken Mattingly (2)
PLT: Loren Shriver (1)

Table (cont)

#	Designation	Orbiter	Launch	Landing	Runway	Days	Incl
MS1:	Ellison Onizuka (1, EV)						
MS2:	James Buchli (1,EV)						
SFE:	Gary Payton (1)						
Payload:	ARC						
Activity:	Deployed Magnum/IUS (USA-8) on 25 Jan.						
STS-16/51D							
16	1985-28A	Discovery (4)	0859 EST 12 Apr	0855 EST 19 Apr	KSC (5)	6.99	28
CDR:	Karol Bobko (2)						
PLT:	Don Williams (1)						
MS1:	Rhea Seddon (1)						
MS2:	David Griggs (1,EV)						
MS3:	Jeff Hoffman (1,EV)						
PS:	Charles Walker (2)						
CO:	Senator Jake Garn (1)						
Payload:	CFES, AFE, PPE, SSIP, GAS						
Activity:	Deployed Anik C1/PAM on 12 Apr and Leasat 3 on 13 Apr.						
Spacewalks:	Griggs and Hoffman						
STS-17/51B							
17	1985-34A	Challenger (7)	1202 EDT 29 Apr	0911 PDT 6 May	EAFB (11)	7.00	57
CDR:	Bob Overmyer (2)						
PLT:	Fred Gregory (1,EV)						
MS1:	Don Lind (1)						
MS2:	Norm Thagard (2,EV)						
MS3:	Bill Thornton (2)						
PS1:	Lodewijk van den Berg (1)						
PS2:	Taylor Wang (1)						
Payload:	Spacelab 3 (AHF), GAS(Nusat)						
Activity:	Deployed Nusat on 29 Apr.						
STS-18/51G							
18	1985-48A	Discovery (5)	0733 EDT 17 Jun	0611 PDT 24 Jun	EAFB (12)	7.07	28
CDR:	Dan Brandenstein (2)						
PLT:	John Creighton (1)						
MS1:	Steve Nagel (1,EV)						
MS2:	John Fabian (2,EV)						
MS3:	Shannon Lucid (1)						
PS1:	Patrick Baudry (1,France)						
PS2:	Prince Sultan Salman Abdul Aziz al-Saud (1)						
Payload:	FPE, ADSF, GAS, HPTE						
Activity:	Deployed Morelos 1/PAM on 17 Jun, Arabsat 1B/PAM on 18 Jun and Telstar 3D/PAM on 19 Jun. Deployed SPARTAN 101 on 22 Jun and retrieveed it on 22 Jun.						
STS-19/51F							
19	1985-63A	Challenger (8)	1700 EDT 29 Jul	1245 PDT 6 Aug	EAFB (13)	7.95	50
CDR:	Gordon Fullerton (2)						
PLT:	Roy Bridges (1)						
MS1:	Karl Henize (1)						
MS2:	Anthony England (1,EV)						
MS3:	Story Musgrave (2,EV)						
PS1:	Loren Acton (1)						
PS2:	John-David Bartoe (1)						
Payload:	Spacelab 2, SAREX, PGU(LIGNIN)						
Activity:	Deployed PDP on 1 Aug and retrieved it on 2 Aug.						

#	Designation	Orbiter	Launch	Landing	Runway	Days	Incl

STS-20/51I

20	1985-76A	Discovery (6)	0658 EDT 27 Aug	0616 PDT 3 Sep	EAFB (14)	7.10	28

CDR: Joe Engle (2)
PLT: Dick Covey (1)
MS1: James van Hoften (2,EV)
MS2: Mike Lounge (1)
MS3: Bill Fisher (1,EV)
Payload: PVTOS
Activity: Deployed Aussat 1/PAM on 27 Aug, ASC 1/PAM on 27 Aug and Leasat 4 on 29 Aug. Captured, repaired and released Leasat 3 on 31 Aug.
Spacewalks: Fisher and van Hoften

STS-21/51J

21	1985-92A	Atlantis (1)	1115 EDT 3 Oct	1000 PDT 7 Oct	EAFB (15)	4.07	28

CDR: Karol Bobko (3)
PLT: Ron Grabe (1)
MS1: Dave Hilmers (1,EV)
MS2: Robert Stewart (2,EV)
SFE: William Pailes (1)
Activity: Deployed DSCS III-2 and DSCS III-3 (both on the same IUS) on 4 Oct.

STS-22/61A

22	1985-104A	Challenger (9)	1200 EST 30 Oct	0945 PST 6 Nov	EAFB (16)	7.03	57

CDR: Hank Hartsfield (3)
PLT: Steve Nagel (2)
MS1: Bonnie Dunbar (1)
MS2: James Buchli 2,EV)
MS3: Guion Bluford (2,EV)
PS1: Earnst Messerschmid (1, West Germany)
PS2: Reinhard Furrer (1, West Germany)
PS3: Wubbo Ockels (1,ESA)
Payload: Spacelab D1, GAS (GLOMR)
Activity: Deployed GLOMR on 31 Oct.

STS-23/61B

23	1985-109A	Atlantis (2)	1929 EST 26 Nov	1334 PST 3 Dec	EAFB (17)	6.88	28

CDR: Brewster Shaw (2)
PLT: Bryan O'Connor (1)
MS1: Sherwood Spring (1,EV)
MS2: Mary Cleave (1)
MS3: Jerry Ross (1,EV)
PS1: Rodolfo Neri Vela (1,Mexico)
PS2: Charles Walker (3)
Payload: EASE, ACCESS, CFES, IMAX, GAS, DMOS, OEX
Activity: Deployed Morelos 2/PAM on 27 Nov, Aussat 2/PAM on 27 Nov and Satcom K2/PAM-D2 on 28 Nov.
Spacewalks: Ross and Spring

STS-24/61C

24	1986-3A	Columbia (7)	0655 EST 12 Jan	0558 PST 18 Jan	EAFB (18)	6.08	28

CDR: Hoot Gibson (2)
PLT: Charles Bolden (1)
MS1: George Nelson (2,EV)
MS2: Steven Hawley (2)
MS3: Franklin Chang-Diaz (1,EV)

Table (cont)

#	Designation	Orbiter	Launch	Landing	Runway	Days	Incl
PS:	Robert Cenker (1)						
CO:	Representative Bill Nelson (1)						
Payload:	MSL, CHAMP, SSIP, GBA, HH-PCG						
Activity:	Deployed Satcom K1/PAM-D2 on 12 Jan.						
STS-25/51L							
25	1986	Challenger (10)	1138 EST 28 Jan	–	–	–	28
CDR:	Dick Scobee (2)						
PLT:	Mike Smith (1)						
MS1:	Judy Resnik (2)						
MS2:	Ellison Onizuka (2,EV)						
MS3:	Ron McNair (2,EV)						
PS:	Greg Jarvis (1)						
TO:	Christa McAuliffe (1)						
Payload:	CHAMP, RME, SSIP, PPE						
Activity:	TDRS-B/IUS, Halley/SPARTAN 203						
Note:	Vehicle destroyed 73 seconds after launch.						
STS-26							
26	1988-91A	Discovery (7)	1137 EDT 29 Sep	0937 PDT 3 Oct	EAFB (19)	4.04	28
CDR:	Rick Hauck (3)						
PLT:	Dick Covey (2)						
MS1:	Mike Lounge (2)						
MS2:	Dave Hilmers (2)						
MS3:	George Nelson (3)						
Payload:	PVTOS, PCG, IRCFE, ARC, IFE, MLE, PPE, ELRAD, ADSF, SSIP, OASIS						
Activity:	Deployed TDRS-C/IUS on 29 Sep.						
STS-27							
27	1988-106A	Atlantis (3)	0930 EST 2 Dec	1536 PST 6 Dec	EAFB (20)	4.38	57
CDR:	Hoot Gibson (3)						
PLT:	Guy Gardner (1)						
MS1:	Mike Mullane (2)						
MS2:	Jerry Ross (2,EV)						
MS3:	Bill Shepherd (1,EV)						
Activity:	Deployed a Lacrosse radar reconnaissance satellite (USA-34) on 2 Dec.						
STS-29							
28	1989-21A	Discovery (8)	0957 EST 13 Mar	0635 PST 18 Mar	EAFB (21)	4.98	28
CDR:	Michael Coats (2)						
PLT:	John Blaha (1)						
MS1:	James Buchli (3)						
MS2:	Robert Springer (1,EV)						
MS3:	James Bagian (1,EV)						
Payload:	IMAX, SHARE, OASIS, PCG, PGU(CHROMEX), SSIP, AMOS, AEM						
Activity:	Deployed TDRS-D/IUS on 13 Mar.						
STS-30							
29	1989-33A	Atlantis (4)	1447 EDT 4 May	1243 PDT 8 May	EAFB (22)	4.04	28
CDR:	David Walker (2)						
PLT:	Ron Grabe (2)						
MS1:	Norm Thagard (3,EV)						
MS2:	Mary Cleave (2)						
MS3:	Mark Lee (1,EV)						
Payload:	MLE, FEA, AMOS						
Activity:	Deployed Magellan/PAM/IUS on 4 May.						

#	Designation	Orbiter	Launch	Landing	Runway	Days	Incl
STS-28							
30	1989-61A	Columbia (8)	0837 EDT 8 Aug	0637 PDT 13 Aug	EAFB (23)	5.04	57
CDR:	Brewster Shaw (3)						
PLT:	Dick Richards (1)						
MS1:	Jim Adamson (1)						
MS2:	David Leestma (2)						
MS3:	Mark Brown (1)						
Payload:	L3, EDS, SILTS, 'the Head'						
Activity:	Deployed an SDS communications relay satellite (USA-40).						
STS-34							
31	1989-84A	Atlantis (5)	1254 EDT 18 Oct	0933 PDT 23 Oct	EAFB (24)	4.99	34
CDR:	Don Williams (2)						
PLT:	Mike McCulley (1)						
MS1:	Shannon Lucid (2)						
MS2:	Ellen Baker (1,EV)						
MS3:	Franklin Chang-Diaz (2,EV)						
Payload:	IMAX, GAS(SSBUV), GHCD, PM, MLE, IMAX, SSIP, AMOS						
Activity:	Deployed Galileo/IUS on 18 Oct.						
STS-33							
32	1989-90A	Discovery (9)	1923 EST 22 Nov	1631 PST 27 Nov	EAFB (25)	5.00	28
CDR:	Fred Gregory (2)						
PLT:	John Blaha (2)						
MS1:	Story Musgrave (3,EV)						
MS2:	Sonny Carter (1,EV)						
MS3:	Kathryn Thornton (1)						
Activity:	Deployed Magnum/IUS (USA-48).						
STS-32							
33	1990-02A	Columbia (9)	0735 EST 9 Jan	0135 PST 20 Jan	EAFB (26)	10.87	28
CDR:	Dan Brandenstein (3)						
PLT:	Jim Wetherbee (1)						
MS1:	Bonnie Dunbar (2,EV)						
MS2:	Marsha Ivins (1)						
MS3:	David Low (1,EV)						
Payload:	IMAX, PCG, FEA, AFE, L3, MLE, AMOS, LBNP, SILTS						
Activity:	Deployed Leasat 5 on 10 Jan. Retrieved LDEF on 12 Jan.						
STS-36							
34	1990-19A	Atlantis (6)	0250 EST 28 Feb	1009 PST 4 Mar	EAFB (27)	4.43	62
CDR:	John Creighton (2)						
PLT:	John Casper (1)						
MS1:	Mike Mullane (3)						
MS2:	Dave Hilmers (3)						
MS3:	Pierre Thuot (1)						
Activity:	Deployed a DoD 'stealthy' reconnaissance satellite (USA-53; released by SPDS) on 1 Mar.						
STS-31							
35	1990-37A	Discovery (10)	0833 EDT 24 Apr	0649 PDT 29 Apr	EAFB (28)	5.05	28
CDR:	Loren Shriver (2)						
PLT:	Charles Bolden (2)						
MS1:	Bruce McCandless (2,EV1)						
MS2:	Steven Hawley (3)						
MS3:	Kathryn Sullivan (2,EV2)						

Table (cont)

#	Designation	Orbiter	Launch	Landing	Runway	Days	Incl
Payload:	IMAX, APM, PCG, RME, PMP, SSIP, AMOS						
Activity:	Deployed HST on 25 Apr.						
STS-41							
36	1990-90A	Discovery (11)	0747 EDT 6 Oct	0657 PDT 10 Oct	EAFB (29)	4.09	28
CDR:	Dick Richards (2)						
PLT:	Bob Cabana (1)						
MS1:	Bruce Melnick (1,EV)						
MS2:	Bill Shepherd (2)						
MS3:	Tom Akers (1,EV)						
Payload:	GAS(SSBUV), ISAC, PGU(CHROMEX), VCS, SSCE, PMP, PSE/AEM, RME, SSIP, AMOS						
Activity:	Deployed Ulysses/PAM/IUS on 6 Oct.						
STS-38							
37	1990-97A	Atlantis (7)	1848 EST 15 Nov	1643 EST 20 Nov	KSC (6)	4.91	28
CDR:	Dick Covey (3)						
PLT:	Frank Culbertson (1)						
MS1:	Robert Springer (2,EV)						
MS2:	Carl Meade (1,EV)						
MS3:	Sam Gemar (1)						
Activity:	Deployed an SDS communications relay satellite (USA-67).						
STS-35							
38	1990-106A	Columbia (10)	0149 EST 2 Dec	2154 PST 10 Dec	EAFB (30)	8.96	28
CDR:	Vance Brand (3)						
PLT:	Guy Gardner (2)						
MS1:	Jeff Hoffman (2,EV1)						
MS2:	Mike Lounge (3,EV2)						
MS3:	Bob Parker (2)						
PS1:	Samuel Durrance (1)						
PS2:	Ronald Parise (1)						
Payload:	ASTRO 1, BBXRT, SAREX, AMOS						
STS-37							
39	1991-27A	Atlantis (8)	0922 EST 5 Apr	0555 PST 11 Apr	EAFB (31)	5.98	28
CDR:	Steve Nagel (3)						
PLT:	Ken Cameron (1)						
MS1:	Linda Godwin (1)						
MS2:	Jerry Ross (3,EV)						
MS3:	Jay Apt (1,EV)						
Payload:	CETA, CLIP, APM, SAREX, PCF, BIMDA, RME, AMOS, SHARE						
Activity:	Deployed CGRO on 7 Apr.						
Spacewalks:	Ross and Apt						
STS-39							
40	1991-31A	Discovery (12)	0733 EDT 28 Apr	1455 EDT 6 May	KSC (7)	8.30	57
CDR:	Mike Coats (3)						
PLT:	Blaine Hammond (1)						
MS1:	Greg Harbaugh (1,EV)						
MS2:	Donald McMonagle (1,EV)						
MS3:	Guion Bluford (3)						
MS4:	Lacy Veach (1)						
MS5:	Richard Hieb (1)						
Payload:	CIRRIS, FUV, HUP, URA, QINMS, CIV, STP (UVLIM, ALFE, SKIRT, APM, DSE), RME, CLOUDS						
Activity:	Deployed CRO-B/C on 1 May, MPEC (USA-70) on 6 May and CRO-A on 3 May. Deployed SPAS 2-01/IBSS on 1 May and retrieved it on 2 May.						

#	Designation	Orbiter	Launch	Landing	Runway	Days	Incl

STS-40

41	1991-40A	Columbia (11)	0925 EDT 5 Jun	0839 PDT 14 Jun	EAFB (32)	9.09	39

CDR: Bryan O'Connor (2)
PLT: Sid Gutierrez (1)
MS1: James Bagian (2,EV)
MS2: Tamara Jernigan (1,EV)
MS3: Rhca Seddon (2)
PS1: Drew Gaffney (1)
PS2: Millie Hughes-Fulford (1)
Payload: SLS 1, GBA , MODE, AEM, ARE, RAHF, GPWS

STS-43

42	1991-54A	Atlantis (9)	1102 EDT 2 Aug	0823 EDT 11 Aug	KSC (8)	8.89	28

CDR: John Blaha (3)
PLT: Mike Baker (1)
MS1: Shannon Lucid (3)
MS2: David Low (2,EV)
MS3: Jim Adamson (2,EV)
Payload: GAS(SSBUV), SHARE, OCTW, GAS(TPCE), APE, PCG, BIMDA, PMP, SAMS, SSCE, UVPI, AMOS, LBNP
Activity: Deployed TDRS-E/IUS on 2 Aug.

STS-48

43	1991-63A	Discovery (13)	1911 EDT 12 Sep	0039 PDT 18 Sep	EAFB (33)	5.35	57

CDR: John Creighton (3)
PLT: Ken Reightler (1)
MS1: Sam Gemar (2,EV)
MS2: James Buchli (4,EV)
MS3: Mark Brown (2)
Payload: AMOS, APM, MODE, SAM, CREAM, PARE, PCG, PMP, RME, ESC
Activity: Deployed UARS on 15 Sep.

STS-44

44	1991-80A	Atlantis (10)	1844 EST 24 Nov	1434 PST 1 Dec	EAFB (34)	6.95	28

CDR: Fred Gregory (3)
PLT: Tom Henricks (1)
MS1: Jim Voss (1,EV)
MS2: Story Musgrave (4)
MS3: Mario Runco (1,EV)
SFE: Tom Hennen (1)
Payload: IOCM, MODE, AMOS, M88-1 (MMIS), CREAM, SAM, RME, VFT, UVPI, BFPT, EDO-MP, Terra Scout
Activity: Deployed DSP-16/IUS (USA-75) on 25 Nov.

STS-42

45	1992-02A	Discovery (14)	0952 EST 22 Jan	0807 PST 30 Jan	EAFB (35)	8.05	57

CDR: Ron Grabe (3)
PLT: Stephen Oswald (1)
MS1: Norm Thagard (4,EV,PC)
MS2: Bill Readdy (1,EV)
MS3: Dave Hilmers (4)
PS1: Roberta Bondar (1,Canada)
PS2: Ulf Merbold (2,ESA)
Payload: IML 1, IMAX, GBA, SSIP, GOSAMR, PMP, RME, PCG, PCF

Table (cont)

#	Designation	Orbiter	Launch	Landing	Runway	Days	Incl
STS-45							
46	1992-15A	Atlantis (11)	0814 EST 24 Mar	0623 EST 2 Apr	KSC (9)	8.93	57
CDR:	Charles Bolden (3)						
PLT:	Brian Duffy (1)						
MS1:	Kathryn Sullivan (3,PC,EV)						
MS2:	David Leestma (3,EV)						
MS3:	Mike Foale (1)						
PS1:	Dirk Frimout (1,ESA/Belgium)						
PS2:	Byron Lichtenberg (2)						
Payload:	ATLAS 1, GAS(SSBUV), STL, PMP, SAREX, VFT, RME, CLOUDS, GAS, APE						
STS-49							
47	1992-26A	Endeavour (1)	1940 EDT 7 May	1357 PDT 16 May	EAFB (36)	8.88	28
CDR:	Dan Brandenstein (4)						
PLT:	Kevin Chilton (1)						
MS1:	Pierre Thuot (2,EV1)						
MS2:	Kathryn Thornton (2,EV3)						
MS3:	Richard Hieb (2,EV2)						
MS4:	Tom Akers (2,EV4)						
MS5:	Bruce Melnick (2)						
Payload:	ASEM, PCF, UVPI, AMOS						
Activity:	Retrieved Intelsat 603 on 13 May, fitted it with an Orbus motor and released it on 14 May.						
Spacewalks:	Thuot, Thornton, Hieb and Akers						
STS-50							
48	1992-34A	Columbia (12)	1212 EDT 25 Jun	0743 EDT 9 Jul	KSC (10)	13.81	28
CDR:	Dick Richards (3)						
PLT:	Ken Bowersox (1)						
MS1:	Bonnie Dunbar (3,PC)						
MS2:	Ellen Baker (2)						
MS3:	Carl Meade (2)						
PS1:	Larry DeLucas (1)						
PS2:	Gene Trinh (1)						
Payload:	USML 1, EDO 1, PCF, GBX, PMP, SAREX, UVPI, AERIS						
STS-46							
49	1992-49A	Atlantis (12)	0956 EDT 31 Jul	0911 EDT 8 Aug	KSC (11)	7.97	28
CDR:	Loren Shriver (3)						
PLT:	Andy Allen (1)						
MS1:	Claude Nicollier (1,ESA/Switzerland)						
MS2:	Marsha Ivins (2)						
MS3:	Jeff Hoffman (3,EV)						
MS4:	Frankin Chang-Diaz (3,EV)						
PS:	Franco Malerba (1,Italy)						
Payload:	TSS 1, LDCE, UVPI, IMAX, EOIM, CONCAP, IMAX, AMOS						
Activity:	Deployed Eureca 1 on 2 Aug.						
STS-47							
50	1992-61A	Endeavour (2)	1023 EDT 12 Sep	0853 EDT 20 Sep	KSC (12)	7.94	57
CDR:	Hoot Gibson (4)						
PLT:	Curt Brown (1)						
MS1:	Mark Lee (2,PC)						
MS2:	Jay Apt (2,EV)						
MS3:	Jan Davis (1)						
MS4:	Mae Jemison (1)						

#	Designation	Orbiter	Launch	Landing	Runway	Days	Incl
PS:	Mamoru Mohri (1,Japan)						
Payload:	Spacelab J1, GEF, SAMS, PCG, CHF, GBA, ISAIAH, SSCE, SAREX, AMOS, UVPI						
STS-52							
51	1992-71A	Columbia (13)	1309 EDT 22 Oct	0905 EST 1 Nov	KSC (13)	9.86	28
CDR:	Jim Wetherbee (2)						
PLT:	Mike Baker (2)						
MS1:	Lacy Veach (2)						
MS2:	Bill Shepherd (3)						
MS3:	Tamara Jernigan (2)						
PS:	Steven MacLean (1,Canada)						
Payload:	USMP 1, CANEX 2, CMIX, SPIFEX, GAS(TPCE), CVTE, HPP, ASP, PCG, PSE						
Activity:	Deployed Lageos 2/IRIS on 23 Oct and CTA on 31 Oct.						
STS-53							
52	1992-86A	Discovery (15)	0824 EST 2 Dec	1243 PST 9 Dec	EAFB (37)	7.30	57
CDR:	David Walker (3)						
PLT:	Bob Cabana (2)						
MS1:	Guion Bluford (4)						
MS2:	Rich Clifford (1)						
MS3:	Jim Voss (2)						
Payload:	GAS(ODERACS), GLO, CRYOHP, MIS, STL, VFT, CREAM, RME, FARE, HERCULES, BLAST, CLOUDS						
Activity:	Deployed an SDS communications relay satellite. (USA-87) on 2 Dec.						
STS-54							
53	1993-03A	Endeavour (3)	0859 EST 13 Jan	0838 EST 19 Jan	KSC (14)	5.97	28
CDR:	John Casper (2)						
PLT:	Donald McMonagle (2)						
MS1:	Mario Runco (2,EV2)						
MS2:	Greg Harbaugh (2,EV1)						
MS3:	Susan Helms (1)						
Payload:	DXS, CGBA, PGU(CHROMEX), PARE, SAMS, SSCE						
Activity:	Deployed TDRS-F/IUS on 12 Jan.						
Spacewalks:	Runco and Harbaugh						
STS-56							
54	1993-23A	Discovery (16)	0129 EDT 8 Apr	0737 EDT 17 Apr	KSC (15)	9.25	57
CDR:	Ken Cameron (2)						
PLT:	Stephen Oswald (2)						
MS1:	Mike Foale (2,PC)						
MS2:	Ken Cockrell (1)						
MS3:	Ellen Ochoa (1)						
Payload:	ATLAS 2, SAREX, GAS(SUVE, SSBUV), CMIX, PARE, STL, CREAM, HERCULES, RME, AMOS						
Activity:	Deployed SPARTAN 201-1 on 11 Apr and retrieved it on 13 Apr.						
STS-55							
55	1993-27A	Columbia (14)	1050 EDT 26 Apr	0730 PDT 6 May	EAFB (38)	9.98	28
CDR:	Steve Nagel (4)						
PLT:	Tom Henricks (2)						
MS1:	Jerry Ross (4,PC)						
MS2:	Charles Precourt (1)						
MS3:	Barnard Harris (1)						
PS1:	Ulrich Walter (1, Germany)						
PS2:	Hans Schlegel (1, Germany)						
Payload:	Spacelab D2 (Anthrorack, MOMS, ROTEX, HOL, CFZF), SAREX						

Table (cont)

#	Designation	Orbiter	Launch	Landing	Runway	Days	Incl
STS-57							
56	1993-37A	Endeavour (4)	0907 EDT 21 Jun	0852 EDT 1 Jul	KSC (16)	9.99	28
CDR:	Ron Grabe (4)						
PLT:	Brian Duffy (2)						
MS1:	David Low (3,PC,EV1)						
MS2:	Nancy Sherlock-Currie (1)						
MS3:	Jeff Wisoff (1,EV2)						
MS4:	Janice Voss (1)						
Payload:	Spacehab 1, SHOOT, CONCAP, GBA, FARE, BLAST, SAREX						
Activity:	Retrieved Eureca 1 on 24 Jun.						
Spacewalks:	Low and Wisoff						
STS-51							
57	1993-58A	Discovery (17)	0745 EDT 12 Sep	0356 EDT 22 Sep	KSC (17)	9.84	28
CDR:	Frank Culbertson (2)						
PLT:	Bill Readdy (2)						
MS1:	Jim Newman (1,EV2)						
MS2:	Dan Bursch (1)						
MS3:	Carl Walz (1,EV1)						
Payload:	IMAX, PCG, PGU(CHROMEX), APE, PMP, RME, LDCE, AMOS						
Activity:	Deployed ACTS/TOS on 12 Sep. Deployed ORFEUS-SPAS 1 on 13 Sep and retrieved it on 19 Sep.						
Spacewalks:	Walz and Newman						
STS-58							
58	1993-65A	Columbia (15)	1053 EDT 18 Oct	0706 PST 1 Nov	EAFB (39)	14.00	39
CDR:	John Blaha (4)						
PLT:	Rick Searfoss (1)						
MS1:	Rhea Seddon (3,PC)						
MS2:	Bill McArthur (1)						
MS3:	David Wolf (1,EV2)						
MS4:	Shannon Lucid (4,EV1)						
PS:	Martin Fettman (1)						
Payload:	SLS 2, EDO 2, OARE, SAREX, PILOT, LBNP						
STS-61							
59	1993-75A	Endeavour (5)	0427 EST 2 Dec	0026 EST 13 Dec	KSC (18)	10.83	28
CDR:	Dick Covey (4)						
PLT:	Ken Bowersox (2)						
MS1:	Kathryn Thornton (3,EV3)						
MS2:	Claude Nicollier (2,ESA/Switzerland)						
MS3:	Jeff Hoffman (4,EV1)						
MS4:	Story Musgrave (5,PC,EV2)						
MS5:	Tom Akers (3,EV4)						
Payload:	HST apparatus (COSTAR, WFPC 2), IMAX						
Activity:	HST Service 1. Retrieved HST on 4 Dec, serviced it and released it on 10 Dec.						
Spacewalks:	Thornton, Hoffman, Musgrave and Akers						
STS-60							
60	1994-06A	Discovery (18)	0710 EST 3 Feb	1419 EST 11 Feb	KSC (19)	8.30	57
CDR:	Charles Bolden (4)						
PLT:	Ken Reightler (2)						
MS1:	Jan Davis (2,EV2)						
MS2:	Ron Sega (1)						
MS3:	Franklin Chang-Diaz (4,PC,EV1)						
MS4:	Sergei Krikalev (1,Russia)						

#	Designation	Orbiter	Launch	Landing	Runway	Days	Incl

Payload: Spacehab 2, WSF 1, SAREX, APE, GBA (CAPL), GAS(ODERACS, Bremsat)
Activity: Deployed ODERACS and Bremsat on 9 Feb.

STS-62

#	Designation	Orbiter	Launch	Landing	Runway	Days	Incl
61	1994-15A	Columbia (16)	0853 EST 4 Mar	0810 EST 18 Mar	KSC (20)	13.97	39

CDR: John Casper (3)
PLT: Andy Allen (2)
MS1: Pierre Thuot (3,EV1)
MS2: Sam Gemar (3,EV2)
MS3: Marsha Ivins (3)
Payload: USMP 2, OAST 2, EDO 3, DEE, GAS(SSBUV), LDCE, APCG, CPCG, PSE, CGBA, BDS, MODE, AMOS, LBNP, PILOT

STS-59

#	Designation	Orbiter	Launch	Landing	Runway	Days	Incl
62	1994-20A	Endeavour (6)	0705 EDT 9 Apr	0954 PDT 20 Apr	EAFB (40)	11.24	57

CDR: Sid Gutierrez (2)
PLT: Kevin Chilton (2)
MS1: Jay Apt (3)
MS2: Rich Clifford (2)
MS3: Linda Godwin (2,PC,EV1)
MS4: Ton Jones (1,EV2)
Payload: SRL 1 (SIR-C, X-SAR), MAPS, CONCAP, SAREX, STL, VFT, GAS, TUFI

STS-65

#	Designation	Orbiter	Launch	Landing	Runway	Days	Incl
63	1994-39A	Columbia (17)	1243 EDT 8 Jul	0638 EDT 23 Jul	KSC (21)	14.75	28

CDR: Bob Cabana (3)
PLT: James Halsell (1)
MS1: Richard Hieb (3,PC)
MS2: Carl Walz (2,EV2)
MS3: Leroy Chiao (1,EV1)
MS4: Don Thomas (1)
PS: Chiaki Naito-Mukai (1,Japan)
Payload: IML 2, EDO 4, EDO-MP, APCF, CPCG, AMOS, OARE, MAST, SAREX

STS-64

#	Designation	Orbiter	Launch	Landing	Runway	Days	Incl
64	1994-59A	Discovery (19)	1823 EDT 9 Sep	1413 PDT 20 Sep	EAFB (41)	10.95	57

CDR: Dick Richards (4)
PLT: Blaine Hammond (2)
MS1: Jerry Linenger (1)
MS2: Susan Helms (2)
MS3: Carl Meade (3,EV)
MS4: Mark Lee (3,EV)
Payload: LITE, SAFER, ROMPS, TCS, GBA, SSCE, BRIC, RME, MAST, SAREX, AMOS, SPIFEX
Activity: Deployed SPARTAN 201-2 on 13 Sep and retrieved it on 15 Sep.
Spacewalks: Lee and Meade

STS-68

#	Designation	Orbiter	Launch	Landing	Runway	Days	Incl
65	1994-62A	Endeavour (7)	0716 EDT 30 Sep	1002 PDT 11 Oct	EAFB (42)	11.24	57

CDR: Mike Baker (3)
PLT: Terry Wilcutt (1)
MS1: Steve Smith (1,EV2)
MS2: Dan Bursch (2)
MS3: Jeff Wisoff (2,EV1)
MS4: Tom Jones (2,PC)
Payload: SRL 2 (SIR-C, X-SAR, MAPS), CPCG, BRIC, PGU(CHROMEX), CREAM, MAST, GAS

Table (cont)

#	Designation	Orbiter	Launch	Landing	Runway	Days	Incl
STS-66							
66	1994-73A	Atlantis (13)	1159 EST 3 Nov	0734 PST 14 Nov	EAFB (43)	10.94	57
CDR:	Donald McMonagle (3)						
PLT:	Curt Brown (2)						
MS1:	Ellen Ochoa (2,PC)						
MS2:	Joe Tanner (1)						
MS3:	Jean-Francois Clervoy (1,ESA/France)						
MS4:	Scott Parazynski (1)						
Payload:	ATLAS 3, GAS(SSBUV), ESCAPE 2, NIH-R, PCG-TES, PCG-STES, NIH-C, SAMS, HPP						
Activity:	Deployed CRISTA-SPAS 1 on 4 Nov and retrieved it on 12 Nov.						
STS-63							
67	1995-04A	Discovery (20)	0022 EST 3 Feb	0650 EST 11 Feb	KSC (22)	8.27	51
CDR:	Jim Wetherbee (3)						
PLT:	Eileen Collins (1)						
MS1:	Bernard Harris (2,EV2,PC)						
MS2:	Mike Foale (3,EV1)						
MS3:	Janice Voss (2)						
MS4:	Vladimir Titov (1,Russia)						
Payload:	Spacehab 3, CONCAP, GLO, CRYOSYS, IMAX, SSCE, AMOS, MSX, TCS, GAS(ODERACS)						
Activity:	Deployed ODERACS on 4 Feb. Closed within 10 metres of Mir on 6 Feb. Deployed SPARTAN 204 on 7 Feb and retrieved it on 9 Feb.						
Spacewalks:	Foale and Harris						
STS-67							
68	1995-07A	Endeavour (8)	0138 EST 2 Mar	1347 PST 18 Mar	EAFB (44)	16.63	28
CDR:	Stephen Oswald (3)						
PLT:	Bill Gregory (1)						
MS1:	John Grunsfeld (1,EV2)						
MS2:	Wendy Lawrence (1)						
MS3:	Tamara Jernigan (3,PC,EV1)						
PS1:	Samuel Durrance (2)						
PS2:	Ronald Parise (2)						
Payload:	ASTRO 2, EDO 5, GAS, PCG-TES, PCG-STES, SAREX, CMIX, MSX, MACE						
STS-71							
69	1995-30A	Atlantis (14)	1532 EDT 27 Jun	1054 EDT 7 Jul	KSC (23)	9.80	51
CDR:	Hoot Gibson (5)						
PLT:	Charles Precourt (2)						
MS1:	Ellen Baker (3,EV2)						
MS2:	Greg Harbaugh (3,EV1)						
MS3:	Bonnie Dunbar (4)						
UP:	Anatoli Solovyev (1)						
UP:	Nikolai Budarin (1)						
DN:	Vladimir Dezhurov (1)						
DN:	Gennadi Strekalov (1)						
DN:	Norm Thagard (5)						
Payload:	Spacelab, IMAX, SAREX, ODS						
Activity:	SMM 1 docked with Mir on 29 Jun and departed on 4 Jul.						
STS-70							
70	1995-35A	Discovery (21)	0942 EDT 13 Jul	0802 EDT 22 Jul	KSC (24)	8.93	28
CDR:	Tom Henricks (3)						
PLT:	Kevin Kregel (1)						
MS1:	Don Thomas (2)						

#	Designation	Orbiter	Launch	Landing	Runway	Days	Incl
MS2:	Nancy Sherlock-Currie (2)						
MS3:	Mary Ellen Weber (1)						
Payload:	MSX, NIH-R, BDS, CPCG, NIH-C, BRIC, SAREX, VFT, MIS, AMOS, HERCULES, WINDEX, RME, MAST						
Activity:	Deployed TDRS-G/IUS on 13 Jul.						

STS-69

#	Designation	Orbiter	Launch	Landing	Runway	Days	Incl
71	1995-48A	Endeavour (9)	1109 EDT 7 Sep	0737 EDT 18 Sep	KSC (25)	10.85	28
CDR:	David Walker (4)						
PLT:	Ken Cockrell (2)						
MS1:	Jim Voss (3,EV1,PC)						
MS2:	Jim Newman (2)						
MS3:	Michael Gernhardt (1,EV2)						
Payload:	GBA(CAPL, IEH, UVSTAR, GLO, CONCAP)						
Activity:	Deployed SPARTAN 201-3 on 8 Sep and retrieved it on 10 Sep. Deployed WSF 2 on 11 Sep and retrieved it on 14 Sep.						
Spacewalks:	Voss and Gernhardt						

STS-73

#	Designation	Orbiter	Launch	Landing	Runway	Days	Incl
72	1995-56A	Columbia (18)	0953 EDT 20 Oct	0745 EDT 5 Nov	KSC (26)	15.92	39
CDR:	Ken Bowersox (3)						
PLT:	Kent Rominger (1)						
MS1:	Cady Coleman (1,EV2)						
MS2:	Michael Lopez-Alegria (1,EV1)						
MS3:	Kathryn Thornton (4,PC)						
PS1:	Fred Leslie (1)						
PS2:	Albert Sacco (1)						
Payload:	USML 2, EDO 6						

STS-74

#	Designation	Orbiter	Launch	Landing	Runway	Days	Incl
73	1995-61A	Atlantis (15)	0730 EST 12 Nov	1201 EST 20 Nov	KSC (27)	8.18	51
CDR:	Ken Cameron (3)						
PLT:	James Halsell (2)						
MS1:	Chris Hadfield (1,Canada)						
MS2:	Jerry Ross (5,EV1)						
MS3:	Bill McArthur (2,EV2)						
Payload:	SVS, DM, ODS, GLO						
Activity:	SMM 2. The DM was mounted on the ODS on 14 Nov. Docked with Mir on 15 Nov and departed on 18 Nov, leaving the DM on Mir's Kristall module.						

STS-72

#	Designation	Orbiter	Launch	Landing	Runway	Days	Incl
74	1996-01A	Endeavour (10)	0441 EST 11 Jan	0242 EST 20 Jan	KSC (28)	8.91	28
CDR:	Brian Duffy (3)						
PLT:	Brent Jett (1)						
MS1:	Leroy Chiao (2, EV1)						
MS2:	Winston Scott (1,EV3)						
MS3	Koichi Wakata (1,Japan)						
MS4:	Daniel Barry (1,EV2)						
Payload:	GAS(SSBUV), PCG						
Activity:	Deployed OAST/SPARTAN 206 on 14 Jan and retrieved it on 16 Jan. Retrieved the Japanese SFU on 13 Jan.						
Spacewalks:	Chiao, Scott and Barry						

STS-75

#	Designation	Orbiter	Launch	Landing	Runway	Days	Incl
75	1996-12A	Columbia (19)	1518 EST 22 Feb	0858 EST 9 Mar	KSC (29)	15.73	28

Table (cont)

#	Designation	Orbiter	Launch	Landing	Runway	Days	Incl
CDR:	Andy Allen (3)						
PLT:	Scott Horowitz (1)						
MS1:	Jeff Hoffman (5)						
MS2:	Maurizio Cheli (1,ESA/Italy)						
MS3:	Claude Nicollier (3,EV2,ESA/Switzerland)						
MS4:	Franklin Chang-Diaz (5,PC,EV1)						
PS:	Umberto Guidoni (1,ESA/Italy)						
Payload:	USMP 3, TSS 2 EDO 7, MGBX						
Activity:	An attempt to unreel TSS was foiled when the tether broke on 25 Feb, inadvertently turning the payload into a satellite.						
STS-76							
76	1996-18A	Atlantis (16)	0313 EST 22 Mar	0529 PST 31 Mar	EAFB (45)	9.22	51
CDR:	Kevin Chilton (3)						
PLT:	Rick Searfoss (2)						
MS1:	Ron Sega (2,PC)						
MS2:	Rich Clifford (3,EV2)						
MS3:	Linda Godwin (3,EV1)						
UP:	Shannon Lucid (5)						
Payload:	Single Spacehab, ODS, KidSat, SAREX, MEEP						
Activity:	SMM 3. Docked with Mir on 23 Mar and departed on 28 Mar.						
Spacewalks:	Clifford and Godwin						
STS-77							
77	1996-32A	Endeavour (11)	0630 EDT 19 May	0709 EDT 29 May	KSC (30)	10.03	28
CDR:	John Casper (4)						
PLT:	Curt Brown (3)						
MS1:	Andy Thomas (1,PC)						
MS2:	Dan Bursch (3,EV2)						
MS3:	Mario Runco (3,EV1)						
MS4:	Marc Garneau (2,Canada)						
Payload:	Spacehab 4, GBA, TEAMS (GANE, VTRE, LMTE), BETSCE, ARF, BRIC						
Activity:	Deployed IAE/SPARTAN 207 on 20 May; IAE inflated and jettisoned same day and then the SPARTAN carrier was retrieved on 21 May. TEAMS (PAMS/STU) deployed on 22 May.						
STS-78							
78	1996-36A	Columbia (20)	1049 EDT 20 Jun	0837 EDT 7 Jul	KSC (31)	16.91	28
CDR:	Tom Henricks (4)						
PLT:	Kevin Kregel (2)						
MS1:	Richard Linnehan (1,EV2)						
MS2:	Susan Helms (3,EV1,PC)						
MS3:	Charles Brady (1)						
PS1:	Jean-Jacques Favier (1,France)						
PS2:	Bob Thirsk (1,Canada)						
Payload:	LMS (APCF, AGHF, BDPU, AEM, PGF, STL, OARE, SAMS, MMA, TRE, HGD, TVD, LFE, COIS); EDO 8						
STS-79							
79	1996-57A	Atlantis (17)	0454 EDT 16 Sep	0813 EDT 26 Sep	KSC (32)	10.13	51
CDR:	Bill Readdy (3)						
PLT:	Terry Wilcutt (2)						
MS1:	Jay Apt (4,EV2)						
MS2:	Tom Akers (4)						
MS3:	Carl Walz (3,EV1)						
UP:	John Blaha (5)						

#	Designation	Orbiter	Launch	Landing	Runway	Days	Incl
DN:	Shannon Lucid (5)						
Payload:	Double Spacehab, ODS						
Activity:	SMM 4. Docked with Mir on 19 Sep and departed on 24 Sep.						

STS-80

#	Designation	Orbiter	Launch	Landing	Runway	Days	Incl
80	1996-65A	Columbia (21)	1456 EST 19 Nov	0649 EST 7 Dec	KSC (33)	17.67	28
CDR:	Ken Cockrell (3)						
PLT:	Kent Rominger (2)						
MS1:	Tamara Jernigan (4,EV1)						
MS2:	Tom Jones (3,EV2)						
MS3:	Story Musgrave (6)						
Payload:	EDO 9, SVS, CMIX, PARE, BRIC, VIEW-CAPL, SEM, CCM, SVS						
Activity:	Deployed ORFEUS-SPAS 2 on 20 Nov and retrieved it on 4 Dec. Deployed WSF 3 on 22 Nov and retrieved it on 25 Nov.						

STS-81

#	Designation	Orbiter	Launch	Landing	Runway	Days	Incl
81	1997-01A	Atlantis (18)	0427 EST 12 Jan	0922 EST 22 Jan	KSC (34)	10.21	51
CDR:	Mike Baker (4)						
PLT:	Brent Jett (2)						
MS1:	Jeff Wisoff (3,EV1)						
MS2:	John Grunsfeld (2,EV2)						
MS3:	Marsha Ivins (4)						
UP:	Jerry Linenger (2)						
DN:	John Blaha (5)						
Payload:	Double Spacehab, ODS, KidSat, CREAM, SAMS, TVIS						
Activity:	SMM 5. Docked with Mir on 14 Jan and departed on 19 Jan.						

STS-82

#	Designation	Orbiter	Launch	Landing	Runway	Days	Incl
82	1997-4A	Discovery (22)	0355 EST 11 Feb	0332 EST 21 Feb	KSC (35)	9.99	28
CDR:	Ken Bowersox (4)						
PLT:	Scott Horowitz (2)						
MS1:	Joe Tanner (2,EV4)						
MS2:	Steven Hawley (4)						
MS3:	Greg Harbaugh (4,EV3)						
MS4:	Mark Lee (4,PC,EV1)						
MS5:	Steve Smith (2,EV2)						
Payload:	NICMOS, STIS						
Activity:	HST Service 2. Retrieved HST on 13 Feb, serviced it and released it on 19 Feb.						
Spacewalks:	Tanner, Harbaugh, Lee and Smith						

STS-83

#	Designation	Orbiter	Launch	Landing	Runway	Days	Incl
83	1997-13A	Columbia (22)	1420 EST 4 Apr	1433 EDT 8 Apr	KSC (36)	3.97	28
CDR:	James Halsell (3)						
PLT:	Susan Still (1)						
MS1:	Janice Voss (3,PC)						
MS2:	Michael Gernhardt (2,EV1)						
MS3:	Don Thomas (3,EV2)						
PS1:	Roger Crouch (1)						
PS2:	Gregory Linteris (1)						
Payload:	MSL 1, EDO 10						

STS-84

#	Designation	Orbiter	Launch	Landing	Runway	Days	Incl
84	1997-23A	Atlantis (19)	0407 EDT 15 May	0928 EDT 24 May	KSC (37)	10.22	51
CDR:	Charles Precourt (3)						
PLT:	Eileen Collins (2)						

Table (cont)

#	Designation	Orbiter	Launch	Landing	Runway	Days	Incl
MS1:	Jean-Francois Clervoy (2,ESA/France,EV1)						
MS2:	Carlos Noriega (1)						
MS3:	Ed Lu (1,EV2)						
MS4:	Yelena Kondakova (1,Russia)						
UP:	Mike Foale (4)						
DN:	Jerry Linenger (2)						
Payload:	Double Spacehab, ODS						
Activity:	SMM 6. Docked with Mir on 16 May and departed on 21 May.						
STS-94							
85	1997-32A	Columbia (23)	1402 EDT 1 Jul	0646 EDT 17 Jul	KSC (38)	15.70	28
CDR:	James Halsell (4)						
PLT:	Susan Still (2)						
MS1:	Janice Voss (4,PC)						
MS2:	Michael Gernhardt (3,EV1)						
MS3:	Don Thomas (4,EV2)						
PS1:	Roger Crouch (2)						
PS2:	Gregory Linteris (2)						
Payload:	MSL 1R, EDO 11						
STS-85							
86	1997-39A	Discovery (23)	1141 EDT 7 Aug	0708 EDT 19 Aug	KSC (39)	11.85	57
CDR:	Curt Brown (4)						
PLT:	Kent Rominger (3)						
MS1:	Jan Davis (3)						
MS2:	Robert Curbeam (1,EV1)						
MS3:	Steve Robinson (1,EV2)						
PS:	Bjarni Tryggvason (1,Canada)						
Payload:	TAS, MFD/SFA, IEH, SWUIS, MIM, BDS, SVS/RSAD, SEM, PCG-STES, BRIC, SSCE, MSX, SIMPLEX						
Activity:	Deployed CRISTA-SPAS 2 on 7 Aug and retrieved it on 16 Aug.						
STS-86							
87	1997-55A	Atlantis (20)	2234 EDT 25 Sep	1755 EDT 6 Oct	KSC (40)	10.80	51
CDR:	Jim Wetherbee (4)						
PLT:	Mike Bloomfield (1)						
MS1:	Vladimir Titov (2,EV2,Russia)						
MS2:	Scott Parazynski (2,EV1)						
MS3:	Jean-Loup Chretien (1,France)						
MS4:	Wendy Lawrence (2)						
UP:	David Wolf (2)						
DN:	Mike Foale (4)						
Payload:	Double Spacehab, ODS						
Activity:	SMM7. Docked with Mir on 27 Sep and departed on 3 Oct.						
Spacewalks:	Titov and Parazynski						
STS-87							
88	1997-73A	Columbia (24)	1446 EST 19 Nov	0720 EST 5 Dec	KSC (41)	15.66	28
CDR:	Kevin Kregel (3)						
PLT:	Steven Lindsey (1)						
MS1:	Kalpana Chawla (1)						
MS2:	Winston Scott (2,EV1)						
MS3:	Takao Doi (1,EV2,Japan)						
PS:	Leonid Kadenyuk (1,Ukraine)						
Payload:	USMP 4, EDO 12						

#	Designation	Orbiter	Launch	Landing	Runway	Days	Incl
Activity:	Deployed SPARTAN 201-4 on 21 Nov and retrieved it on 24 Nov. Deployed Sprint on 3 Dec and retrieved it same day.						
Spacewalks:	Scott and Doi						

STS-89

#	Designation	Orbiter	Launch	Landing	Runway	Days	Incl
89	1998-3A	Endeavour (12)	2148 EST 22 Jan	1736 EST 31 Jan	KSC (42)	8.82	51
CDR:	Terry Wilcutt (3)						
PLT:	Joe Edwards (1)						
MS1:	Jim Reilly (1,EV1)						
MS2:	Mike Anderson (1,EV2)						
MS3:	Bonnie Dunbar (5,PC)						
MS4:	Salizhan Sharipov (1,Russia)						
UP:	Andy Thomas (2)						
DN:	David Wolf (2)						
Payload:	Double Spacehab, ODS						
Activity:	SMM 8. Docked with Mir on 24 Jan and departed on 29 Jan.						

STS-90

#	Designation	Orbiter	Launch	Landing	Runway	Days	Incl
90	1998-22A	Columbia (25)	1419 EDT 17 Apr	1209 EDT 3 May	KSC (43)	15.91	28
CDR:	Rick Searfoss (3)						
PLT:	Scott Altman (1)						
MS1:	Richard Linnehan (2,PC,EV1)						
MS2:	Kathryn Hire (1)						
MS3:	Dave Williams (1,EV2,Canada)						
PS1:	Jay Buckey (1)						
PS2:	James Pawelczyk (1)						
Payload:	NeuroLab, EDO 13						

STS-91

#	Designation	Orbiter	Launch	Landing	Runway	Days	Incl
91	1998-34A	Discovery (24)	1806 EDT 2 Jun	1400 EDT 12 Jun	KSC (44)	9.83	51
CDR:	Charles Precourt (4)						
PLT:	Dominic Gorie (1)						
MS1:	Franklin Chang-Diaz (6,PC,EV1)						
MS2:	Wendy Lawrence (3)						
MS3:	Janet Kavandi (1,EV2)						
MS4:	Valeri Ryumin (1,Russia)						
DN:	Andy Thomas (2)						
Payload:	Single Spacehab, ODS						
Activity:	SMM 9. Docked with Mir on 4 Jun and departed on 8 Jun.						

STS-95

#	Designation	Orbiter	Launch	Landing	Runway	Days	Incl
92	1998-64A	Discovery (25)	1419 EST 29 Oct	1204 EST 7 Nov	KSC (45)	8.90	28
CDR:	Curt Brown (5)						
PLT:	Steven Lindsey (2)						
MS1:	Steve Robinson (2,EV1)						
MS2:	Scott Parazynski (3)						
PS3:	Pedro Duque (1,EV2,ESA/Spain)						
PS1:	Chiaki Mukai (2,Japan)						
PS2:	John Glenn (1)						
Payload:	Spacehab, IEH, HOST						
Activity:	Deployed PANSAT on 30 Oct. Deployed SPARTAN 201-5 on 1 Nov and retrieved it on 3 Nov.						

STS-88

#	Designation	Orbiter	Launch	Landing	Runway	Days	Incl
93	1998-69A	Endeavour (13)	0335 EST 4 Dec	2254 EST 15 Dec	KSC (46)	11.80	51
CDR:	Bob Cabana (4)						

Table (cont)

#	Designation	Orbiter	Launch	Landing	Runway	Days	Incl
PLT:	Fred Sturckow (1)						
MS1:	Jerry Ross (6,EV1)						
MS2:	Nancy Sherlock-Currie (3)						
MS3:	James Newman (3,EV2)						
MS4:	Sergei Krikalev (2,Russia)						
Payload:	ODS, ICBX, SEM 07, GAS						
Activity:	SSAF 2A. Lifted the Unity package (Node-1 with PMA-1/2) from the bay and mated with the ODS on 5 Dec. Zarya was grappled, soft-docked and hard-docked on 6 Dec. The stack was released on 13 Dec. Deployed SAC-A on 13 Dec and MightySat 1 on 14 Dec.						
Spacewalks:	Ross and Newman						
STS-96							
94	1999-30A	Discovery (26)	0650 EDT 27 May	0203 EDT 6 Jun	KSC (47)	9.55	51
CDR:	Kent Rominger (4)						
PLT:	Rick Husband (1)						
MS1:	Tamara Jernigan (5,EV1)						
MS2:	Ellan Ochoa (3,FE)						
MS3:	Daniel Barry (2,EV2)						
MS4:	Julie Payette (1, Canada)						
MS5:	Valeri Tokarev (1, Russia)						
Payload:	ODS, Spacehab DM, ICC (with OTD and Strela 1 base), SVFE						
Activity:	SSAF 2A.1. Docked with PMA-2 on Unity's forward CBM on 29 May and departed on 3 Jun. Deployed STARSHINE 1 on 5 Jun.						
Spacewalks:	Jernigan and Barry						
STS-93							
95	1999-40A	Columbia (26)	0031 EDT 23 Jul	2320 EDT 27 Jul	KSC (48)	4.95	28
CDR:	Eileen Collins (3)						
PLT:	Jeffrey Ashby (1)						
MS1:	Cady Coleman (2,EV2)						
MS2:	Steven Hawley (5,FE)						
MS3:	Michel Tognini (1,EV1,ESA/Frnance)						
Payload:	MSX, SIMPLEX, SWUIS, GOSAMR, STL-B, LFSAH, CCM, SAREX-II, EarthKAM, PGIM, CGBA (GBA-ICM), MEMS, BRIC						
Activity:	Deployed Chrandra/IUS on 23 Jul.						
STS-103							
96	1999-69A	Discovery (27)	1950 EST 19 Dec	1901 EST 27 Dec	KSC (49)	7.97	28
CDR:	Curt Brown (6)						
PLT:	Scott Kelly (1)						
MS1:	Steven Smith (3,PC,EV1)						
MS2:	Jean-Francois Clervoy (3,ESA/France)						
MS3:	John Grunsfeld (3,EV2)						
MS4:	Michael Foale (5,EV3)						
MS5:	Claude Nicollier (4,EV4,ESA/Switzerland)						
Payload:	ORC, FSS						
Activity:	HST Service 3A. Retrieved HST on 21 Dec, serviced it and released at on 25 Dec.						
Spacewalks:	Foale, Grunsfeld, Nicollier and Smith						
STS-99							
97	2000-10A	Endeavour (14)	1244 EST 11 Feb	1822 EST 22 Feb	KSC (50)	11.23	57
CDR:	Kevin Kregel (4)						
PLT:	Dom Gorie (2)						
MS1:	Gerhard Thiele (1,EV2,ESA/Germany)						
MS2:	Janet Kavandi (2,FE,EV1)						

#	Designation	Orbiter	Launch	Landing	Runway	Days	Incl
MS3:	Janice Voss (5,PC)						
MS4:	Mamoru Mohri (2,Japan)						
Payload:	SSRTM, EarthKAM						

STS-101

98	2000-27A	Atlantis (21)	0611 EDT 19 May	0220 EDT 29 May	KSC (51)	9.84	51
CDR:	James Halsell (5)						
PLT:	Scott Horowitz (3)						
MS1:	Mary Ellen Weber (2)						
MS2:	Jeff Williams (1,EV1)						
MS3:	James Voss (4,EV2)						
MS4:	Susan Helms (4)						
MS5:	Yuri Usachev (1,Russia)						
Payload:	ODS, Spacehab DM, ICC, BTPE, SEM 6, SEEDS-III						
Activity:	SSAF 2A.2a. Docked with PMA-2 on Unity's forward CBM on 21 May and departed on 26 May.						
Spacewalks:	Voss and Williams						

STS-106

99	2000-53A	Atlantis (22)	0846 EDT 8 Sep	0356 EDT 20 Sep	KSC (52)	11.46	51
CDR:	Terry Wilcutt (4)						
PLT:	Scott Altman (2)						
MS1:	Ed Lu (2,EV1)						
MS2:	Rick Mastracchio (1)						
MS3:	Dan Burbank (1)						
MS4:	Yuri Malenchenko (1,EV2,Russia)						
MS5:	Boris Morukov (1,Russia)						
Payload:	ODS, Spacehab DM, ICC, SEM 8, GAS, CGBA, PCG-EGND						
Activity:	SSAF 2A.2b. Docked with PMA-2 on Unity's forward CBM on 10 Sep and departed on 17 Sep.						
Spacewalk:	Lu and Malenchenko						

STS-92

100	2000-62A	Discovery (28)	1917 EDT 11 Oct	1700 EDT 24 Oct	EAFB (46)	12.90	51
CDR:	Brian Duffy (4)						
PLT:	Pamela Melroy (1)						
MS1:	Leroy Chiao (3,EV1)						
MS2:	Bill McArthur (3,EV2)						
MS3:	Jeff Wisoff (4,EV3)						
MS4:	Michael Lopez-Alegria (3,EV4)						
MS5:	Koichi Wakata (1,Japan)						
Payload:	Z1 Truss Assembly (with CMGs, HPGT and Ku-Band), PMA 3, IMAX, DDCU (heat pipe)						
Activity:	SSAF 3A. Docked with PMA-2 on Unity's forward CBM on 13 Oct and departed on 20 Oct. Meanwhile the Z1 was mounted on Unity's zenith on 14 Oct and PMA-3 on its nadir on 16 Oct.						
Spacewalks:	Chiao, McArthur, Wisoff and Lopez-Alegria						

STS-97

101	2000-78A	Endeavour (15)	2206 EST 30 Nov	1803 EST 11 Dec	KSC (53)	10.83	51
CDR:	Brent Jett (3)						
PLT:	Michael Bloomfield (2)						
MS1:	Joseph Tanner (3,EV1)						
MS2:	Marc Garneau (3,Canada)						
MS3:	Carlos Noriega (2,EV2)						
Payload:	P6 segment of the ITS						
Activity:	SSAF 4A. Docked with PMA-3 on Unity's nadir CBM on 2 Dec and departed on 9 Dec. Meanwhile P6 was mounted on the Z1 on 3 Dec.						
Spacewalks:	Tanner and Noriega						

Table (cont)

#	Designation	Orbiter	Launch	Landing	Runway	Days	Incl
STS-98							
102	2001-6A	Atlantis (23)	1813 EST 7 Feb	1533 EST 20 Feb	EAFB (47)	12.88	51
CDR:	Kenneth Cockrell (4)						
PLT:	Mark Polansky (1)						
MS1:	Robert Curbeam (2,EV2)						
MS2:	Marsha Ivins (5)						
MS3:	Tom Jones (4,EV1)						
Payload:	US Laboratory (Destiny), SIMPLEX						
Activity:	SSAF 5A. Docked with PMA-3 on Unity's nadir CBM on 9 February and departed on 16 Feb. Meanwhile, PMA-2 was detached from Unity's forward CBM and temporarily set on the MBM on Z1, Destiny was mounted in its place on 10 Feb, and PMA-2 was retrieved and placed on the end of Destiny.						
Spacewalks:	Jones and Curbeam						
STS-102							
103	2001-10A	Discovery (29)	0642 EST 8 Mar	0231 EST 21 Mar	KSC (54)	12.82	51
CDR:	Jim Wetherbee (5)						
PLT:	James Kelly (1)						
MS1:	Andy Thomas (3,EV3)						
MS2:	Paul Richards (1,EV4)						
UP:	Yuri Usachev (2,Russia, ISS-2 CDR/PLT)						
UP:	James Voss (5,EV1,ISS-2 FE1)						
UP:	Susan Helms (5,EV2,ISS-2 FE2)						
DN:	Bill Shepherd (4,ISS-1 CDR)						
DN:	Yuri Gidzenko (1,Russia,ISS-1 PLT/FE)						
DN:	Sergei Krikalev (3,Russia,ISS-1 FE)						
Payload:	MPLM (Leonardo), ICC with LCA, APCU, EAS						
Activity:	SSAF 5A.1. Docked with Destiny/PMA-2 on 10 Mar and departed on 18 Mar. Meanwhile PMA-3 was detached from Unity's nadir and placed on its port CBM on 11 Mar. Leonardo was berthed on Unity's nadir on 12 Mar for cargo transfer, and retrieved on 18 Mar.						
Spacewalks:	Helms, Voss, Richards and Thomas						
STS-100							
104	2001-16A	Endeavour (16)	1441 EDT 19 Apr	1211 EDT 1 May	EAFB (48)	11.90	51
CDR:	Kurt Rominger (5)						
PLT:	Jeffrey Ashby (2)						
MS1:	Chris Hadfield (2,EV1,Canada)						
MS2:	John Phillips (1)						
MS3:	Scott Parazynski (4,EV2)						
MS4:	Umberto Guidoni (2,ESA/Italy)						
MS5:	Yuri Lonchakov (1,Russia)						
Payload:	MPLM (Raffaello), SSRMS, UHF Antenna						
Activity:	SSAF 6A. Docked with Destiny/PMA-2 on 21 Apr and departed on 29 Apr. Meanwhile Raffaello was berthed on Unity's nadir on 23 Apr for cargo transfer, and retrieved on 27 Apr. The SSRMS engaged the grapple fixture on Destiny and 'walked' off its pallet on 23 Apr and the pallet was retrieved in a 'handover' from SSRMS to RMS on 28 Apr.						
Spacewalks:	Hadfield and Parazynski						
STS-104							
105	2001-28A	Atlantis (24)	0504 EDT 12 Jul	2339 EDT 24 Jul	KSC (55)	12.77	51
CDR:	Steven Lindsey (3)						
PLT:	Charles Hobaugh (1)						
MS1:	Michael Gernhardt (4,EV1)						
MS2:	Janet Kavandi (3)						
MS3:	James Reilly (2,EV2)						
Payload:	Joint Airlock(Quest) with three HPGT, SIMPLEX, IMAX, EarthKAM						

#	Designation	Orbiter	Launch	Landing	Runway	Days	Incl
Activity:	SSAF 7A. Docked with Destiny/PMA-2 on 13 Jul and departed on 22 Jul. Meanwhile Quest was mounted on Unity's starboard CBM on 15 Jul.						
Spacewalks:	Gernhardt and Reilly						

STS-105

106	2001-35A	Discovery (30)	1710 EDT 10 Aug	1423 EDT 22 Aug	KSC (56)	11.88	51
CDR:	Scott Horowitz (4)						
PLT:	Rick Sturckow (2)						
MS1:	Patrick Forrester (1,EV2)						
MS2:	Dan Barry (3,EV1)						
UP:	Frank Culbertson (3,ISS-3 CDR)						
UP:	Vladimir Dezhurov (2,Russia,ISS-3 PLT/FE)						
UP:	Mikhail Tyurin (1,Russia,ISS-3 FE)						
DN:	Yuri Usachev (2,Russia,ISS-2 CDR/PLT)						
DN:	James Voss (5,ISS-2 FE1)						
DN:	Susan Helms (5,ISS-2 FE2)						
Payload:	MPLM (Leonardo), MISSE, HEAT, SEM, GAS, EAS, ICC						
Activity:	SSAF 7A.1. Docked with Destiny/PMA-2 on 12 Aug and departed on 20 Aug. Meanwhile Leonardo was berthed on Unity's nadir on 13 Aug for cargo transfer, and retrieved on 19 Aug. Deployed SIMPLESAT 1 on 20 Aug.						
Spacewalks:	Barry and Forrester						

STS-108

107	2001-54A	Endeavour (17)	1719 EST 5 Dec	1255 EST 17 Dec	KSC (57)	11.82	51
CDR:	Dominic Gorie (3)						
PLT:	Mark Kelly (1)						
MS1:	Linda Godwin (4,EV1)						
MS2:	Daniel Tani (1,FE,EV2)						
UP:	Yuri Onufrienko (1,Russia,ISS-4 CDR/PLT)						
UP:	Carl Walz (4,ISS-4 FE)						
UP:	Daniel Bursch (4,ISS-4 FE)						
DN:	Frank Culbertson (3,ISS-3 CDR)						
DN:	Mikhail Tyurin (1,Russia,ISS-3 FE)						
DN:	Vladimir Dezhurov (2,Russia,ISS-3 PLT/FE)						
Payload:	MPLM (Raffaello), ADF, MACH 1, LMC, CPLE-3, CBTM-AEM						
Activity:	SSUF 1. Docked with Destiny/PMA-2 on 7 Dec and departed on 15 Dec. Meanwhile Raffaello was berthed on Unity's nadir CBM on 8 Dec for cargo transfer, and retrieved on 15 Dec. Deployed STARSHINE 2 on 16 Dec.						
Spacewalks:	Godwin and Tani						

STS-109

108	2002-10A	Columbia (27)	0622 EST 1 Mar	0432 EST 12 Mar	KSC (58)	10.92	28
CDR:	Scott Altman (3)						
PLT:	Duane Carey (1)						
MS1:	John Grunsfeld (4,PC,EV1)						
MS2:	Nancy Sherlock-Currie (4)						
MS3:	Richard Linnehan (3,EV2)						
MS4:	James Newman (4,EV3)						
MS5:	Michael Massimino (1,EV4)						
Payload:	HST Solar Arrays, ACS, NICMOS cryocooler, PCU, RWA						
Activity:	HST Service 3B. Retrieved HST on 3 Mar, serviced it and released it on 9 Mar.						
Spacewalks:	Grunsfeld, Linnehan, Newman and Massimino						

Table (cont)

#	Designation	Orbiter	Launch	Landing	Runway	Days	Incl
STS-110							
109	2002-18A	Atlantis (25)	1644 EDT 8 Apr	1227 EDT 19 Apr	KSC (59)	10.82	51
CDR:	Michael Bloomfield (3)						
PLT:	Stephen Frick (1)						
MS1:	Rex Walheim (1,EV2)						
MS2:	Ellen Ochoa (4)						
MS3:	Lee Morin (1,EV4)						
MS4:	Jerry Ross (7,EV3)						
MS5:	Steven Smith (4,EV1)						
Payload:	S0 segment of the ITS, MT						
Activity:	SSAF 8A. Docked with Destiny/PMA-2 on 10 Apr and departed on 17 Apr. Meanwhile the S0 was mounted on the Lab Cradle Assembly on 11 Apr.						
Spacewalks:	Smith, Walheim, Ross and Morin						
STS-111							
110	2002-28A	Endeavour (18)	1723 EDT 5 Jun	1358 EDT 19 Jun	EAFB (49)	13.85	51
CDR:	Kenneth Cockrell (5)						
PLT:	Paul Lockhart (1)						
MS1:	Franklin Chang-Diaz (7,EV1)						
MS2:	Philippe Perrin (1,EV2,France)						
UP:	Valeri Korzun (1,Russia,ISS-5 CDR/PLT)						
UP:	Peggy Whitson (1,ISS-5 FE1/SO)						
UP:	Sergei Treschev (1,Russia,ISS-5 FE2)						
DN:	Yuri Onufrienko (1,Russia,ISS-4 CDR/PLT)						
DN:	Carl Walz (4,ISS-4 FE)						
DN:	Dan Bursch (4,ISS-4 FE)						
Payload:	MPLM (Leonardo), MBS						
Activity:	SSUF 2. Docked with Destiny/PMA-2 on 7 June and departed on 15 Jun. Meanwhile Leonardo was berthed on Unity's nadir on 8 Jun for cargo transfer, and retrieved on 14 Jun.						
Spacewalks:	Chang-Diaz and Perrin						
STS-112							
111	2002-47A	Atlantis (26)	1546 EDT 7 Oct	1144 EDT 18 Oct	KSC (60)	10.83	51
CDR:	Jeff Ashby (3)						
PLT:	Pamela Melroy (2)						
MS1:	David Wolf (3,EV1)						
MS2:	Sandra Magnus (1)						
MS3:	Piers Sellers (1,EV2)						
MS4:	Fyodor Yurchikhin (1,Russia)						
Payload:	S1 segment of the ITS with CETA 1						
Activity:	SSAF 9A. Docked with Destiny/PMA-2 on 9 Oct and departed on 16 Oct. Meanwhile the S1 was mounted it on the starboard end of S0 on 10 Oct.						
Spacewalks:	Wolf and Sellers						
STS-113							
112	2002-52A	Endeavour (19)	1950 EST 23 Nov	1437 EST 7 Dec	KSC (61)	13.78	51
CDR:	Jim Wetherbee (6)						
PLT:	Paul Lockhart (2)						
MS1:	Michael Lopez-Alegria (3,EV1)						
MS2:	John Herrington (1,EV2)						
UP:	Ken Bowersox (5,ISS-6 CDR)						
UP:	Nikolai Budarin (2,Russia,ISS-6 PLT/FE1)						
UP:	Don Pettit (1,ISS-6 FE2/SO)						
DN:	Valeri Korzun (1,Russia,ISS-5 CDR/PLT)						
DN:	Peggy Whitson (1,ISS-5 FE1/SO)						

#	Designation	Orbiter	Launch	Landing	Runway	Days	Incl

DN: Sergei Treschev (1,Russia,ISS-5 FE2)
Payload: P1 segment of ITS with CETA 2
Activity: SSAF 11A. Docked with Destiny/PMA-2 on 25 Nov and departed on 2 Dec. Meanwhile the P1 was mounted on the port end of S0 on 26 Nov. Deployed MEPSI on 2 Dec.
Spacewalks: Lopez-Alegria and Herrington

STS-107

#	Designation	Orbiter	Launch	Landing	Runway	Days	Incl
113	2003-3A	Columbia (28)	1039 EST 16 Jan	–	–	–	39

CDR: Rick Husband (2)
PLT: William McCool (1)
MS1: Dave Brown (1)
MS2: Kalpana Chawla (2,FE)
MS3: Mike Anderson (2,PC)
MS4: Laurel Clark (1)
PS: Ilan Ramon (1,Israel)
Payload: Spacehab RDM, EDO, FREESTAR, OARE, OEX
Note: The de-orbit burn at 0815 EST 1 Feb was nominal, but the spacecraft was destroyed at 0900 EST, in re-entry at an altitude of 200,000 feet over Texas, some 16 minutes before the scheduled 0916 EST landing at KSC

STS-114

#	Designation	Orbiter	Launch	Landing	Runway	Days	Incl
114	??	Atlantis (27)	–	–	()	–	51

CDR: Eillen Collins (4)
PLT: James Kelly (2)
MS-: Soicho Noguchi (1,Japan)
MS-: Steve Robinson (3)
MS-: Wendy Lawrence (4)
MS-: Andy Thomas (4)
MS-: Charles Camarda (1)
Payload: MPLM (Raffaello), ESP 2
Activity: SSULF 1

Acronyms

AADSF	Advanced ADSF
AAEU	Aquatic Animal Experiment Unit
ACCESS	Assembly Concept for Construction of Erectable Space Structures
ACIP	Aerodynamic Coefficient Identification Package
ACRIM	Active Cavity Radiometer Irradiance Monitor
ACRV	Assured Crew-Return Vehicle
ACS	Advanced Camera System
ACS	Advanced Camera for Surveys
ACTS	Advanced Communications Technology Satellite
ADF	Avian Development Facility
ADSEP	ADvanced SEparation Process
ADSF	Automated Directional Solidification Furnace
AEM	Animal Enclosure Module
AEPI	Atmospheric Emissions Photometric Imager
AERIS	American Echocardiograph Research Imaging System
AFE	American Flight Echocardiograph
AGHF	Advanced Gradient Heating Furnace
AHF	Animal Holding Facility
AKM	Apogee Kick-Motor
ALAE	Atmospheric Lyman-Alpha Emission
ALFE	Advanced Liquid Feed Experiment
ALT	Approach and Landing Test
AMEE	Advanced Materials Exposure Experiment
AMOS	Air Force Maui Optical Site
AOA	Abort Once Around
APA	Anticipatory Postural Activity
APAS	Androgynous Peripheral Attachment System
APCF	Advanced Protein Crystallisation Facility
APCG	Advanced PCG
APCU	Assembly Power Converter Unit
APDS	Androgynous Peripheral Docking System

APE	Auroral Photography Experiment
APM	Ascent Particle Monitor
APU	Auxiliary Power Unit
ARC	Aggregation of Red blood Cells
ARCU	American–Russian Converter Unit
ARF	Aquatic Research Facility
ARIS	Active Rack Isolation System
ASE	Airborne Support Equipment
ASEM	Assembly of Station by EVA Methods
ASP	Attitude Sensor Package
ASTP	Apollo-Soyuz Test Programme
ATLAS	ATmospheric Laboratory for Applications and Science
ATMOS	Atmospheric Trace MOolecule Spectroscope
ATO	Abort To Orbit
AXAF	Advanced X-ray Astronomy Facility
BATSE	Burst And Transient Source Experiment
BBXRT	Broad-Band X-Ray Telescope
BCAT	Binary-Colloid Alloy Test
BDPU	Bubble, Drop and Particle Unit
BDS	Bioreactor Demonstration System
BETSCE	Brilliant Eyes Ten K Sorption Cryocooler Experiment
BFPT	Bioreactor Flow and Particle Trajectory
BGA	Beta Gimbal Assembly
BIMDA	Bioserve ITA MDA
BLAST	Battlefield Laser Acquisition Sensor Test
BPA	Back Pain in Astronauts
BPL	Bioserve Pilot Laboratory
BR	BioRack
BREMSAT	BREMen SATellite
BRIC	Biological Research In Canisters
BSK	BioStacK
BTS	BioTechnology System
BTPE	BioTube Precursor Experiment
CAPL	CApillary Pumped Loop
CBTME	Commercial Biomedical Testing Module Experiment
CBTM-AEM	Commercial Biomedical Testing Module – AEM
CCAFS	Cape Canaveral Air Force Station
CCD	Charge-Coupled Device
CCDS	Center for the Commercial Development of Space
CCM	Cell Culture Module
CDR	Shuttle Commander (left seat)
CETA	Crew and Equipment Translation Aids
CFCs	ChloroFluoroCarbons
CFE	Candle Flame Experiment
CFES	Continuous Flow Electrophoresis System

CFLSE	Critical Fluid Light Scattering Experiment
CFZF	Commercial Floating-Zone Furnace
CGBA	Commercial Generic Bioprocessing Apparatus
CGBA-ICM	CGBA – Isothermal Containment Module
CGF	Crystal Growth Furnace
CHAMP	Comet Halley Active Monitoring Programme
CHF	Continuous Heating Furnace
CIRRIS	Cryogenic InfraRed Radiance Instrument in Space
CIV	Critical Ionisation Velocity
CLAES	Cryogenic Limb Array Etalon Spectrometer
CLIP	Crew Loads Instrumented Pallet
CLOUDS	Cloud Logic to Optimise Use of Defence Systems
CT	Crawler Transporter
DMSP	Defence Meteorological Satellite Programme
CMIX	Commercial MDA ITA Experiments
COIS	Canal and Otolith Integration Studies
COMPTEL	COMpton TELescope
CONCAP	CONsortium for materials development in space, Complex Autonomous Payload)
COS	Cosmic Origins Spectrograph
COSTAR	Corrective Optics Space Telescope Axial Replacement
CPCG	Commercial PCG
CPF	Critical Point Facility
CPLE	Capillart Pumed Loop Experiment
CREAM	Cosmic Radiation Effects and Activation Monitor
CRISTA	CRyogenic Infrared Spectrometer and Telescope for the Atmosphere
CRO	Chemical Release Observation
CRYOHP	CRYOgenic HeatPipe
CSA	Cooperative Solar Array
CTA	Canadian Target Assembly
CTPE	Cryogenic Two-Phase Experiment
CVTE	Chemical Vapour Transport Experiment
DATE	Dynamic, Acoustic and Thermal Environment
DCAM	Diffusion Crystallisation Apparatus for Microgravity
DDCU	DC-to-DC Converter Unit
DDU	Data Display Unit
DEE	Dexterous End-Effector
DFI	Development Flight Instrumentation
DM	Docking Module
DMOS	Diffuse Mixing of Organic Solutions
DoD	Department of Defense
DPM	Drop Physics Module
DSCS	Defense Satellite Communications System
DSE	Data Systems Experiment

DSP	Defense Support Programme
DVOS	Direct-View Optical System
DXS	Diffuse X-ray Spectrometer
EAS	Early Ammonia Servicer
EASE	Experimental Assembly of Structures in EVA
EATCS	External Active Thermal Control System
ECLIPSE	Equipment for Controlled LIquid-Phase Sintering Experiments
ECT	Emulsion Chamber Technology
EDO	Extended Duration Orbiter
EDO-MP	EDO Medical Programme
EDS	Energy Deposition Spectrometer
EES	Energy Expenditure in Spaceflight
EEVT	Electrophoresis Equipment Verification Test
EGRET	Energetic Gamma Ray Experiment Telescope
EISG	Experimental Investigation of Spacecraft Glow
ELRAD	Earth Limb RAdiance Experiment
EMU	Extravehicular Mobility Unit
EOIM	Evaluation of Oxygen Interaction with Materials
EORF	Enhanced Orbiter Refrigerator and Freezer
EOS	Earth Observing System
EOS	Electrophoresis Operations in Space
EOSAT	Earth Observation Satellite Corporation
ERB	Earth's Radiation Budget
ERBS	Earth's Radiation Budget Satellite
ESC	Electronic Still Camera
ESOC	European Space Operations Centre
ESP	External Stowage Platform
ET	External Tank
ETR	Eastern Test Range
ETTF	Extreme-Temperature Translation Furnace
EURECA	EUropean REtrievable CArrier
eV	electron Volt
EVA	ExtraVehicular Activity
EXPRESS	EXpedite PRocessing of Experiments to Space Station
FARE	Fluid Acquisition and Resupply Experiment
FAUST	FAr Ultraviolet Space Telescope
FEA	Fluids Experiment Apparatus
FES	Fluids Experiment System
FES	Flash Evaporator System)
FFEU	Free-Flow Electrophoresis Unit
FFFT	Forced Flow Flame Test
FGBA	Fluid Generic Bioprocessing Apparatus
FGS	Fine Guidance Sensor
FILE	Feature Identification and Location Experiment
Fltsatcom	Fleet satellite communications

FME	Foil Microabrasion Package
FOC	Faint Object Camera
FOS	Faint Object Spectrograph
FPE	French Posture Experiment
FPP	Floating Potential Probe
FREESTAR	Fast Reaction Experiments Enabling Science, Technology, Applications and Research
FRF	Flight Readiness Firing
FRGF	Flight-Releasable Grapple Fixture
FSDCE	Fibre-Supported Droplet Combustion Experiment
FSE	Flight Support Equipment
FSS	Flight Support Structure
FUV	Far-UltraViolet
GANE	GPS Attitude and Navigation Experiment
GAS	Get-Away-Special
GBA	GAS Bridge Assembly
GBX	GloveBoX
GEF	Gas Evaporation Facility
GEO	GEostationary Orbit
GFFC	Geophysical Fluid Flow Cell
GFVT	GAS Flight Verification Test
GHCD	Growth Hormone Concentration and Distribution
GHRS	Goddard High-Resolution Spectrograph
GLOMR	Global Low Orbiting Message Relay
GOSAMR	Gelation Of Sols, Applied Microgravity Research
GPC	General Purpose Computer
GPPF	Gravitational Plant Physiology Facility
GPPM	Gas-Permeable Polymeric Membrane
GPS	Global Positioning System
GPWS	General Purpose WorkStation
GRO	Gamma Ray Observatory
GS	Grille Spectrometer
GSC	Grab Sample Container
GSO	GeoSynchronous Orbit
GTO	Geostationary Transfer Orbit
HALOE	HALogen Occultation Experiment
HBT	Hyflex Bioengineering Test
HEAT	Hitchhiker Experiments Advanced Technology
HERCULES	Hand-held Earth-oriented Real-time Cooperative User-friendly, Location-targeting and Environmental System
HGD	Hand Grip Dynamometer
HH-DTC	Hand-Held Diffusion Test Cell
HH-PCG	Hand-Held PCG
HOL	Holographic Optics Laboratory
HOST	HST Orbital Systems Test

HPGT	High-Pressure Gas Tank
HPP	Heat Pipe Performance
HPTE	High-Precision Tracking Experiment
HRDI	High-Resolution Doppler Imager
HRS	High-Resolution Spectrograph
HSP	High-Speed Photometer
HST	Hubble Space Telescope
HTLPE	High-Temperature Liquid Phase Experiment
HUP	Horizon Ultraviolet Programme
HUT	Hopkins Ultraviolet Telescope
IAE	Inflatable Antenna Experiment
IBSS	Infrared Background Signature Survey
ICBE	IMAX Cargo Bay Camera
ICC	Integrated Cargo Carrier
ICE	Interface Configuration Experiment
IDGE	Isothermal Dendritic Growth Experiment
IECM	Induced Environment Contamination Monitor
IEH	International Extreme-ultraviolet Hitchhiker
IFE	Isoelectric Focusing Experiment
IFM	In-Flight Maintenance
IGY	International Geophysical Year
IMEWS	Integrated Missile Early Warning System
IML	International Microgravity Laboratory
IMS	Inventory Management System
IMU	Inertial Measurement Unit
IOCM	Interim Operational Contamination Monitor
IPS	Instrument Pointing System
IRCFE	InfraRed Communications Flight Experiment
IRIS	InfraRed Imagery of the Shuttle
IRIS	Italian Research Interim Stage
IRT	Integrated Rendezvous Target
ISAC	Intelsat Solar Array Coupon
ISAIAH	Israeli Space Agency Investigation About Hornets
ISAMS	Improved Stratospheric And Mesospheric Sounder
ISIS	Infrared Spectral Imaging Radiometer
ISO	Imaging Spectrometric Observatory
ISS	International Space Station
ISY	International Space Year
ITA	Instrumentation Technologies Associates
ITS	Integrated Truss Structure
IUE	International Ultraviolet Explorer
IUS	Inertial Upper Stage
JPL	Jet Propulsion Laboratory
JSC	Johnson Space Center
KSC	Kennedy Space Center

L3	Latitude and Longitude Locator
LACE	Laser Atmospheric Compensation Experiment
LAGEOS	LAser GEOS
LBNP	Lower Body Negative Pressure
LC 39	Launch Complex 39
LCA	Lab Cradle Assembly
LCC	Launch Control Centre
LDCE	Limited Duration space environment Candidate materials Exposure
LDEF	Long Duration Exposure Facility
LEE	Latching End-Effector
LEMZ	Liquid Encapsulated Melt Zone
LEO	Low Earth Orbit
LFC	Large Format Camera
LFE	Lung Function Experiment
LFSAH	Lightweight Flexible Solar Array Hinge
LIF	Large Isothermal Furnace
LIGNIN	Gravity-Influenced Lignification In Higher Plants
LiOH	Lithium hydroxide
LITE	Lidar-In-space Technology Experiment
LMC	Lightweight MPESS Carrier
LMTE	Liquid Metal Thermal Experiment
LPE	Lambda Point Experiment
LRU	Line Replacement Units
LSRF	Life Sciences Refrigerator and Freezer
LTA	Lower Torso Assembly
MACE	Middeck Active Control Experiment
MACH	Multiple Application Cusomised Hitchhiker
MAHRSI	Middle-Atmosphere High-Resolution Spectrograph Instrument
MAPS	Measurement of Air Pollution from Space
MAS	Microbial Air Sampler
MAS	Millimetre Atmospheric Sounder
MAST	Military Applications of Ship Tracks
MBM	Manual Berthing Mechanism
MBS	Mobile Base System
MCC	Mission Control Center
MDA	Materials Dispersion Apparatus
MEA	Materials Experiments Assembly
MECO	Main Engine Cut-Off
MEDS	Multifunction Electronic Display System
MEEP	Mir Environmental Effects Package
MEFCE	Mir Electric Fields Characterisation Experiment
MELEO	Materials Exposure in LEO
MEMS	Micro-Electrical Mechanical System
MET	Mission Elapsed Time

MFD	Manipulator Flight Demonstration
MFR	Manipulator Foot Restraint
MGBX	Middeck GBX
MGME	Mechanics of Granular Materials Experiment
MICG	Mercury Iodide Crystal Growth
MIDAS	Materials In Devices As Semiconductors
MIDAS	MIssile Defence Alarm System
MIM	Microgravity Isolation Mount
MIS	Microencapsulation In Space
MISDE	MIr Structural Dynamics Experiment
MISSE	Materials ISS Experiment
MLE	Mesoscale Lightning Experiment
MLP	Mobile Launch Platform
MLR	Monodisperse Latex Reactor
MLS	Microwave Limb Sounder
MMA	Microgravity Measurement Assembly
MMIS	Military-Man-In-Space
MMS	Multiple Mission Spacecraft
MMU	Manned Manoeuvring Unit
MODE	Middeck 0-gravity Dynamics Experiment
MOMS	Modular Opto-electronic Multispectral Scanner
MPEC	Multipurpose Experiment Canister
MPESS	Mission-Peculiar Equipment Support Structure
MPLM	Multi-Purpose Logistics Module
MSG	Microgravity Science GBX
MSL	Materials Science Laboratory
MSL	Materials Science Laboratory
MSRE	Mir Sample Return Experiment
MSX	Midcourse Space Experiment
MT	Mobile Transporter
MTPE	Mission To Planet Earth
MVC	Measurement of Venous Compliance
MVI	Microgravity Vestibular Investigations
MWPE	Mental Workload and Performance Experiment
NASA	National Aeronautics and Space Administration
NICMOS	Near-Infrared Camera and Multi-Object Spectrometer
NIH	National Institutes of Health
NIH-C	NIH-Chicken
NIH-R	NIH-Rodent
NOAA	National Oceanic and Atmospheric Administration
NOSL	Night-day Optical Survey of Lightning
NSTS	National Space Transportation System
OAIMT	Oxygen Atom Interaction with Materials Test
OARE	Orbiter Acceleration Research Equipment
OASIS	Orbiter experiments Autonomous Supporting Instrumentation

	System
OAST	Office of Aeronautics and Space Technology
O&C	Operations & Checkout
OCE	Ocean Colour Experiment
OCGF	Organic Crystal Growth Facility
OCS	Operations Control Software
OCTW	Optical Communication Through Window
ODERACS	Orbital DEbris RAdar Calibration Spheres
ODS	Orbiter Docking System
OEX	Orbiter EXperiment
OFT	Orbital Flight Test
OGLOW	Orbiter GLOW
OLIPSE	Optizon LIquid Phase Sintering Experiment
OMS	Orbital Manoeuvring System
OMV	Orbital Manoeuvring Vehicle
OPF	Orbiter Processing Facility
ORC	Orbital Replacement Carrier
ORFEUS	Orbiting Retrievable Far- and Extreme-Ultraviolet Spectrometer
ORS	Orbital Refuelling System
ORU	Orbital Replacement Unit
OSS	Office of Space Sciences
OSSE	Oriented Scintillation Spectrometer Experiment
OSTA	Office of Space and Terrestrial Applications
OTD	ORU Transfer Device
PAM	Payload Assist Module
PAMS	Passive Aerodynamically-stabilised Magnetically-damped Satellite
PANSAT	Petite Amateur Naval Satellite
PAO	Public Affairs Office
PARE	Physiological and Anatomical Rodent Experiment
PAS	Passive Accelerometer System
PAWS	Performance Assessment Workstation
PCF	Protein Crystallisation Facility
PCG	Protein Crystal Growth
PCG-EGND	Protein Crystal Growth Enhanced Gaseous Nitrogen Dewar
PCU	Plasma Contactor Unit
PCU	Power Control Unit
PDE	Particle Dispersion Experiment
PDGF	Power and Data Grapple Fixture
PDP	Plasma Diagnostics Package
PEM	Particle Environment Monitor
PEP	Power Extension Package
PFR	Portable Foot Restraint
PFTA	Payload Flight Test Article
PGBA	Plant Generic Bioprocessing Apparatus
PGF	Plant Growth Facility
PGIM	Plant Growth in Microgravity

PGU	Plant Growth Unit
PIE	Particle Impact Experiment
PILOT	Portable In-flight Landing Operations Trainer
PKM	Perigee Kick-Motor
PLT	Shuttle Pilot (right seat)
PM	Polymer Morphology
PMP	Polymer Membrane Processing
POA	Payload ORU Accommodation
POCC	Payload Operations Control Centre
PPE	Phase Partitioning Experiment
PSE	Physiological Systems Experiment
PSN	Positional and Spontaneous Nystagmus
PTI	Programmed Test Input
PVTOS	Physical Vapour Transport Organic Solid experiment
QINMS	Quadropole Ion Neutral Mass Spectrometer
QSAM	Quasi-Steady Acceleration Measurement
QUELD	Queen's University Experiment in Liquid Diffusion
RACU	Russian-American Converter Unit
RAHF	Research Animal Holding Facility
RCC	Reinforced Carbon–Carbon
RCRS	Regenerative Carbon dioxide Removal System
RCS	Reaction Control System
RITSI	Radiative Ignition and Transition to Spread Investigation
RME	Radiation Monitoring Equipment
RMS	Remote Manipulator System
ROMPS	Robot Operated Materials Processing System
ROSAT	ROentgen SATellite
ROTEX	RObоT EXperiment
RPU	Relay Power Unit
RRMD	Real-time Radiation Monitoring Device
RSA	Russian Space Agency
RSAD	RMS Situational Awareness Display
RSI	Reusable Surface Insulation
RSP	Resupply Stowage Platform
RSS	Rotating Service Structure
RSU	Rate Sensor Unit
RTAS	Rocketdyne Truss Attachment System
RTG	Radioisotope Thermoelectric Generator
RTLS	Return To Landing Site
RWA	Reaction Wheel Assembly
RWS	Robotic Workstation
S4	Spacehab Soft Stowage System
SAC	Satelite de Aplicaciones
SAFER	Simplified Aid For EVA Rescue
SAGE	Stratospheric Aerosol and Gas Experiment

SAM	Shuttle Activation Monitor
SAMPIE	Solar Array Module Plasma Interaction Experiment
SAMS	Space Acceleration Measurement System
SAR	Synthetic Aperture Radar
SAREX	Shuttle Amateur Radio EXperiment
SAS	Space Adaptation Syndrome
SASE	SAS Experiment
SATO	Space Adaptation Tests and Observations
SBU	Spun-Bypass Unit
SBUV	Solar and Backscattered UltraViolet
SCA	Shuttle Carrier Aircraft
SCE	Solar Constant Experiment
SCE	Smoldering Combustion Experiment
SCM	Spinal Changes in Microgravity
SDI	Strategic Defense Initiative
SDOC	Space Defense Operations Center
SDS	Satellite Data System
SEF	Space Experiment Facility
SEH	Solar Extreme-ultraviolet Hitchhiker
SEM	Student Experiment Module
SEPAC	Space Experiments with a Particle ACcelerator
SEPRE	Shuttle Electric Potential and Return Experiment
SETS	Shuttle Electrodynamic Tether System
SFA	Small Fine Arm
SFOG	Solid Fuel Oxygen Generator
SFU	Space Flyer Unit
SHARE	Station Heatpipe Advanced Radiator Experiment
SHEAL	Spacelab High Energy Astrophysics Laboratory
SHOOT	Superfluid Helium On-Orbit Transfer
SIMPLEX	Shuttle Ionospheric Modification with Pulsed Local Exhaust Experiment
SIR	Spaceborne Imaging Radar
SIR	Standard Interface Rack
SKIRT	Spacecraft Kinetic InfraRed Test
SLA	Shuttle Laser Altimeter
SLF	Shuttle Landing Facility
SLP	Spacelab Logistics Pallet
SMIRR	Shuttle Multispectral InfraRed Radiometer
SOLCON	SOLar CONstant
SOLSPEC	SOLar SPECtrum
SOLSTICE	SOLar-STellar Irradiance Comparison Experiment
SOSP	SOlar SPectrum
SOVA	SOlar VAriability
SPARTAN	Shuttle Pointed Autonomous Research Tool for AstroNomy
SPAS	Shuttle Pallet Applications Satellite

SPD	Spool Positioning Devices
SPDS	Stabilised Payload Deployment System
SPE	Space Physiology Experiments
SPEAM	Sun Photospectrometer Earth Atmosphere Measurement
SPIFEX	Shuttle Plume Impingement Flight EXperiment
SRB	Solid Rocket Booster
SSAF	Space Station Assembly Flight
SSAS	Solid Sorbent Air Sampler
SSBUV	Shuttle SBUV
SSCE	Solid Surface Combustion Experiment
SSCP	Small Self-Contained Payload
SSIP	Shuttle Student Involvement Programme
SSME	Space Shuttle Main Engine
SSR	Solid State Recorder
SSRMS	Space Station Remote Manipulator System
SSUF	Space Station Utility Flight
SS-RMS	Space Station Remote Manipulator System
STA	Shuttle Training Aircraft
STALE	Suppression of Transient Accelerations by Levitation Experiment
STARSHINE	Student Tracked Atmospheric Research Satellite for Heuristic International Networking Equipment
STDCE	Surface Tension Driven Convection Experiment
STIS	Space Telescope Imaging Spectrograph
STL	Space Tissue Loss
STLV	Slow-Turning Lateral Vessel
STP	Space Test Programme
STS	Space Transportation System
STScI	Space Telescope Science Institute
STU	Satellite Test Unit
SUSIM	Solar Ultraviolet Spectral Irradiance Monitor
SUVE	Solar UltraViolet Experiment
SVFE	Shuttle Vibration Forces Experiment
SVS	Space Vision System
SWUIS	SouthWest Ultraviolet Imaging System
3DMA	Three-Dimensional Microgravity Accelerometer
TAGS	Text And Graphics System
TAL	Transatlantic Abort Landing
TAS	Technology Applications and Science
TC	Trash Compactor
TCDT	Terminal Countdown Demonstration Test
TCS	Trajectory Control Sensor
TDRS	Tracking and Data Relay System
TEAMS	Technology Experiments for Advancing Missions in Space
TEHM	Thermo-Electric Holding Module
TEI	ThermoElectric Incubator

TEPC	Tissue-Equivalent Proportional Counter
TIROS	TV InfraRed Operational System
TLD	ThermoLuminescent Dosimeter
TOMS	Total Ozone Mapping Spectrometer
TOS	Transfer Orbit Stage
TPAD	Trunnion Pin Attachment Device
TPCE	Tank Pressure Control Experiment
TPS	Thermal Protection System
TRE	Torso Rotation Experiment
TSS	Tethered Satellite System
TSS	Temporary Sleep Station
TUFI	Toughened Unipiece Fibrous Insulation
TVD	Torque Velocity Dynamometer
TVIS	Treadmill with Vibration Isolation and Stabilisation
UARS	Upper Atmosphere Research Satellite
UFO	UHF Follow-On
UHS	Ultraviolet Horizon Scanner
UIT	Ultraviolet Imaging Telescope
URA	Uniformly Redundant Array
USAR	Ultraviolet Spectrograph for Astronomical Research
USML	US Microgravity Laboratory
USMP	US Microgravity Payload
UTA	Upper Torso Assembly
UVLIM	UltraViolet LIMb
UVPI	UltraViolet Plume Instrument
UVSTAR	UltraViolet Spectrograph Telescope for Astronomical Research
VAB	Vehicle Assembly Building
VAFB	Vandenberg AFB
VCAP	Vehicle Charging and Potential
VCGS	Vapour Crystal Growth System
VCS	Voice Command System
VFT	Visual Function Tester
VIBES	Vibration Isolation Box Experiment System
VIEW-CAPL	Visualisation In Experimental Water CAPL
VIK	Voltage/Temperature Improvement Kit
VOA	Volatile Organic Analyser
VTRE	Vented Tank Resupply Experiment
WAMDII	Wide Angle Michelson Doppler Imaging Interferometer
WCS	Waste Containment System
WETA	Wireless video system External Transceiver Assembly
WETF	Weightless Environment Training Facility
WF/PC	Wide Field/Planetary Camera
WIF	Work Interface Fixture
WIFE	Wire Insulation Flammability Experiment
WINDEX	WINdow EXperiment

WINDII	WIND Imaging Interferometer
WSF	Wake Shield Facility
WUPPE	Wisconsin Ultraviolet Photo-Polarimeter Experiment
X-SAR	X-band SAR
ZCGF	Zeolite Crystal Growth Facility

Index